Recent Advances in Biological and Technological Research of Fresh Fruit and Vegetable

Recent Advances in Biological and Technological Research of Fresh Fruit and Vegetable

Guest Editor

Zhongqi Fan

MDPI

Basel • Beijing • Wuhan • Barcelona • Belgrade • Novi Sad • Cluj • Manchester

Guest Editor
Zhongqi Fan
College of Food Science
Fujian Agriculture and
Forestry University
Fuzhou
China

Editorial Office
MDPI AG
Grosspeteranlage 5
4052 Basel, Switzerland

This is a reprint of the Special Issue, published open access by the journal *Foods* (ISSN 2304-8158), freely accessible at: https://www.mdpi.com/journal/foods/special_issues/IA8UA14903.

For citation purposes, cite each article independently as indicated on the article page online and as indicated below:

Lastname, A.A.; Lastname, B.B. Article Title. *Journal Name* **Year**, *Volume Number*, Page Range.

ISBN 978-3-7258-7448-4 (Hbk)
ISBN 978-3-7258-7449-1 (PDF)
https://doi.org/10.3390/books978-3-7258-7449-1

Contents

foods MDPI

Editorial

Recent Advances in Biological and Technological Research of Fresh Fruit and Vegetable

Zhongqi Fan

College of Food Science, Fujian Agriculture and Forestry University, Fuzhou 350002, China; fanzqi@fafu.edu.cn

Fresh fruit and vegetables are sources of vitamins, minerals, and dietary fiber; however, due to their short postharvest life, a large portion of the produce is lost. It is, therefore, necessary to investigate the biological changes underlying ripening and senescence and to find appropriate ways to control the perishability of fresh fruit and vegetables, thereby increasing their shelf life [1,2]. The basic science of the ripening and senescence of fruit and vegetables has progressed rapidly in recent years. This is largely due to a series of breakthrough discoveries regarding key factors and signaling pathways that activate ripening-related and senescence-associated genes [3,4]. On the other hand, advanced postharvest technologies, including chemical treatment, physical methods, and biotechnology, have been widely applied to minimize postharvest loss and maintain the commercial quality of fresh fruit and vegetables [5,6]. This Special Issue, "Recent Advances in Biological and Technological Research of Fresh Fruit and Vegetable", features 15 articles. These articles represent cutting-edge research on the biological mechanisms, physiology disorders, and preservation technologies of fresh fruit and vegetables.

Proteomics and metabolomics technologies are important omics approaches to detect the nutrients in fruit and vegetables [7]. In this Special Issue, an integrating metabolomics and proteomics approach was utilized to investigate the flavor precursor in the maturity stage of *coffea arabica* L. cherries. A total of 456 differently expressed metabolites (DEMs) and 402 differently expressed proteins (DEPs) were identified. Among them, 45 DEPs were revealed to modulate 40 flavor precursors of primary amino acids. The results reveal the amino acid and organic acid metabolic pathways are important for flavor formation during the ripening of coffee cherries [contribution 1].

Additionally, HPLC-MS was widely used to determine the chemical substances in various food materials [8]. Regarding the quality of apple juice, HPLC-MS was used to detect changes in phenolic compounds, organic acids, and sugar content after processing and storage. A total of 26 phenolic compounds were identified and quantified in both of red-fleshed cultivar 'Baya Marisa' and white-fleshed cultivar 'Golden Delicious'. The results suggest that phenolic compound preservation of apple juice during storage depends on the fruit cultivar, and refrigeration is the best storage condition for apple juice [contribution 2].

Quality deterioration in fresh fruit and vegetables occurs due to the vigorous respiratory metabolism postharvest. Advanced postharvest technologies, including chemical treatment, physical methods, and biotechnology, have been utilized to control product loss and maintain the commercial quality of fresh fruit and vegetables. For physical treatment, a cooling method was used to delay core browning in postharvest 'Yali' pear fruit. Slow cooling and rapid cooling treatments were performed to store the pears. The respiration intensity and ethylene production in slow-cooled fruits were higher than that of rapid-cooled fruit. The rapid cooling group exhibited better fruit quality than the slow cooling group. Furthermore, RNA-seq analysis reveal pear browning was associated with hormone signal transduction pathways, especially ethylene metabolism. The results suggest that rapid cooling inhibits core browning in postharvest 'Yali' pear by partly regulating ethylene biosynthesis and signal transduction [contribution 3].

Controlled atmosphere (CA) packaging has been widely used to prolong the shelf-life of fresh produce. A gas environment with O_2 (5%), CO_2 (5%), and N_2 (90%) was used to treat fresh-cut *Dictyophora rubrovolvata*. This CA treatment protected membrane integrity, prevented oxidation, and inhibited the browning of fresh-cut *Dictyophora rubrovolvata* during 8 days storage at 4 ± 1 °C. Moreover, the quality, taste, and volatile components of fresh-cut *Dictyophora rubrovolvata* were maintained by the CA treatment [contribution 4].

Regarding chemical treatment, ε-poly-L-lysine (ε-PL) combined with a chitosan coating effectively inhibits diseases development and weight loss in fresh *Tremella fuciformis* during storage. ε-PL is a homopolymer composed of lysine residues; it is non-toxic and harmless to humans, displaying efficient endotoxin removal capabilities. Chitosan is a high-molecular-weight cationic polysaccharide that is safe and non-toxic, soluble in various organic acids, and has certain antibacterial, antioxidant, and film-forming adsorption properties. This chemical treatment (ε-PL + chitosan) suppressed the respiration rate and activities of respiratory-metabolism-related enzymes and energy-metabolism-related enzymes. This work demonstrated that the application of ε-PL and chitosan coating has the potential to maintain the quality of postharvest, fresh *Tremella fuciformis* [contribution 5].

Another edible coating technology, comprising carboxymethyl cellulose (CMC) and gelatin (Gel) integrated with melatonin (MT), mitigated pericarp browning and chilling injury in longkong fruit. The results showed that the CMC-Gel-MT coating alleviated chilling injury by reducing the activities of cellular-degrading enzymes and improving antioxidant ability [contribution 6].

Furthermore, pullulan coating (a bio-polysaccharide produced by *Aureobasidium pullulans*) combined with potassium metabisulfite could reduce the decay rate of kiwifruit and maintain fruit quality during storage, as evidenced by higher levels of soluble solid content and firmness, increased accumulation of phenols and flavonoids, as well as higher activities and genes expression of antioxidant enzymes. Overall, this study suggested that the use of pullulan-based active coating combining with potassium metabisulfite treatment inhibited soft rot and maintained the quality of postharvest kiwifruit [contribution 7].

Another study revealed that exogenous triterpenoids, including oleanolic acid (OA) and ursolic acid (UA), effectively reduced water loss and improved the quality of blueberry fruit during storage [contribution 8].

Moreover, calcium chloride ($CaCl_2$) treatment maintained litchi fruit quality by preventing the activities of cell-wall-degrading enzymes, thereby stabilizing the cell wall's architecture [contribution 9].

Melatonin treatment prolonged postharvest quality and controlled disease development in strawberry fruit by enhancing the reactive oxygen species (ROS) scavenging system [contribution 10].

Dry walnut and hazelnut leaves used for composting could affect the yield and phenolic composition of *Lactuca sativa* L. This study suggested that dry walnut leaves and cut grass could be used for composting; however, the dry hazelnut leaves still contained some unknown allelochemicals after two years, which inhibited plant growth and, thus, yield. Therefore, dry hazelnut leaves should not be recommended for composting [contribution 11].

More interestingly, a combined chemical–physical method was explored. Near-freezing temperature (NFT) with jujube polysaccharide treatment could maintain the fruit quality of *Prunus armeniaca* L. Further proteomic analysis identified 1054 differentially expressed proteins (DEPs), which were enriched in carbohydrate and energy metabolism, cell wall degradation, stress response, and antioxidant system [contribution 12].

Regarding biocontrol treatment technologies, fruit-isolated *Metschnikowia* yeasts were effective against three major citrus pathogens, including *Penicillium digitatum*, *Penicillium italicum*, and *Geotrichum citri-aurantii*; further investigations revealed that the *Metschnikowia* yeasts' biocontrol efficacy was mainly attributable to their ability to competitively consume iron ions in a shared environment. Moreover, the size of their pigment halo was directly related to their antagonistic ability [contribution 13].

Additionally, the antifungal effects of the volatile organic compounds (VOCs) produced by *Scheffersomyces spartinae* W9 (a marine biocontrol yeast) were evaluated against *Botrytis cinerea* on strawberry fruit. The findings displayed a biocontrol strategy to prevent disease development in strawberry fruit [contribution 14].

Regarding molecular studies, a cold-induced WRKY transcription factor, MaWRKY70, was associated with banana chilling injury development by directly binding to the promoters of four lipoxygenase (LOX) genes and activating their transcriptions. This study suggests MaWRKY70 plays a positive role in chilling injury development in banana fruit [contribution 15].

This editorial has summarized all research included in this Special Issue. The authors investigate the biologic and chemical changes underlying ripening and senescence, as well as the postharvest technologies of fresh, postharvest products for quality maintenance during storge time. Advanced physical approaches, including rapid cooling and CA packaging; chemical methods, such as ε-PL and chitosan coating, CMC-Gel-MT coating, pullulan coating, triterpenoids, and $CaCl_2$; and biocontrol treatment technologies are recorded in this Special Issue. Thus, this Special Issue expands our knowledge of nutrition changes and molecular mechanisms and contributes advanced technologies for maintaining the shelf life and commercial value of fresh fruit and vegetables.

Funding: This research received no external funding.

Conflicts of Interest: The author declares no conflicts of interest.

List of Contributions:

1. Li, Z.; Zhou, B.; Zheng, T.; Zhao, C.; Shen, X.; Wang, X.; Qiu, M.; Fan, J. Integrating Metabolomics and Proteomics Technologies Provides Insights into the Flavor Precursor Changes at Different Maturity Stages of Arabica Coffee Cherries. *Foods* **2023**, *12*, 1432.
2. Juhart, J.; Medic, A.; Jakopic, J.; Veberic, R.; Hudina, M.; Stampar, F. Use of HPLC-MS to Determine the Loss of Metabolites in Apple Juices under Different Storage Conditions. *Foods* **2023**, *12*, 2822.
3. Zhang, H.; Han, Y.; Liang, L.; Deng, B. Rapid Cooling Delays the Occurring of Core Browning in Postharvest 'Yali' Pear at Advanced Maturity by Inhibiting Ethylene Metabolism. *Foods* **2024**, *13*, 1072.
4. Xia, Z.; Wang, R.; Ma, C.; Li, J.; Lei, J.; Ji, N.; Pan, X.; Chen, T. Effect of Controlled Atmosphere Packaging on the Physiology and Quality of Fresh-Cut *Dictyophora rubrovolvata*. *Foods* **2023**, *12*, 1665.
5. Sun, J.; Wei, Y.; Li, L.; Tang, B.; Yang, Y.; Xiao, Z.; Chen, J.; Lai P. Investigating the Respiratory and Energy Metabolism Mechanisms behind ε-Poly-L-lysine Chitosan Coating's Improved Preservation Effectiveness on *Tremella fuciformis*. *Foods* **2024**, *13*, 707.
6. Venkatachalam, K.; Charoenphun, N.; Lekjing, S.; Noonim, P. Investigation of Melatonin Incorporated CMC-Gelatin Based Edible Coating on the Alleviation of Chilling Injury Induced Pericarp Browning in Longkong. *Foods* **2023**, *13*, 72.
7. Tian, Y.; Li, L.; Wang, R.; Ji, N.; Ma, C.; Lei, J.; Guan, W.; Zhang, X. Pullulan-Based Active Coating Incorporating Potassium Metabisulfite Maintains Postharvest Quality and Induces Disease Resistance to Soft Rot in Kiwifruit. *Foods* **2023**, *12*, 3197.
8. Kong, Q.; Liu, R.; Wu, W.; Fang, X.; Chen, H.; Han, Y.; Chen, J. Cuticular Wax Triterpenes Maintain Storage Quality of Blueberries by Reducing Water Loss. *Foods* **2023**, *12*, 2643.
9. Guo, X.; Li, Q.; Luo, T.; Han, D.; Zhu, D.; Wu, Z. Postharvest Calcium Chloride Treatment Strengthens Cell Wall Structure to Maintain Litchi Fruit Quality. *Foods* **2023**, *12*, 2478.
10. Promyou, S.; Raruang, Y.; Chen, Z. Melatonin Treatment of Strawberry Fruit during Storage Extends Its Post-Harvest Quality and Reduces Infection Caused by Botrytis cinerea. *Foods* **2023**, *12*, 1445.
11. Medic, A.; Solar, A.; Hudina, M.; Veberic, R.; Zamljen, T. Effect of Different Walnut and Hazelnut Leaf Compost Treatments on Yield and Phenolic Composition of *Lactuca sativa* L. *Foods* **2023**, *12*, 2738.
12. Wang, Z.; Wang, W.; Li, W.; Yang, R.; Li, Y.; Zhang, L.; Zhang, M.; Li, X. Effects of Near-Freezing Temperature Combined with Jujube Polysaccharides Treatment on Proteomic Analysis of 'Diaogan' Apricot (*Prunus armeniaca* L.). *Foods* **2023**, *12*, 4504.

13. Wang, S.; Tan, Z.; Wang, C.; Liu, W.; Hang, F.; He, X.; Ye, D.; Li, L.; Sun, J. Iron Competition as an Important Mechanism of Pulcherrimin-Producing *Metschnikowia* sp. Strains for Controlling Postharvest Fungal Decays on Citrus Fruit. *Foods* **2023**, *12*, 4249.
14. Zou, X.; Wei, Y.; Zhu, J.; Sun, J.; Shao, X. Volatile Organic Compounds of Scheffersomyces spartinae W9 Have Antifungal Effect against *Botrytis cinerea* on Strawberry Fruit. *Foods* **2023**, *12*, 3619.
15. Lin, H.; Bai, L.; Wei, W.; Su, W.; Wu, Y.; Wu, R.; Wang, H.; Chen, J.; Lin, H.; Fan, Z. The Role of MaWRKY70 in Regulating Lipoxygenase Gene Transcription during Chilling Injury Development in Banana Fruit. *Foods* **2024**, *13*, 854.

References

1. Guo, Y.; Ren, G.; Zhang, K.; Li, Z.; Miao, Y.; Guo, H. Leaf senescence: Progression, regulation, and application. *Mol. Hortic.* **2021**, *1*, 5. [CrossRef] [PubMed]
2. Seymour, G.B.; Østergaard, L.; Chapman, N.H.; Knapp, S.; Martin, C. Fruit development and ripening. *Annu. Rev. Plant Biol.* **2013**, *64*, 219–241. [CrossRef] [PubMed]
3. Figueroa, C.R.; Jiang, C.Z.; Torres, C.A.; Fortes, A.M.; Alkan, N. Editorial: Regulation of Fruit Ripening and Senescence. *Front. Plant Sci.* **2021**, *12*, 711458. [CrossRef] [PubMed]
4. Gambhir, P.; Raghuvanshi, U.; Kumar, R.; Sharma, A.K. Transcriptional regulation of tomato fruit ripening. *Physiol. Mol. Biol. Plants* **2024**, *30*, 289–303. [CrossRef] [PubMed]
5. Liu, T.; Albornoz, K.; Deltsidis, A.; Beckles, D.M. Editorial: Postharvest ripening, senescence, and technology. *Front. Genet.* **2022**, *13*, 920584. [CrossRef] [PubMed]
6. Wu, J.; Tang, R.; Fan, K. Recent advances in postharvest technologies for reducing chilling injury symptoms of fruits and vegetables: A review. *Food Chem. X* **2023**, *21*, 101080. [CrossRef] [PubMed]
7. Habibi, F.; Boakye, D.A.; Chang, Y.; Casorzo, G.; Hallman, L.M.; Madison, M.; Clavijo-Herrera, J.; Sarkhosh, A.; Liu, T. Molecular mechanisms underlying postharvest physiology and metabolism of fruit and vegetables through multi-omics technologies. *Sci. Hortic.* **2024**, *324*, 112562. [CrossRef]
8. Feng, Y.; Xu, Y.; Li, W.; Chen, S.; Su, Z.; Xi, L.; Li, G. Improved enrichment and analysis of heterocyclic aromatic amines in thermally processed foods by magnetic solid phase extraction combined with HPLC-MS/MS. *Food Control* **2022**, *137*, 108929. [CrossRef]

foods

MDPI

Article

Rapid Cooling Delays the Occurring of Core Browning in Postharvest 'Yali' Pear at Advanced Maturity by Inhibiting Ethylene Metabolism

Hongyan Zhang [1], Yunyun Han [2], Liya Liang [1] and Bing Deng [3,*]

[1] College of Food Science and Biological Engineering, Tianjin Agricultural University, Tianjin 300392, China; 18789420684@163.com (H.Z.); liangliya@126.com (L.L.)
[2] College of Horticulture, Shanxi Agricultural University, Taigu 030801, China; hanyunyun_x@163.com
[3] College of Food Science and Engineering, Shanxi Agricultural University, Taigu 030801, China
* Correspondence: dengbing_y@163.com

Abstract: During the storage and transportation processes, the occurrence of core browning in 'Yali' pear fruit due to adversity injury can be easily mitigated by implementing different cooling methods, especially in advanced maturity fruits. In this study, 'Yali' pears at an advanced maturity stage were subjected to slow cooling and rapid cooling treatment. The quality-related physiological percentage and severity, and the rate of good fruits were determined, and RNA-seq was used to explore the effects of different cooling methods on pathways related to core browning in advanced-maturity pears at the transcriptional level. The results indicated that, compared with slow cooling treatment, rapid cooling significantly inhibited core browning in advanced-maturity 'Yali' pears. Measurements of quality-related physiological indexes suggested that rapid cooling treatment led to higher SSC content, firmness, L* value, and b* value, indicating better brightness, coloration, and higher soluble solid content, which are desirable for commercial sale. Rapid cooling effectively suppressed the physiological metabolism of 'Yali' pears, delaying fruit senescence compared with slow-cooling treatment. Furthermore, the RNA-Seq sequencing results revealed that pathways related to browning are involved in hormone signal transduction pathways, which are associated with resistance and aging processes of pear fruit. In summary, rapid cooling treatment delayed the core browning of advanced maturity of 'Yali' pears, indicating that the core browning of 'Yali' pears is related to the cooling method, and the mechanism of rapid cooling in reducing the core browning of advanced maturity of 'Yali' pears was by delaying the aging process of the fruit. This provides a new perspective for alleviating the core browning of advanced-maturity 'Yali' pears during storage and transportation, and provides a theoretical reference for studying the mechanism of core browning of 'Yali' pears.

Keywords: 'Yali' pear; core browning; rapid cooling; respiration intensity; ethylene; senescence

1. Introduction

The 'Yali' pear (*Pyrus bretschneideri* Rehd cv. Yali) is one cultivar of white pear, which has long been a staple in Hebei Province, China. As the primary exported pear fruit, it is well-liked by customers [1,2]. When stored and transported properly, the 'Yali' pear can be kept until February or March of the next year after being harvested in September [3].

Core browning of the pear, which has been documented in cases of 'Yali pears', 'Nanguo pear', 'Whangkeumbae pear', 'Bartlett', 'Korla pears', and other pear kinds, is a common physiological disorder [4–8]. According to studies, the core browning of the 'Yali' pear occurs in a pattern that starts in the core of the fruit and then spreads to the flesh [3]. Insights into the mechanism of fruit browning have been studied extensively. The Vc protection concept, the regionalized distribution of phenols and phenol oxidase, and the oxygen radical theory are the three primary hypotheses [1,9]. The most extensively researched theory is the regionalized distribution theory, which argues that the core browning of the

'Yali' pear during storage is attributed to damage to the membrane function, interfering with the compartmentalization of metabolism [1]. Moreover, there are some reports arguing that the core browning of the 'Yali' pear was related to the activity of enzymes. For example, hypobaric treatment has the potential to manage core browning of the 'Yali' pear by regulating enzymes such as polyphenol oxidase (PPO) activity, succinate dehydrogenase (SDH), and glucose-6-phosphate dehydrogenase (G-6-PDH) [10]. The results of Liu et al. also indicated that the suppressive impact of fumigation with 20 μL of L^{-1} NO on core browning was linked to its ability to reduce PPO activity [11]. Atmosphere packaging with a 10 μm thickness reduced the core browning of 'Yali' pears and delayed the peak appearance of PPO activity [1]. The core browning under various cooling methods was found to be linked to the actions of superoxide dismutase (SOD), catalase (CAT), peroxidase (POD), glutathione reductase (GR), and lipoxygenase (LOX). These results suggest that PPO, SOD, SDH, CAT, POD, et al., might be involved in the core browning of 'Yali' pears.

Pear, as a typical respiratory climacteric fruit, undergo an after ripening process to achieve the optimal edible quality [12]. A variety of fruits can release ethylene, a gaseous hormone that is crucial for the ripening and senescence of respiratory climacteric fruits [13]. As ethylene inhibitors, 1-MCP and aminoethoxyvinylglycine (AVG) dramatically reduced browning of pear fruit after application. This suggests that by delaying the senescence of pear fruit, 1-MCP and AVG can prevent the development of browning in pear fruit [14].

The time of harvest, fruit quality, storage temperature, CO_2 concentration, storage method, and exogenous preservative treatment are numerous variables that contribute to core browning in 'Yali' pears [15], and make it more susceptible to temperature changes; 'Yali' pears can be stored at 2–3 °C for 12 weeks and 0.5 °C (0.5 °C) for 30 weeks [3]. Our previous research demonstrated that the core browning of 'Yali' pears at different maturity levels can be alleviated by various cooling methods [3,16]. Therefore, we consider various cooling techniques for managing 'Yali' pears at advanced ripeness, assess the most effective cooling method for 'Yali' pears at advanced maturity, pinpoint physiological indicators related to the quality of 'Yali' pears and the percentage of good fruit, and utilize RNA-seq technology to investigate the impact of the cooling method on ethylene-related pathways of 'Yali' pears at the transcriptional level. This is to elucidate the mechanisms through which different cooling methods regulate the core browning of 'Yali' fruits at advanced maturity.

2. Materials and Methods

2.1. Fruit and Treatments

2.1.1. Fruit

The 'Yali' pears in this experiment were harvested from Xinji City, Hebei Province, China. After careful selection, fruits with similar size, free from mechanical injuries, diseases, and pests were transported in cartons to the cold store of Tianjin Agricultural University. A total of 20 boxes, each containing 70 fruits were used, and there were two sets of 10 boxes for each cooling treatment.

2.1.2. Cooling Treatment and Sampling

'Yali' pears were divided into two treatments: rapid cooling and slow cooling. In the rapid cooling treatment, the pears were pre-cooled at 0 °C for 24 h and then directly placed into cold store which maintained 0 ± 1 °C; for the slow cooling treatment, the pears were pre-cooled at 12 °C for 24 h and then placed into a 12 °C cold store, where the temperature was gradually reduced to 0 ± 1 °C in 30 days, by dropping 2 °C over a period of 5 days. Physiological indexes were measured every 30 days for the samples. The core parts of the pear were snapped frozen in liquid nitrogen and stored in a −80 °C refrigerator for subsequent experiments.

2.2. Core Browning Index of 'Yali' Pears

The core browning index of 'Yali' pears was determined by the method of Yan Shijie et al. [3]. In each storage time, 30 pears were randomly chosen from each treatment

(10 boxes), and taken out of the cold store. These were then divided into 3 repetitions, with 10 pears in each repetition. The core browning of 'Yali' pears was observed and recorded at intervals of 0, 30, 60, 90, 120, 150, 180, and 210 days of storage. All the fruits were cut horizontally, and the degree of core browning was assessed based on the level of browning in the cross-section and the extent of browning in the core or flesh tissue. No browning was rated as 0; slight browning with brown spots in the core was rated as 1; slight to 20% browning was rated as 2; 21–50% browning was rated as 3; and greater than 50% browning was rated as 4. Subsequently, the core browning index was calculated according to the following formula:

$$\text{Core browning index} = \frac{\sum(\text{Browning level} \times \text{The number of fruits of this level})}{\text{The highest browning level} \times \text{total number of fruits}}$$

2.3. *Physiological Indexes Assessments*

2.3.1. Respiratory Intensity of Fruit

The measurement of fruit respiration intensity followed the method described by Li Ling et al. [16] with some modification. The respiratory intensity and ethylene release of 'Yali' pears were measured using fixed fruit and measured in cold store. For the selection of fixed fruits, 12 fruits were randomly selected, with 4 fruits in each group, and a total of 3 groups before storage. Next, the grouped pears were placed in a modified Lekou box, sealed for 1 h, and the gas inside the box was extracted using a 1 mL medical syringe. A total of 4 needles were extracted from each box, with 3 for repetition and 1 left as a backup. The respiration intensity of the 'Yali' pears was measured using a GXH-3051 infrared CO_2 analyzer (Beijing Junfang Physics & Chemical Science & Technology Research Institute, Beijing, China), and calculated using the calculation formula as follows (the unit was CO_2 mg kg^{-1} h^{-1}):

$$\text{Fruit respiration intensity} = \frac{F \times 60 \times C}{22.4} \times \frac{44}{W} \times 10^{-6} \times \frac{273}{273 + T}$$

in which F is the flow rate of gas, the unit was expressed in mL min^{-1};
C is the CO_2 concentration, the unit was expressed in μL L^{-1};
W is the weight of the measured fruit, the unit was expressed in kg; and
T is the temperature at the time of measurement, the unit was expressed in °C.

2.3.2. Respiratory Intensity of Seed

For each treatment, a total of 30 relatively full seeds were selected and divided into 3 groups. During the storage period, the seeds were placed in 10 mL glass vials with rubber caps, which were then sealed for 1 h, and 0.5 mL of gas was extracted from each vial. The calculation of respiratory intensity was the same as Section 2.3.1.

2.3.3. CO_2 Content of Core

A total of 10 'Yali' pears were selected for the experiment. By using a syringe, the core chamber of 'Yali' pear was pierced from the center part of the calyx, then 0.5 mL of gas was extracted for analysis, following the same method as described in Section 2.3.1, and the unit was μL L^{-1}.

2.3.4. Ethylene Production

The ethylene gas was obtained in the same way as the respiratory intensity and was measured by GC-14 gas chromatograph (Shimadzu, Kyoto, Japan) and repeated three times. The setup parameters and calculation formula were referred from Li Ling et al. [16] with some modification. The experimental parameters were set as follows: the chromatographic column was a GDX-502 stainless steel packed column; the detector used was a hydrogen flame ionization detector; and the carrier gas was N_2. The inlet temperature, column temperature box, and detector temperature were all set at 60 °C, and the unit was expressed in μL kg^{-1}·h^{-1}.

$$\text{Ethylene Production} = \frac{C \times V}{M \times T \times 1000}$$

in which C is the ethylene content in the sample gas, with the unit as μL L^{-1};
V is the volume of the enclosed space, with the unit as mL;
M is the mass of fruits, with the unit as kg; and
T is the smothering time, with the unit as h.

2.3.5. Soluble Solids Content and Titratable Acids

Refer to the method of Li Jian et al. [7] with some modification. At each storage time for testing, 15 fruits were randomly selected from each treatment (10 boxes), taken out of the cold store, and divided into 3 groups. Fruits were peeled and cut to extract juice from the pulp part, and the soluble solids content (SSC) of the juice was measured by PAL-10 (Atago, Tokyo, Japan). The titratable acid (TA) was measured by GMK-835N acidimeter (Beijing Shangdetong Technology Co., Ltd., Beijing, China).

2.3.6. Firmness

The firmness of 'Yali' pears were determined by referring to the method of Li Jian et al. [7]. At each storage time for testing, 15 fruits were randomly selected from each treatment (10 boxes), taken out of the cold store, and divided into 3 groups. The firmness of 'Yali' pear fruits were determined using a TA-XT plus physical property analyzer (Stable Micro System, Godalming, UK). The unit was expressed in kg cm^{-2}.

2.3.7. L*, a* and b* Value

The color of 'Yali' pears was determined with reference to the method of Li Jian et al. [7]. For the determination of L*, a* and b* values, 15 fixed 'Yali' pears were used and tested in cold store, and divided into 3 groups. The fruits were marked in the middle with a circle of approximately 2.0 cm in diameter. Measurements were carried out using a SC-10 colorimeter (Kunshan Henggang Electronic Technology Co., Suzhou, China) to record the values of L*, a* and b* of the 'Yali' pear fruit during storage.

2.3.8. Good Fruit Rate

After 210 days of storage, the remaining 'Yali' pears were counted to determine the rate of good fruit.

$$\text{Good fruit rate} = \frac{\text{Number of good fruits per treatment}}{\text{total number of statistics per treatment}}$$

2.4. RNA-Seq

The core of 'Yali' pears was sampled at 0 days and 60 days after slow cooling and rapid cooling treatment. Total RNA was extracted using the Trizol reagent kit following the manufacturer's protocol. The quality of RNA extraction was detected by 1% agarose gel electrophoresis, and the concentration and purity of RNA were detected by NanoDrop 2000 spectrophotometer (Thermo, Waltham, MA, USA). Subsequently, mRNA was enriched with Oligo (dT) beads. The enriched mRNA was then fragmented into short fragments using fragmentation buffer and reverse transcribed into cDNA with random primers. Second strand cDNA synthesis was carried out using DNA polymerase I, RNase H, dNTPs, and buffer. The cDNA fragments were purified with the QiaQuick PCR extraction kit, end-repaired, poly(A)-tailed, and ligated to Illumina sequencing adapters. The ligation products were size-selected via agarose gel electrophoresis, PCR amplified, and sequenced using Illumina HiSeq2500 (Illumina, San Diego, CA, USA) by Beijing Novogene (Beijing, China).

2.5. Determination of ACO Activity

The determination of ACC oxidase (ACO) activity was referred from Fonseca et al. [17] with some modification. The sample from the core of a 'Yali' pear was taken, and enzyme extract and enzyme activity determination and computing method were carried out as in references. After the experiment, 1 mL of gas was extracted from the headspace, and the ethylene release was measured by Shimadsu GC-14. The ACO activity was expressed as the production rate of ethylene per hour, and the ACO activity was expressed as nmol $h^{-1}g^{-1}$.

2.6. Statistical Analysis

Each trial was executed with three independent biological duplicates. Data reorganization and standard errors were computed using Excel 2019. All statistical assessments were carried out utilizing SPSS 22 (IBM Inc., Armonk, NY, USA). Differences between the groups on a single storage day were appraised through *t*-test. Mean separations were determined using the least significant difference (LSD) test. A *p*-value of 0.05 was established as the threshold for significance. Figures were generated using GraphPad Prism 8.

3. Results

3.1. Development Process of Browning

Figure 1A shows the core of 'Yali' pears completely in a non-browning state; Figure 1B illustrates the initial browning in the locule; Figure 1C shows the core of 'Yali' pears completely browned, and Figure 1D shows the browning extended from the core to the pulp. The progression from Figure 1A–D represented the development of fruit browning. It was evident that the browning of 'Yali' pears originally started from the core chamber, and gradually spread to the pulp.

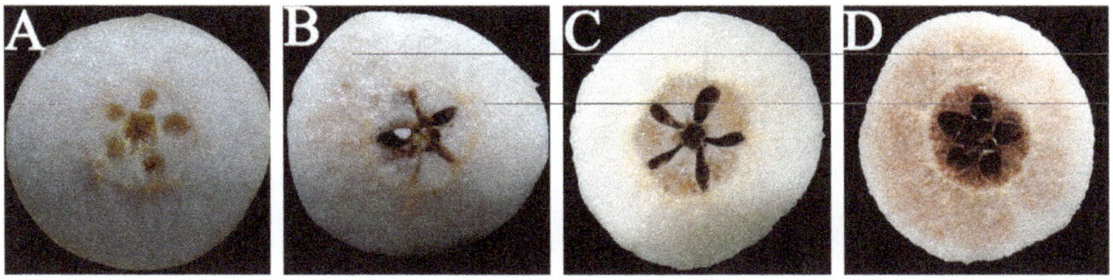

Figure 1. Development process of core browning in 'Yali' pears. (**A**): the core of 'Yali' pears completely in a non-browning state; (**B**): the initial browning in the locule; (**C**): the core of 'Yali' pears completely browned, (**D**): the browning extended from the core to the pulp.

3.2. Effects of Different Cooling Treatments on Browning Index of 'Yali' Pears at Advanced Maturity

As shown in Figure 2A, the browning index of 'Yali' pears increases gradually as the storage period extended. The occurrence time of browning in 'Yali' pears differed between the two cooling treatments. With the slow cooling treatment, slight browning was observed at 30 days, while it occurred at 60 days with the rapid cooling treatment. During the storage period of 60 to 150 days, there was a sharp increase in the browning index. Notably, the browning was more pronounced in the slow cooling treatment. Towards the end of storage, the browning index of the slow cooling treatment exceeded that of the rapid cooling treatment, with the rapid cooling treatment exhibiting a browning index of 0.5 and the slow cooling treatment reaching as high as 0.75. Figure 2B depicts the browning state of 'Yali' pears after 210 days of storage.

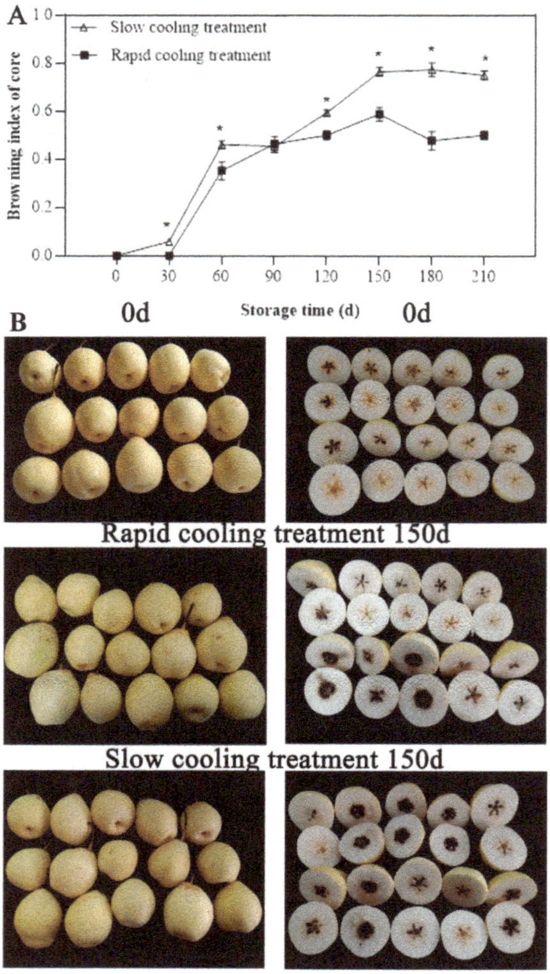

Figure 2. Core browning index (**A**) and the browning phenotype (**B**) of 'Yali' pears. Each value in (**A**) is the mean ± SE, n = 3. Asterisks (*) indicate significant differences between the two fruit treatments ($p < 0.05$).

3.3. *Effects of Different Cooling Treatments on Physiological Indexes of 'Yali' Pears at Advanced Maturity*

As shown in Figure 3A, the respiratory intensity of freshly harvested 'Yali' pears was 26.48 mg CO_2 $kg^{-1}h^{-1}$. With the extension of storage duration, the respiration intensity initially decreased and then exhibited an upward trajectory, reaching its peak at 60 days. In general, the respiration intensity of the rapid cooling treatment surpassed that of the slow cooling treatment at 30, 60, 90, 120, and 180 days of storage.

Figure 3B illustrates the changes in the respiratory intensity of seeds. Initially, the respiratory intensity of seeds was high when the fruit was freshly harvested. With the extension of the storage, the respiratory intensity first decreased and then peaked, and subsequently decreased again. Towards the end of the storage, there was a tendency for an increase. The respiratory intensity of slow cooling treatment was higher than that of rapid cooling treatment at 30 days and 120–180 days.

The CO_2 accumulation in the core showed a similar pattern as the respiration intensity in fruits. As demonstrated in Figure 3C, the CO_2 accumulation in the core first decreases, then increases to a peak, followed by another decrease. The CO_2 accumulation in the core peaked at 60 days, which aligned with the results of respiration intensity in seeds as well as in fruits. The CO_2 accumulation was slightly higher in the slow cooling treatment than in the rapid cooling treatment.

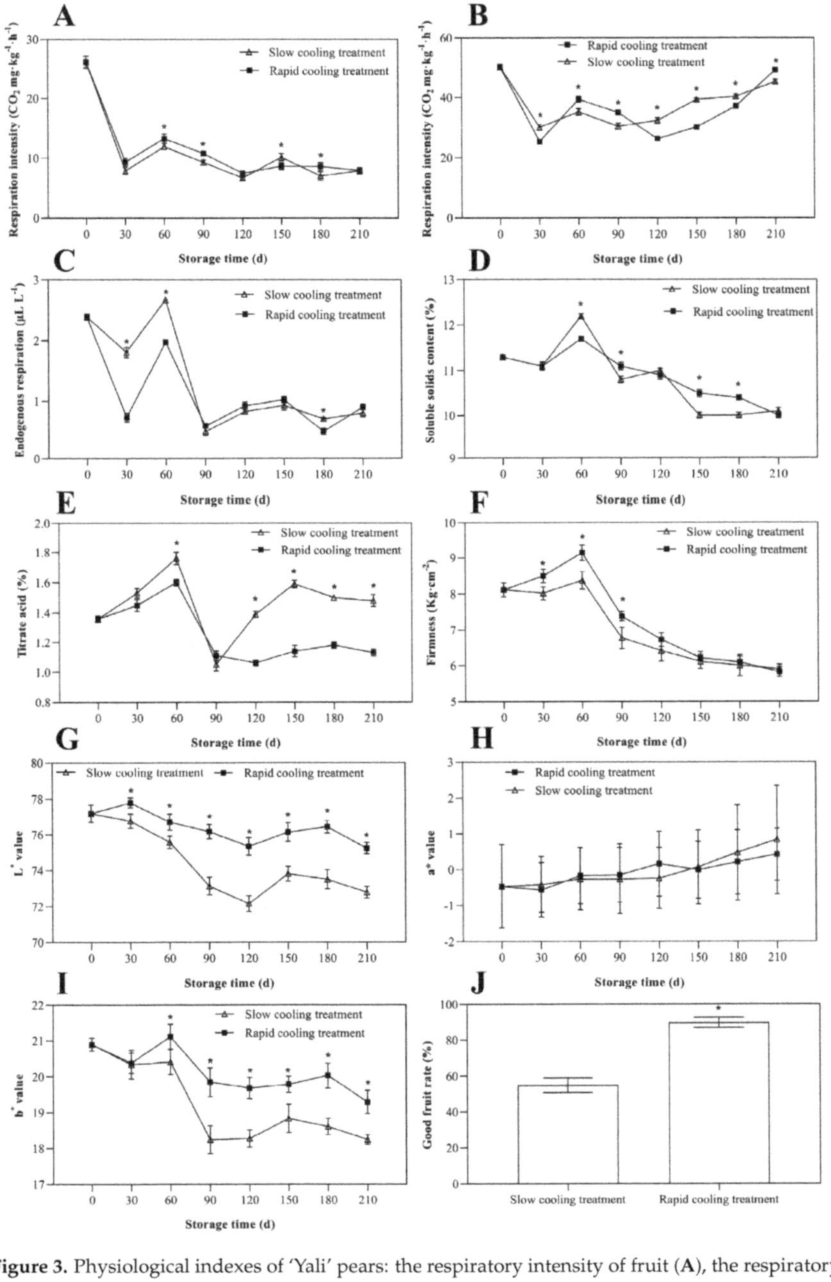

Figure 3. Physiological indexes of 'Yali' pears: the respiratory intensity of fruit (**A**), the respiratory intensity of seed (**B**), endogenous respiration (**C**), soluble solids content (**D**), titratable acids (**E**), firmness (**F**), L* value (**G**), a* value (**H**), b* value (**I**), and the good fruit rate (**J**). Each value is the mean ± SE, *n* = 3. Asterisks (*) indicate significant differences between the two treatments.

Figure 3D depicts that as storage time increases, the SSC first increases to the maximum value and then declines. Towards the end of storage (150~180 days), SSC decreased rapidly, with the SSC of the slow cooling treatment being lower than that of the rapid cooling treatment.

As shown in Figure 3E, the TA reaches the peak at 60 days and then decreases as the storage extends, and finally plateaus. Up to the middle of storage (60~90 days), the content of titratable acid decreased sharply, and in the late storage period (120–210 days), the TA of the slow cooling treatment was greater than that of the rapid cooling treatment.

The initial firmness of 'Yali' pears was 8.12 kg cm^{-2} as shown in Figure 3F, with the increasing storage time, the firmness initially peaks at 60 days, then decreases and finally plateaus. At the beginning of storage (30–120 days), rapid cooling significantly kept the firmness of 'Yali' pears, and towards the end of storage (120–210 days), there was no difference between the firmness of the slow cooling treatment and that of the rapid cooling treatment.

L* indicates the degree of brightness of 'Yali' pears, and the larger the L*, the brighter the color of the peel. Figure 3G reveals that the L* value of the 'Yali' pear after picking decreases, and the L* value of the slow cooling treatment is significantly lower than that of the rapid cooling treatment.

The a* value represents the redness (+a*) or greenness (−a*) of a color. The a*-values of both treatments showed an increase with the prolongation of storage time, and no notable discrepancy in the a-values was observed between the two treatments at each measurement time point (Figure 3H).

Figure 3I presents the changing of b* values. A b* value indicates the change between yellow (+b*) and blue (−b*), which decreases with extended time in the storage, and b* values of the slow cooling treatment were lower than that of the rapid cooling treatment.

As shown in Figure 3J, good fruit rate wis 89.76% in the rapid cooling treatment group, and 54.84% in the slow cooling treatment group. This indicated that the rapid cooling treatment was better for 'Yali' pears at advanced maturity.

3.4. GO and KEGG Enrich Analysis of DEGs

Figure 4 illustrates the GO and KEGG enrichment analysis of all differentially expressed genes (DEGs) from core samples of 'Yali' pear treated with various cooling methods for 60 days compared to 0 days. The subclasses of response to metal ion, response to endogenous stimulus, and cellular amino acid metabolic process were connected to the pear's stress response to the environment; other subcategories included amino acid metabolism, and other life activities, according to GO enrichment analyses (Figure 4A,B). The top 20 primary important KEGG pathways were presented, with amino acid biosynthesis, plant hormone signal transduction, and carbon metabolism ranking as the top three, which were found to be linked to resistance and senescence in the 'Yali' pear (Figure 4C,D).

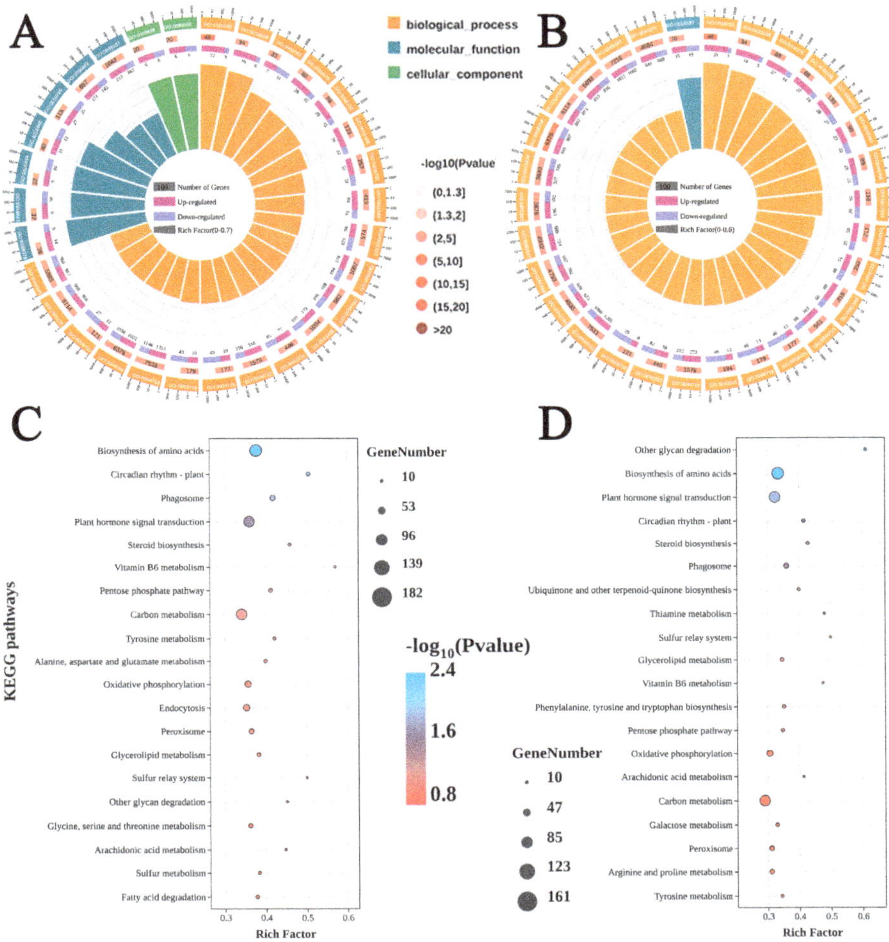

Figure 4. The GO enrichment analysis of DEGs from 0 days and 60 days after slow cooling treatment (**A**); 0 days and 60 days after rapid cooling treatment (**B**); the KEGG enrichment analysis of DEGs from 0 days and 60 days after slow cooling treatment (**C**); and 0 days and 60 days after rapid cooling treatment (**D**). *p*-value: statistically significant level of enrichment analysis; the color represents the level of *p*-value, where the larger the $-\log_{10}$ (*p* value), the smaller the *p*-value, indicating that the pathway is more significant.

3.5. Ethylene Metabolism in 'Yali' Pears during Slow Cooling and Rapid Cooling Treatment

Figure 5 shows the biosynthesis and signal transduction of internal ETH in 'Yali' pears after slow cooling and rapid cooling treatment at 60 days. The expression levels of genes related to ethylene biosynthesis (*SAMS*, *ACS*, *ACO*) as well as genes related to ethylene signal transduction pathway (*ETR*, *MPK6*, *EIN2*, *EIN3*, *EBF1/2* and *ERF1/2*) were higher in 'Yali' pear treated with the slow cooling treatment compared to those treated with the rapid cooling treatment. Additionally, the fruit and core exhibited extremely low ethylene release in the rapid cooling treatment—nearly zero. As storage time was extended, the ethylene release of fruit, core, and seed gradually rose until it reached its peak at 60 days, and then declined before showing a tendency to increase again. In general, the ethylene release of fruit was higher in 'Yali' pear treated with the slow cooling treatment compared to those treated with the rapid cooling treatment, which was consist with the expression of related genes.

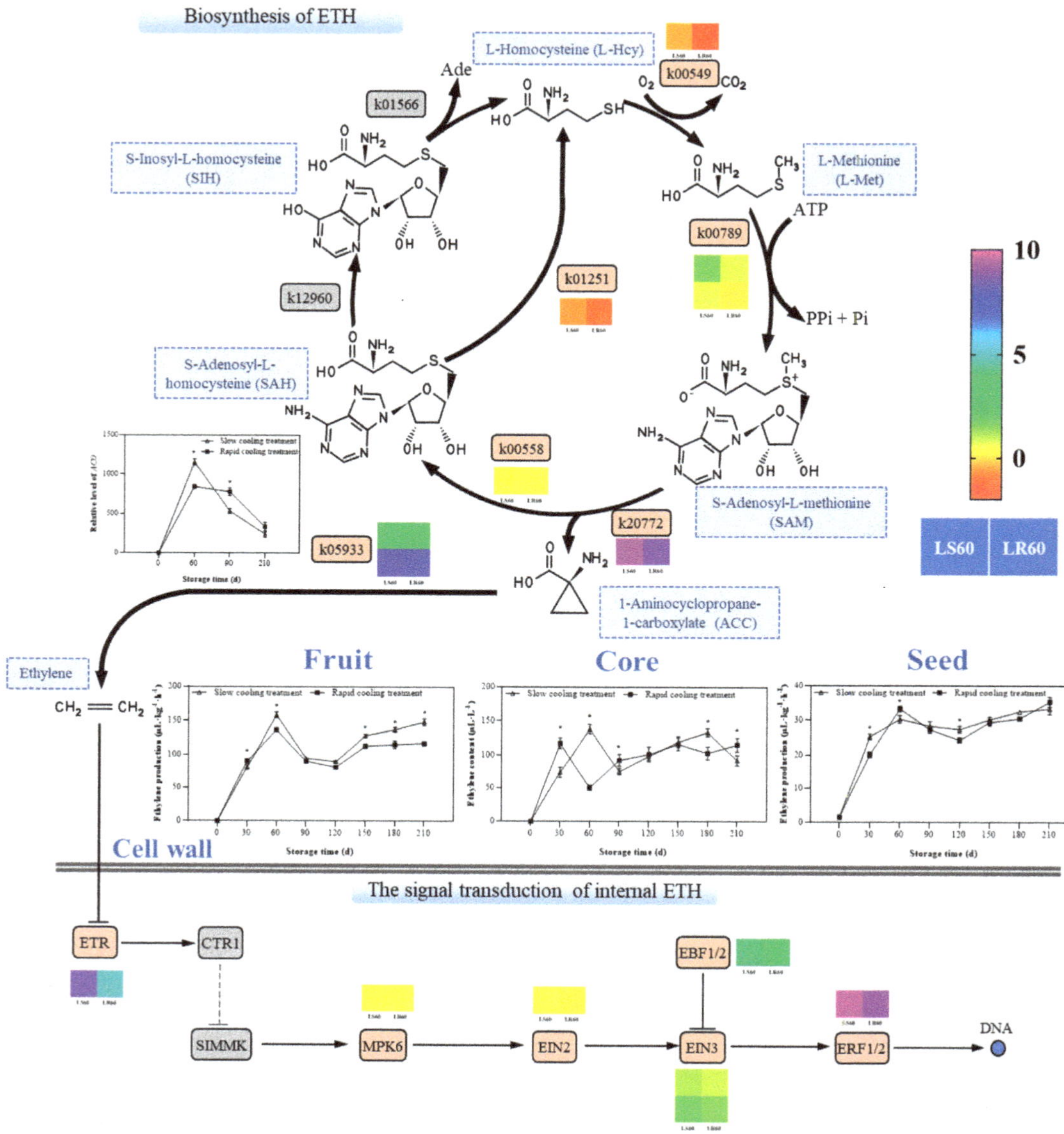

Figure 5. Biosynthesis and signal transduction of internal ETH in 'Yali' pears after the slow cooling and rapid cooling treatment at 60 days. Note: The heat map next to the gene in the figure represents the FPKM of the gene in the slow cooling and rapid cooling treatment at 60 days. K00789: S-adenosylmethionine synthetase, SAMS; K00558: DNA (cytosine-5)-methyltransferase 1, DNMT1; K20772: 1-aminocyclopropane-1-carboxylate synthase, ACS; K12960: 5-methylthioadenosine/S-adenosylhomocysteine deaminase, mtaD; K01251: aminocyclopropanecarboxylate oxidase, AHCY; K01566: S-inosyl-L-homocysteine hydrolase, SIHH; K00549: 5-methyltetrahydropteroyltriglutamate—homocysteine methyltransferase, metE; and K05933: aminocyclopropanecarboxylate oxidase, ACO. Each value is the mean ± SE, n = 3. Asterisks (*) indicate significant differences between the two treatments.

4. Discussion

Core browning reduces the commodity value of the fruit and shortens its market supply time and sales [7]. The browning process in 'Yali' pear fruits progress through multiple stages, from the core to the flesh (Figure 1). During the observations, we classified the fruits based on the degree of browning, which ultimately determined the browning index [3]. According to our research, the core browning signs in advanced-maturity 'Yali' pears appeared after 30 days when treated with the slow cooling method, and at 60 days when treated with the rapid cooling method. As storage was prolonged, the core browning index of 'Yali' pears treated with the rapid cooling were significantly lower than that of the slow cooling treatment group (Figure 2), indicating that the rapid cooling treatment can inhibit core browning incidence and achieve a higher rate of good fruits for advanced-maturity pear fruits.

Respiratory intensity and ethylene production indicate the plant's vitality and are closely linked to the quality and senescence process of pears during storage [18]. The cooling methods have an effect on the respiratory intensity of 'Yali' pear fruits. In the early storage period (0–90 days), the respiration intensity of the rapid cooling treatment group is higher than that of the slow cooling treatment group, indicating that rapid cooling promoted physiological metabolism of the 'Yali' pear. In the late storage period (150–210 days), there is almost no difference in respiration intensity between the slow cooling and rapid cooling groups (Figure 3A). The regularity in respiration intensity of seed is consistent with the results of fruit, indicating that the internal physiological metabolism of seeds affects the physiological metabolism of fruits. At the late stage, as fruits tend to senesce and seeds begin to germinate, there is an increase in respiration intensity (Figure 3B). There are numerous variables that contribute to core browning in the 'Yali' pear, with CO_2 concentration being one of them [2,3]. We examined CO_2 accumulation in the core using the pinprick method to investigate the correlation between CO_2 accumulation in the core and core browning in 'Yali' pears. The peak of CO_2 buildup occurred at 60 days, which corresponded with the respiration intensity of both the fruit and the seeds. At the same time, 'Yali' pears exhibited varying degrees of core browning at 60 days due to the rapid accumulation of CO_2 during the early storage period, resulting in damage to the core tissues. The core browning in the slow cooling treatment was higher than that in the rapid cooling treatment, likely due to slightly higher CO_2 accumulation in the slow cooling treatment (Figure 3C).

High-quality fruits have been increasingly popular in recent years, and TA and SSC are two significant determinants of fruit quality [7]. SSC and TA changes could be attributed to the conversion of sugars and acids during fruit ripening [19]. The type and content of soluble sugars in fruits not only determine the flavor and quality of the fruit, but they are also important energy substances, intermediates, and important substrates of many metabolic pathways, and they play a role in the regulation of plant growth, ripening, and senescence as signaling molecules in many processes; for example, the delay of fruit senescence was related to the maintaining of higher energy status in apple fruit [20]. A higher amount of SSC content can prolong the storage life of 'Yali' pears and prevent core browning and senescence; our results also showed a correlation between core browning and SSC content, as SSC declined rapidly towards the end of the storage period (120–210 days), coinciding with the severe browning of fruit (Figure 3D). At 60 days, TA reached its peak (Figure 3E). This indicated that increased SSC might partially prevent core browning and delay TA degradation. Due to extensive browning in advanced-maturity 'Yali' pears, the firmness of the fruits was at its lowest at the end of the storage period (120–210 days), particularly with the slow cooling treatment, which led to more pronounced core browning. Consequently, the fruit firmness decreased significantly compared to the rapid cooling treatment group (Figure 3F). The brightness of 'Yali' pears also decreased more in the slow cooling treatment group than in the rapid cooling treatment group (Figure 3G). This was due to the prolonged storage time causing severe core browning. Furthermore, the brown index was higher for the slow cooling treatment group compared to the rapid cooling treatment group as it extended from cores to flesh, resulting in reduced peel brightness.

In order to further explore how the rapid cooling treatment alleviated the browning of advanced-maturity 'Yali' pears, we investigated the mechanisms by RNA-Seq sequencing, and the results showed that the GO enrichment related to browning mainly focused on biosynthetic process and molecular function (Figure 4), and the DEGs in GO term focused on the stimulation aspect, single cell process, and metal-binding activity. Regarding KEGG pathway annotation, genes causing browning were found in amino acid metabolism, plant hormone signal transduction, and carbon metabolism pathways. According to a report, maturity and senescence of 'Nanguo' pears are directly related to the core browning of the fruit at room temperature [6]. Advanced-maturity 'Yali' pears underwent various cooling treatments, resulting in browning caused by alterations in metabolic and activity levels of browning-related substances during the storage of 'Yali' pears, and we focused on DEGs associated with fruit senescence.

It is generally recognized that ripening and senescence serve as the causes of browning in 'Yali' pears [15]. One factor that accelerates the senescence of fruit and causes browning is the ethylene released during the storage of fruit [21]. Having reported that the respiratory intensity and ethylene production serve as indicators of the plant's vigor and are intimately related to the quality and senescence process of pears during storage [18], any intervention that delays fruit senescence or prolongs the storage period of fruit can only postpone the peak of ethylene or decrease ethylene production. For instance, the low temperature controlled the ethylene release rate to a certain extent during an 80 day low-temperature storage, but the cold conditions did not change the general trend of increasing ethylene release in 'Meihong' fruits [22]. Furthermore, with the ripening of fruits at RT (25 ± 1 °C), ethylene release increased and peaked at 20 days; for fruits under the LR (2 ± 1 °C for 10 days then 25 ± 1 °C) environment, ethylene release remained almost unchanged during the first 10 days but peaked at 15 days [23]. 'Yali' pears are climacteric fruits with a peak ethylene release at around 60 days of storage, coinciding with their peak respiratory intensity (Figure 5). Throughout the late storage period (120–210 days), especially in the slow cooling treatment group, core browning of fruits became more severe as indicated by an increase in fruit browning index due to continuous ethylene release leading to nutrient consumption. The ethylene release of the slow cooling treatment group was higher than that of the rapid cooling, and at the same time, the browning level of slow cooling was higher than that of rapid cooling.

Ethylene, as a little compound, could regulate many metabolic and developmental processes in plants, including seed germination, seedling development, organ growth and senescence, fruit ripening, and stress response [18]. L-Met is the primary base substance for the synthesis of ethylene in plants. S-adenosylmethionine synthetase (SAM), 1-aminocyclopropane-1-carboxylic acid synthetase (ACS), and 1-aminocyclopropane-1-carboxylic acid oxidase (ACO) regulate this process in turn. Ethylene molecules are produced by the catalysis of three enzymes [13]. The *ACS* and *ACO* gene families are the two most significant genes involved in ethylene production out of these three genes. The alternating expression of the *ACO* and *ACS* genes regulate ethylene production during the developmental and ripening phases of climacteric fruit, completing the progression from development to ripening. The 'Yali' pear is a type of fruit known as a respiratory climacteric fruit. During the postharvest ripening process, there is an obvious accumulation of ethylene that motivates the organism to accelerate maturation. This is accompanied by several changes including an increased respiration rate in fruits, decreased flesh firmness, accumulation of soluble solids content, increase in flavor compounds content, etc.

In this experiment, the rapid cooling treatment dramatically reduced the ethylene release of 'Yali' pears compared to the slow cooling treatment group. Rapid cooling enables 'Yali' pears to adapt more quickly to a low temperature environment. Additionally, it reduces the metabolic rate of the fruit overall than slow cooling and releases less amounts of ethylene, which delays the core browning related to senescence. Thus, it has been established that rapid cooling prevented the development of core browning in 'Yali' pears at advanced maturity.

5. Conclusions

In our study, the application of rapid cooling resulted in a decrease in the browning index of 'Yali' pears. Additionally, the peaks in respiration for 'Yali' pear fruits, seeds, and core chambers treated with the two different cooling methods all occurred at 60 days during the storage. Overall, the fruits treated with slow cooling exhibited higher respiratory intensity and ethylene production compared to those treated with rapid cooling. Moreover, the rapid cooling group showed higher levels of soluble solids, firmness, L-value, b-value, and a higher rate of good fruits, as compared to the slow cooling treatment group. These findings suggest that for advanced-maturity 'Yali' pears, rapid cooling can reduce ethylene production and respiratory intensity, thereby delaying the senescence process and preserving fruit quality. Consequently, this approach could lead to a decrease in the core browning index of the fruit, offering potential benefits for commercialization.

Author Contributions: Conceptualization, Y.H.; methodology, B.D.; software, H.Z.; validation, L.L.; formal analysis, B.D.; investigation, Y.H.; resources, Y.H.; data curation, Y.H.; writing—original draft preparation, H.Z.; writing—review and editing, H.Z.; visualization, H.Z.; supervision, L.L.; project administration, B.D.; funding acquisition H.Z. and B.D. All authors have read and agreed to the published version of the manuscript.

Funding: This research was funded by the National Nature Science Foundation Project of China, Grant/Award Number: 32072278 and the Key project of Tianjin Natural Science Foundation, Grant/Award Number: 20JCZDJC00420.

Institutional Review Board Statement: Not applicable.

Informed Consent Statement: Not applicable.

Data Availability Statement: The original contributions presented in the study are included in the article, further inquiries can be directed to the corresponding author.

Conflicts of Interest: The authors declare no conflicts of interest.

References

1. Cheng, Y.; Liu, L.; Zhao, G.; Shen, C.; Yan, H.; Guan, J.; Yang, K. The effects of modified atmosphere packaging on core browning and the expression patterns of PPO and PAL genes in 'Yali' pears during cold storage. *LWT Food Sci. Technol.* **2015**, *60*, 1243–1248. [CrossRef]
2. Hao, Y.; Li, X.; Zhang, C.; Lei, Z. Online Inspection of Browning in Yali Pears Using Visible-Near Infrared Spectroscopy and Interpretable Spectrogram-Based CNN Modeling. *Biosensors* **2023**, *13*, 203. [CrossRef]
3. Yan, S.; Li, L.; He, L.; Liang, L.; Li, X. Maturity and cooling rate affects browning, polyphenol oxidase activity and gene expression of 'Yali' pears during storage. *Postharvest Biol. Technol.* **2013**, *85*, 39–44. [CrossRef]
4. Zhou, H.; Tian, M.; Huang, W.; Luo, S.; Hu, H.; Zhang, Y.; Zhang, L.; Li, P. Physiological and transcriptomic analysis of 'Whangkeumbae' pear core browning during low-temperature storage. *Gene Expr. Patterns* **2020**, *36*, 119113. [CrossRef] [PubMed]
5. Zhang, H.; Zha, Z.; Kulasiri, D.; Wu, J. Detection of Early Core Browning in Pears Based on Statistical Features in Vibro-Acoustic Signals. *Food Bioprocess Technol.* **2021**, *14*, 887–897. [CrossRef]
6. Sun, Y.; Sun, H.; Luo, M.; Zhou, X.; Zhou, Q.; Wei, B.; Cheng, S.; Ji, S. Membrane lipid metabolism in relation to core browning during ambient storage of 'Nanguo' pears. *Postharvest Biol. Technol.* **2020**, *169*, 111288. [CrossRef]
7. Li, J.; Yao, T.; Xu, Y.; Cai, Q.; Wang, Y. Elevated CO_2 exposure induces core browning in Yali pears by inhibiting the electron transport chain. *Food Chem.* **2022**, *378*, 132101. [CrossRef]
8. Wang, Y.; Sugar, D. Internal browning disorder and fruit quality in modified atmosphere packaged 'Bartlett' pears during storage and transit. *Postharvest Biol. Technol.* **2013**, *83*, 72–82. [CrossRef]
9. Wu, G.X.; Zhou, H.W.; Wang, J.M. Biochemical mechanism and substances determination of enzymic browning of Yali pear (*Pyrus bretschneideri* Rehd). *Acta Hortic. Sin.* **1992**, *19*, 198–202.
10. Li, J.; Bao, X.; Xu, Y.; Zhang, M.; Cai, Q.; Li, L.; Wang, Y. Hypobaric storage reduced core browning of Yali pear fruits. *Sci. Hortic.* **2017**, *225*, 547–552. [CrossRef]
11. Liu, L.Q.; Dong, Y.; Guan, J.F. Effects of nitric oxide fumigation on post-harvest core browning in 'Yali' pear (*Pyrus bretschneideri* Rehd.) during cold storage. *J. Hortic. Sci. Biotechnol.* **2012**, *87*, 571–576. [CrossRef]

12. Peralta-Ruiz, Y.; Grande-Tovar, C.D.; Navia Porras, D.P.; Sinning-Mangonez, A.; Delgado-Ospina, J.; González-Locarno, M.; Maza Pautt, Y.; Chaves-López, C. Packham's Triumph Pears (*Pyrus communis* L.) Post-Harvest Treatment during Cold Storage Based on Chitosan and Rue Essential Oil. *Molecules* **2021**, *26*, 725. [CrossRef] [PubMed]
13. Li, X.; Guo, W.; Xu, M.; Zhao, J.; Wang, G.; Yuan, H.; Wang, A. PuWRKY31 affects ethylene production in response to sucrose signal in pear fruit. *Hortic. Res.* **2022**, *9*, uhac156. [CrossRef] [PubMed]
14. Cheng, Y.; Liu, L.; Feng, Y.; Dong, Y.; Guan, J. Effects of 1-MCP on fruit quality and core browning in 'Yali' pear during cold storage. *Sci. Hortic.* **2019**, *243*, 350–356. [CrossRef]
15. Franck, C.; Lammertyn, J.; Ho, Q.T.; Verboven, P.; Verlinden, B.; Nicolaï, B.M. Browning disorders in pear fruit. *Postharvest Biol. Technol.* **2007**, *43*, 1–13. [CrossRef]
16. Li, L.; Zhang, Y.; Fan, X.; Wang, J.; Liang, L.; Yan, S.; Xiao, L. Relationship between activated oxygen metabolism and browning of "Yali" pears during storage. *J. Food Process Preserv.* **2020**, *44*, e14392. [CrossRef]
17. Fonseca, S.; Monteiro, L.; Barreiro, M.G.; Pais, M.S. Expression of genes encoding cell wall modifying enzymes is induced by cold storage and reflects changes in pear fruit texture. *J. Exp. Bot.* **2005**, *56*, 2029–2036. [CrossRef]
18. Wang, Y.; Mao, H.; Lv, Y.; Chen, G.; Jiang, Y. Comparative analysis of total wax content, chemical composition and crystal morphology of cuticular wax in Korla pear under different relative humidity of storage. *Food Chem.* **2021**, *339*, 128097. [CrossRef]
19. Gao, C.; Zhang, Y.; Li, H.; Gao, Q.; Cheng, Y.; Ogunyemi, S.O.; Guan, J. Fruit bagging reduces the postharvest decay and alters the diversity of fruit surface fungal community in 'Yali' pear. *BMC Microbiol.* **2022**, *22*, 239. [CrossRef]
20. Chen, L.; Wang, M.; Wang, H.; Zhou, C.; Yuan, J.; Li, X.; Pan, Y. Isothermal Storage Delays the Senescence of Post-Harvest Apple Fruit through the Regulation of Antioxidant Activity and Energy Metabolism. *Foods* **2023**, *12*, 1765. [CrossRef]
21. Jung, S.-K.; Watkins, C.B. Involvement of ethylene in browning development of controlled atmosphere-stored 'Empire' apple fruit. *Postharvest Biol. Technol.* **2011**, *59*, 219–226. [CrossRef]
22. Lu, L.; Zuo, W.; Wang, C.; Li, C.; Feng, T.; Li, X.; Wang, C.; Yao, Y.; Zhang, Z.; Chen, X. Analysis of the postharvest storage characteristics of the new red-fleshed apple cultivar 'meihong'. *Food Chem.* **2021**, *354*, 129470. [CrossRef] [PubMed]
23. Dai, X.; Jiang, F.; Li, Q.; Yu, X.; Xu, X.; Cao, W.; Guo, T.; Song, Z.; Zhang, A.; Zhang, H.; et al. Transcriptome analysis for key softening-related genes in 'Docteur Jules Guyot' pear (*Pyrus communis* L.). *Postharvest Biol. Technol.* **2023**, *205*, 112484. [CrossRef]

foods

MDPI

Article

The Role of MaWRKY70 in Regulating Lipoxygenase Gene Transcription during Chilling Injury Development in Banana Fruit

Han Lin [1,2,†], Lijuan Bai [1,†], Wei Wei [3], Wenbing Su [2], Yanting Wu [1], Rong Wu [1], Hui Wang [1], Jianye Chen [3], Hetong Lin [1] and Zhongqi Fan [1,*]

1 Institute of Postharvest Technology of Agricultural Products, College of Food Science, Fujian Agriculture and Forestry University, Fuzhou 350002, China; linhan_haha@163.com (H.L.); bailijuan2024@163.com (L.B.); wuyanting2024@163.com (Y.W.); 15093706039@163.com (R.W.); xiafengxixi@126.com (H.W.); hetonglin@163.com (H.L.)

2 Fruit Research Institute, Fujian Academy of Agricultural Science, Fuzhou 350013, China; suwenbing13@163.com

3 Guangdong Provincial Key Laboratory of Postharvest Science of Fruits and Vegetables, College of Horticulture, South China Agricultural University, Guangzhou 510642, China; weiwei@scau.edu.cn (W.W.); chenjianye@scau.edu.cn (J.C.)

* Correspondence: fanzqi@fafu.edu.cn

† These authors contributed equally to this work.

Abstract: Banana is a typical cold-sensitive fruit; it is prone to chilling injury (CI), resulting in a quality deterioration and commodity reduction. However, the molecular mechanism underlying CI development is unclear. In this study, cold storage (7 °C for 5 days) was used to induce CI symptoms in bananas. As compared with the control storage (22 °C for 5 days), cold storage increased the CI index and cell membrane permeability. Moreover, we found that the expression levels of the WRKY transcription factor *MaWRKY70* were increased consistently with the progression of CI development. A subcellular localization assay revealed that MaWRKY70 was localized in the nucleus. Transcriptional activation analyses showed that MaWRKY70 processed a transactivation ability. Further, an electrophoretic mobility shift assay (EMSA) and dual-luciferase reporter (DLR) assays showed that MaWRKY70 was directly bound to the W-box motifs in the promoters of four lipoxygenase (LOX) genes associated with membrane lipid degradation and activated their transcription. Collectively, these findings demonstrate that MaWRKY70 activates the transcription of *MaLOXs*, thereby acting as a possible positive modulator of postharvest CI development in banana fruit.

Keywords: banana; chilling injury; WRKY; LOX; transcriptional regulation

1. Introduction

Banana (*Musa acuminata*) is one of the most widely grown fruits in tropical and subtropical regions due to its sweet flavor and high nutritional value [1,2]. However, banana is a typical cold-sensitive fruit that is susceptible to chilling injury (CI) symptoms, including pitting, water soaking, and even abnormal fruit ripening, leading to a quality deterioration and commodity loss [3,4]. Therefore, it is important to investigate the underlying molecular mechanism of CI development in postharvest banana fruit.

Plants are regularly exposed to changing environments throughout their life cycle and respond and acclimate to this environmental stress to survive and reproduce [5,6]. Sophisticated signaling cascades are used to induce changes in the expression of temperature-responsive genes that enable plants to withstand temperature stress [7]. The alterations of plant membrane function and structure caused by cold are considered to be the principal events of CI [8]. Under cold stress, the liquid crystal of the cell membrane undergoes a phase transition and transforms into a gel state with poor fluidity. This phase transition leads to an increase in membrane permeability, ultimately resulting in physiology disorder, membrane damage, and CI occurrence [9]. Membrane lipids are the predominant components of plant cell membranes; therefore, the composition of the membrane lipids has an important role in maintaining the stability of the membrane's structure and function [10]. The composition and content of membrane lipids are regulated by a series of enzymes, such as phospholipase A (PLA), phospholipase D (PLD), lipoxygenase (LOX), and lipase. Specifically, phospholipids are degraded by PLD, PLA, and lipase to free fatty acids (FFAs). LOX promotes membrane lipids' peroxidation via catalyzing the conversion of unsaturated fatty acids (USFAs) to saturated fatty acids (SFAs) [11]. Previous reports have shown that the enzymatic activity and the transcription of PLA, PLD, LOX, and lipase in banana fruit were induced by a cold environment during storage [4]. Moreover, the enzyme activity and gene expression of PLD and LOX increased in cold-stored green bell peppers [12]. The activity of PLD and LOX rose in line with CI development in zucchini fruit [13]. Generally, the occurrence of CI is caused by changes in membrane lipid degradation and peroxidation enzyme activity in fresh fruit.

WRKYs are one of the largest transcription factor (TF) families in plants. It is widely known that WRKYs are essential modulators of biotic and abiotic stress [14]. In 1994, Ishiguro and Nakamura cloned the first WRKY transcription factor, SPF1, from sweet potatoes. In recent years, with the rapid development of molecular biology and bioinformatics techniques, as well as the unveiling of numerous plant genome sequences, a large number of *WRKY* genes have been discovered and identified from plants, such as Arabidopsis, corn, rice, potato, tomato, cotton, and cucumber [15]. For example, the overexpression of the *OsWRKY71* and *OsWRKY76* genes in rice enhances its cold resistance, while the overexpression of *OsWRKY63* in rice exacerbates the membrane damage caused by low temperatures, leading to a cold tolerance decrease in the plant. The *oswrky63* mutant plant improved its cold tolerance [16,17]. *Vitis amurensis VaWRKY12* was upregulated by cold and the overexpression of *VaWRKY12* in Arabidopsis could enhance its cold tolerance [18]. The overexpression *KoWRKY40* from *Kandelia obovate* in Arabidopsis improved its cold resistance by regulating the antioxidant system and cold signaling pathway (ICE-CBF-COR) [19]. The overexpression of *PmWKY57* from *Prunus mume* reduced the cold sensitivity in *Arabidopsis* by upregulating the expression of cold-response genes, including *AtCOR6.6*, *AtCOR47*, *AtKIN1*, and *AtRCI2A* [20]. To date, however, little has been reported about the molecular aspects of cold-induced WRKY TFs in cold-sensitive fruit, such as banana.

Previously, we demonstrated that CI development in banana fruit was caused by membrane integrity damage and was associated with the enzymatic and genetic manipulation of membrane lipid metabolism. These activities facilitated the degradation of membrane phospholipids and USFAs in banana fruit during the cold storage period [4]. Nevertheless, the transcriptional regulatory mechanisms underpinning WRKY-TF-modulated membrane lipid metabolism in cold-stored banana fruit remain largely unknown. Here, a cold-induced WRKY TF, MaWRKY70, was isolated and characterized from banana fruit. We demonstrate that MaWRKY70 is positively associated with CI development in postharvest banana fruit by activating four *LOX* genes, namely *MaLOX1*, *MaLOX3.1*, *MaLOX3.2* and *MaLOX4*. Our findings thus provide a novel link between MaWRKY70 and CI development in banana fruit.

2. Materials and Methods

2.1. Plant Material and Treatments

Banana fruits at 70–80% maturation were taken from an orchard in Zhangzhou, Fujian, China. The banana fruits were separated into individual fingers and selected based on the absence of mechanical damage and disease, with a uniform size and color. The selected bananas were washed in water and then soaked in prochloraz solution for one minute. After air drying, the fruits were divided randomly into two groups. One group was stored at 7 °C (cold treatment); the other group was stored at 22 °C (control). Banana peel tissues were sampled at 0, 1, 2, 3, 4, and 5 d; frozen in liquid nitrogen; and preserved at −80 °C for further use [4].

2.2. Assessment of CI Index and Cell Membrane Permeability

The CI index was assessed following the methodology of Li et al. [4]. Banana fruits are graded for cold damage according to the size of external browning areas, where 0 indicates no browning area; 1 indicates areas with 1–25% browning; 2 indicates areas with 25–50% browning; 3 indicates areas with 50–75% browning; and 4 indicates areas with 75–100% browning [21,22].

Cell membrane permeability was determined by using the method of Li et al. [4]; the results were calculated as the ratio of the electrical conductivity in both the non-boiled and boiled state and were expressed in %.

2.3. RNA Extraction and cDNA Synthesis

The total RNA of banana peel was extracted by an RNAprep Pure Plant Kit (Tiangen, Bejing, China). The protocol of RNA extraction followed the manufacturer's instructions, including the following steps: (a) 100 mg fruit tissues were ground into a powder in liquid nitrogen; 450 μL buffer RL was added (β-mercaptoethanol was added before use); and it was vortexed vigorously, shaken, and stirred evenly; (b) all solutions were transferred to the RNAase-free filter column set (CS) and centrifuged at 13,400× *g* for 5 min; the supernatant was aspirated from the collection tube into the RNase-free centrifuge tube, avoiding contact with the cell debris precipitate in the collection tube; (c) 225 μL anhydrous ethanol was added and the obtained solution and precipitation were transferred together into the adsorption column set (CR3); it was centrifuged at 13,400× *g* for 60 s; the waste liquid was discarded from the collection tube, and the adsorption column CR3 was replaced in the collection tube; (d) 350 μL deproteinized solution RW1 was added to the adsorption column CR3 and it was then centrifuged at 13,400× *g* for 60 s; the waste liquid was discarded from the collection tube and the adsorption column CR3 was replaced in the collection tube; (e) 80 μL DNaseI working solution was added to the center of the adsorption column CR3 and placed at room temperature for 15 min; (f) 350 μL deproteinized solution RW1 was added to the adsorption column CR3 and it was then centrifuged at 13,400× *g* for 60 s; (g) 500 μL RW buffer was added (ethyl alcohol was added before use) and it was placed at room temperature for 2 min and then centrifuged at 13,400× *g* for 60 s; the waste liquid was discarded from the collection tube and the adsorption column CR3 was placed into the collection tube; this was repeat once; (h) it was centrifuged at 13,400× *g* for 2 min, the waste liquid was discarded from the collection tube, and it was placed at room temperature for 5 min; (i) the adsorption column CR3 was placed into a new RNase-free centrifuge tube and 50 μL RNase-free ddH_2O was added; it was placed at room temperature for 2 min and centrifuged at 13,400× *g* for 2 min to obtain an RNA solution.

First-strand cDNA was synthesized via reverse transcription by a cDNA synthesis kit (Vazyme, Nanjing, China), following the manufacturer's instructions, including the following steps: (a) the genomic DNA was removed and RNase-free ddH_2O was added, with 4 × gDNA wiper mix and the template RNA; the reaction took place at 42 °C for 2 min; (b) to prepare the reverse transcription reaction system, 5 × HiScript III qRT SuperMix was added to the above solution; (c) the reverse transcription reaction was performed at 37 °C for 15 min and 85 °C for 5 s.

RT-qPCR was carried out using the Taq Pro universal SYBR qPCR Master Mix (Vazyme, China) on a Bio-Rad CFX96 Real-Time PCR System. The required primers were designed using the Integrated DNA Technologies Website (https://sg.idtdna.com/pages) (accessed on 10 June 2022), and all primers are recorded in Table S1. *MaEIF5A-2* was used as the reference gene [23].

2.4. Subcellular Localization

The full-length MaWRKY70 was cloned into the pBE-GFP vector. The pBE-GFP-MaWRKY70 or pBE-GFP (positive control) was transferred to the *Agrobacterium tumefaciens* strain EHA105 and then injected into tobacco leaves. NLS-mCherry was used as the nuclear marker. After 48 h of infiltration, the fluorescent signal of GFP was observed and imaged through a fluorescence microscope (ZEISS, Oberkochen, Germany) [24,25].

2.5. Dual-Luciferase Transient Expression Analysis

For the analyses of MaWRKY70 transactivation, its coding sequence was inserted into the pGreenII BD-62-SK vector as an effector. The pGreenII 0800-LUC vector, which contains firefly luciferase (LUC) and renilla luciferase (REN), was used as the reporter. For the analyses of MaWRKY70 regarding the activation of the *MaLOX1*, *MaLOX3.1*, *MaLOX3.2*, and *MaLOX4* promoters, the MaWRKY70 coding region was cloned into the pGreenII 62-SK vector as an effector, while the promoters of *MaLOX1*, *MaLOX3.1*, *MaLOX3.2*, and *MaLOX4* were cloned into the pGreenII 0800-LUC vector as a reporter. Different pairs of effectors and reporters were co-transferred to tobacco leaves. After 48 h of infiltration, the LUC and REN activity was measured by a Dual Luciferase Assay Kit (Promega, Madison, WI, USA) in a Luminoskan Ascent Microplate Luminometer (Thermo, Waltham, MA, USA). The LUC/REN ratio was calculated to obtain results on the transcriptional activity [26].

2.6. Y2H Assay

To estimate the transcriptional activity of MaWRKY70 in yeast, the full-length MaWRKY70 was inserted into the pGBK7 vector. The pGBK7-MaWRKY70, pGBK7-empty, and pGBK7-p53+pGAD7 plasmids were transformed into the yeast strain Y2H, respectively, following growth on synthetic defined (SD) medium lacking tryptophan (SD/-Trp) or tryptophan, histidine, and adenine (SD/-Trp-His-Ade). The transcriptional activity of MaWRKY70 was determined according to the growth state and α-galactosidase activity in yeast cells [27].

2.7. Electrophoretic Mobility Shift Assay (EMSA)

The full-length MaWRKY70 was inserted into the pGEX4T-1 vector to fuse to the glutathione S-transferase (GST) tag. GST-MaWRKY70 or empty GST (negative control) was expressed in Rosetta (DE3) *E. coli* cells, induced with isopropyl-D-thiogalatopyranoside (IPTG), and purified by the GST Purification Kit (Takara, Tokyo, Japan). Probes containing the W-box recognition sequence (C/T)TGAC(C/T)) from the target promoter regions were synthesized and biotin-labeled by 3′ biotin and HPLC purification [27]. EMSA was conducted with a LightShift Chemiluminescent EMSA Kit (Thermo, Waltham, MA, USA), and biotin-labeled DNA was observed on the ChemiDoc™ MP Imaging System (Bio-Rad, Hercules, CA, USA) [27].

2.8. Statistical Analysis

All experiments were performed at least in triplicate. Data were recorded as the mean $\pm$ standard error (S.E.) of three or six independent biological replicates. Significant differences between samples in cold-stored fruits and control fruits were determined by Student's *t*-test (* $p < 0.05$ or ** $p < 0.01$).

3. Results

3.1. Physiological Changes during Banana Fruit CI Development

Peel browning is an apparent symptom of CI in bananas. As displayed in Figure 1a, the fruits showed an intact peel appearance without any browning. All banana fruits did not exhibit any CI symptoms in the control group (22 °C), but fruit browning was seen after 1 day of storage at 7 °C, and it became increasingly severe (Figure 1a). The CI index and cell membrane permeability are used to evaluate the severity of CI progression [5]. In the control group, the CI index remained 0 during the whole storage period, but it increased greatly in the cold-stored group (Figure 1b). The cell membrane permeability displayed an increasing tendency with the extension of the storage time; it was increased remarkably in cold-stored bananas, and it maintained higher levels compared with control fruits during storage. In cold-stored bananas, the values of cell membrane permeability on day 5 were 4.13-fold higher than on day 0 (Figure 1c).

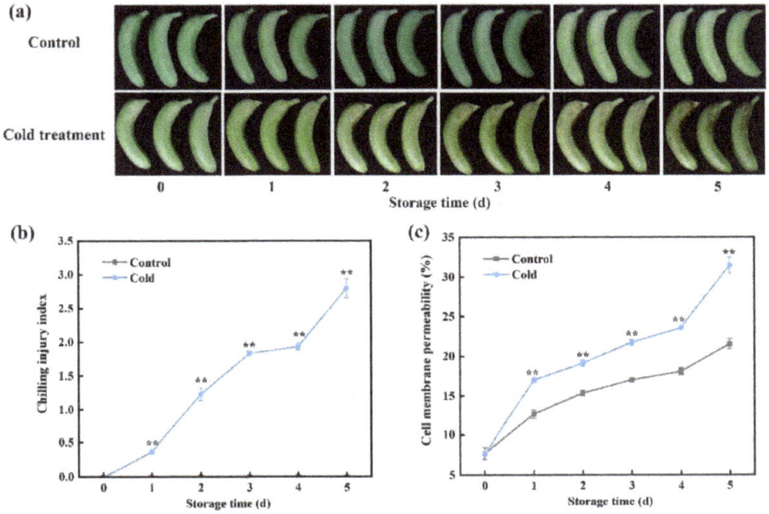

Figure 1. Cold storage induces chilling injury (CI) occurrence in banana fruit. (**a**) Changes in fruit appearance. (**b**) CI index and (**c**) cell membrane permeability in control and cold-stored bananas during storage. Each datum represents the mean ± standard error of three replicates, and the asterisks indicate significant differences by Student's *t*-test (** *p*-value < 0.01).

3.2. Isolation of MaWRKY70

Banana WRKY TFs have been reported to play central roles in regulating CI development [17]. Based on our transcriptome database relevant to CI in banana fruits [2], a *WRKY* gene (Ma06_g17380), which was markedly upregulated by cold, was selected and identified. The open reading frame (ORF) of Ma06_g17380 is 888 bp, encoding a protein of 296 aa, and it was named MaWRKY70 (XP_009404676.1) following the NCBI database. To investigate the phylogeny of MaWRKY70, a phylogenetic tree showed that Ma06_g17380 was clustered with AtWRKY46 (AT2G46400), which belonged to Group III (Figure 2a). Moreover, the protein alignment of MaWRKY70 with AtWRKY46 (AT2G46400), AtWRKY54 (AT2G40750), AtWRKY55 (AT2G40740), and AtWRKY70 (AT3G56400) showed that they possessed a highly conserved WRKY domain and the putative C2HC zinc-finger motif (Figure 2b).

3.3. Molecular Characterization of MaWRKY70

To analyze the potential association between *MaWRKY70* and CI development in banana fruit, the gene expression (RT-qPCR) of *MaWRKY70* was studied during storage. The results showed that the expression of *MaWRKY70* was obviously upregulated in cold-stored banana fruit, but there was no obvious change in control fruit during the storage time (Figure 3a). To investigate the subcellular localization of MaWRKY70, it was fused with GFP and the MaWRKY70-GFP fusion protein was generated and injected into tobacco leaves for transient expression. As exhibited in Figure 3b, the GFP signal of the MaWRKY70-GFP fusion protein was observed in the nuclei of tobacco cells, while GFP fluorescence (positive control) was observed throughout the cell. NLS-mCherry was used as a marker, and it completely overlapped with the green fluorescence in the cell nucleus.

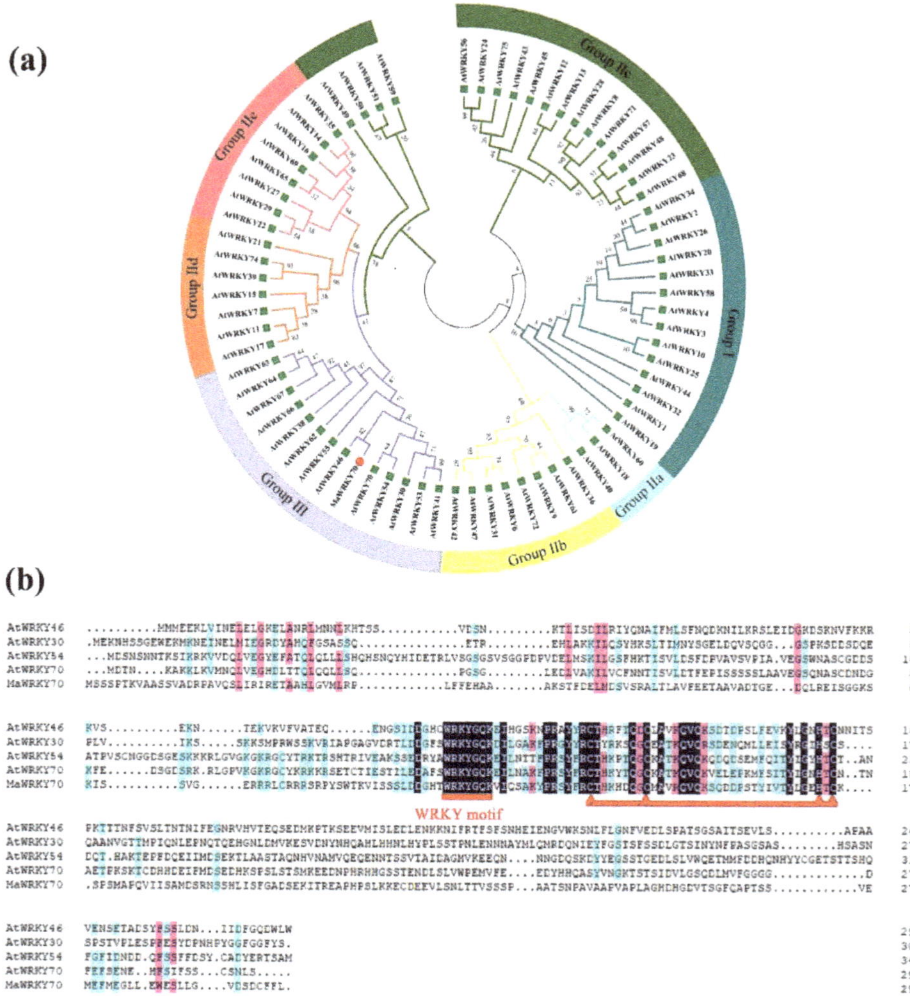

Figure 2. Sequence analysis of MaWRKY70. (**a**) Phylogenic tree of AtWRKYs and MaWRKY70. MaWRKY70, along with AtWRKY46, AtWRKY63, AtWRKY55, and AtWRKY70, was classified into Group III. (**b**) Multiple alignment of MaWRKY70 with AtWRKY46, AtWRKY54, AtWRKY55, and AtWRKY70. WRKY domain and C2HC zinc-finger motif are marked.

To test the transactivation activity of MaWRKY70, we employed the Y2H assay in the yeast system and the DLR assay in tobacco leaves. For the Y2H assay, the yeast cells transformed with pGBKT7-53 + pGADT7 (positive control) and the yeast cells harboring pGBKT7-MaWRKY70 survived well on a selection medium (SD/-Trp-His-Ade) and turned blue in the presence of X-α-gal, but the negative control (pGADT7-T vector only) did not survive (Figure 3c), indicating that MaWRKY70 exhibited transcriptional activation activity in yeast. Furthermore, we studied the transcriptional activation of MaWRKY70 in tobacco leaves. Compared with BD-62SK negative control which was expressed as 1, BD-62SK-MaWRKY70 activities the LUC/REN ratio. Similar to the positive control, BD-62SK-VP16 enhanced this value (Figure 3d), suggesting the transcriptional activation activity of MaWRKY70 in tobacco leaves. Collectively, these results demonstrate that

MaWRKY70 is a nuclear-localized transcriptional activator that is associated with CI development in postharvest banana fruit.

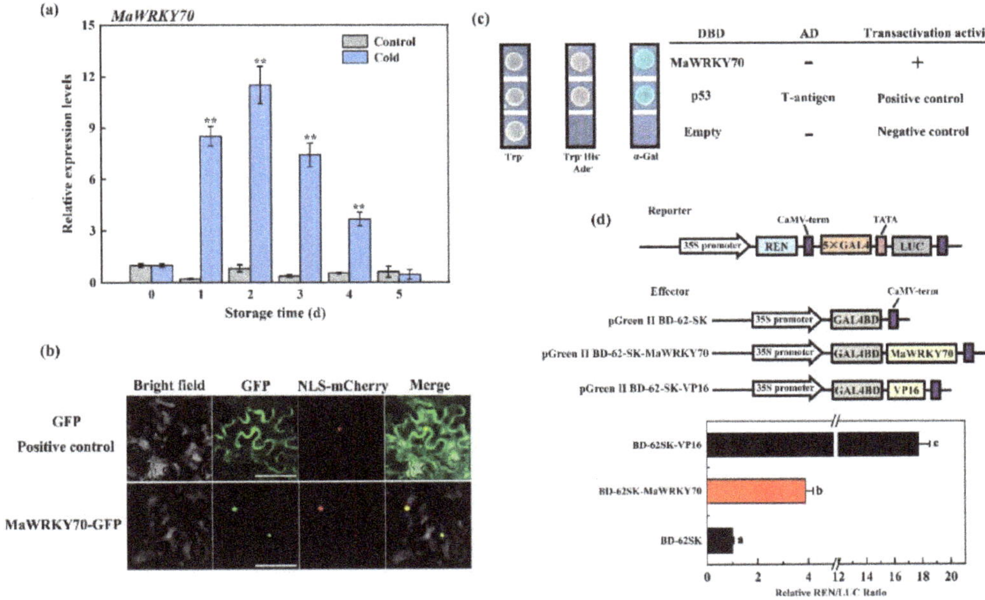

Figure 3. Molecular characterization of MaWRKY70. (**a**) Relative expression levels of MaWRKY70 in control and cold-stored banana fruit during storage, the asterisks indicate significant differences by Student's *t*-test (** *p*-value < 0.01). (**b**) Nuclear localization of MaWRKY70 in tobacco leaves. GFP and mCherry signals were captured by a fluorescence microscope. Bars, 50 μm. (**c**) Transcriptional activation of MaWRKY70 in yeast cells. pGBKT7 and pGBKT7-53 + pGADT7-T were used as negative and positive control, respectively. (**d**) Transcriptional activation capacity of MaWRKY70 in tobacco cells. The ratio of LUC to REN indicates the transactivation ability of MaWRKY70. The ratio of LUC to REN of the BD-62SK vector was used as a calibrator (value set as 1). Each datum represents the mean ± standard error of three replicates, and the different letters indicate significant differences by Student's *t*-test (*p*-value < 0.01).

3.4. Expression Patterns of MaLOXs during Banana Fruit CI Development

LOX promotes membrane lipids' peroxidation, which has an important role in the CI development process. Our previous transcriptome revealed that the *MaLOX* genes were greatly upregulated by cold storage [2]; therefore, an RT-qPCR analysis was conducted to confirm the expression of *MaLOXs*. As shown in Figure 4, the expression of *MaLOX1.1* (Ma06_g26850), *MaLOX3.1* (Ma08_g23400), *MaLOX3.2* (Ma09_g15420), *MaLOX1.2* (Ma09_g19140), and *MaLOX4* (Ma10_g17560) was induced by cold treatment and they maintained higher levels in cold-stored bananas compared to control fruits. At the peak of the upward trend period (day 4 of storage), the mRNA transcripts of *MaLOX1*, *MaLOX3.1*, *MaLOX3.2*, and *MaLOX4* in cold-stored bananas were 3.29-, 3.57-, 5.05-, and 3.03-fold higher than those of control fruits (Figure 4).

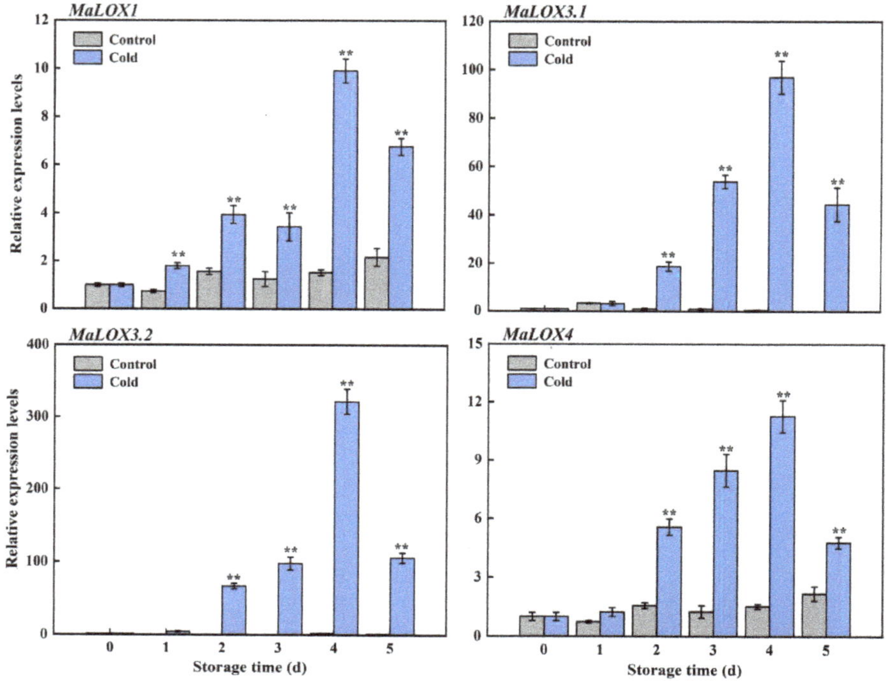

Figure 4. Relative expression levels of *MaLOX1*, *MaLOX3.1*, *MaLOX3.2*, and *MaLOX4* in control and cold-stored banana fruit during storage. Each datum represents the mean ± standard error of three replicates, and the asterisks indicate significant differences by Student's *t*-test (** *p*-value < 0.01).

3.5. *MaWRKY70 Activates Transcription of MaLOXs*

It is widely known that WRKY TFs regulate their target genes by binding to the W-box *cis*-element in the promoter [17]. Here, the promoter regions of *MaLOXs* were scanned, and we found that they all contained putative W-box motifs (Text S1). Moreover, the expression trends of *MaWRKY70*, *MaLOX1*, *MaLOX3.1*, *MaLOX3.2*, and *MaLOX4* were similar during the CI development process (Figures 3a and 4). Therefore, we speculated that these four *MaLOXs* might be the targets of MaWRKY70. An EMSA was performed to verify the direct binding of the MaWRKY70 protein to the promoters of *MaLOX1*, *MaLOX3.1*, *MaLOX3.2*, and *MaLOX4*. An EMSA with the purified recombinant MaWRKY70 protein was performed to investigate whether MaWRKY70 could recognize the W-box. The results showed that the purified GST-MaWRKY70 fusion protein (Figure 5a) directly targeted the DNA biotin probes harboring the W-box cis-acting element in the promoters of *MaLOX1*, *MaLOX3.1*, *MaLOX3.2*, and *MaLOX4* and caused shifted bands, and the band shifting was weakened upon the addition of increasing amounts of the cold probe. No shifted band was observed when the GST protein alone was used with biotin probes (Figure 5b). These results demonstrate that MaWRKY70 targets the W-box motif in the promoters of *MaLOX1*, *MaLOX3.1*, *MaLOX3.2*, and *MaLOX4*.

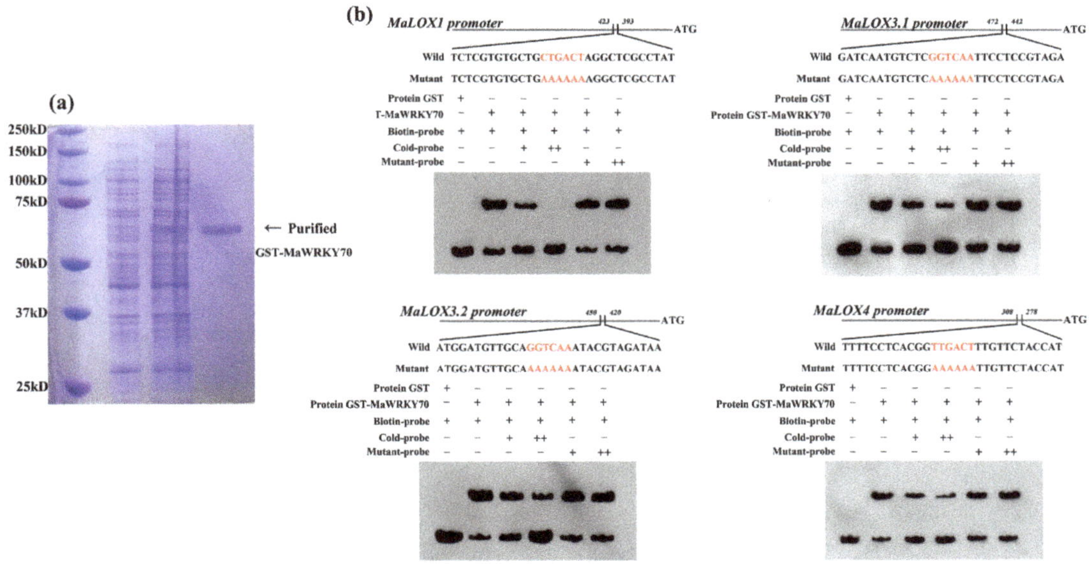

Figure 5. Electrophoretic mobility shift assay (EMSA) of MaWRKY70 binding to the promoters of *MaLOX1*, *MaLOX3.1*, *MaLOX3.2*, and *MaLOX4* containing W-box element. (**a**) Affinity purification of recombinant GST-MaWRKY70 protein as presented by SDS-PAGE gel stained with Coomassie brilliant blue. (**b**) EMSA assays. GST protein was used as negative control. GST-MaWRKY70 fusion protein binds to promoters of *MaLOX1*, *MaLOX3.1*, *MaLOX3.2*, and *MaLOX4*. Unlabeled probes (cold probe) with 50-fold (50×) and 500-fold (500×) were used for competition. Shifted bands suggest the formation of DNA–protein complexes. Symbols − and + represent absence or presence, respectively, and ++ indicates increasing amounts.

Since MaWRKY70 is a transcription activator (Figure 3c,d), the transactivation of *MaLOX1*, *MaLOX3.1*, *MaLOX3.2*, and *MaLOX4* by MaWRKY70 was further examined through the DLR system in tobacco leaves. For this experiment, MaWRKY70 driven by the 35S promoter was used as an effector, and the LUC gene driven by each promoter of *MaLOX1*, *MaLOX3.1*, *MaLOX3.2*, and *MaLOX4* was used as a reporter (Figure 6a). As displayed in Figure 6b, when 62SK-MaWRKY70 was co-transformed with the *MaLOX1*, *MaLOX3.1*, *MaLOX3.2*, or *MaLOX4* promoter, the LUC/REN ratio was obviously enhanced, as compared to the control, which involved co-transformation with the empty 62SK vector. MaWRKY70 significantly enhanced the reporter activity, 3.02-, 3.29-, 3.57-, 5.05-, and 3.03-fold, under the control of the *MaLOX1*, *MaLOX3.1*, *MaLOX3.2*, and *MaLOX4* promoters, respectively, compared with the empty 62SK vector (Figure 6b). These results indicate that MaWRKY70 is a transcriptional activator of *MaLOX1*, *MaLOX3.1*, *MaLOX3.2*, and *MaLOX4* and acts by directly targeting the binding motifs in their promoters.

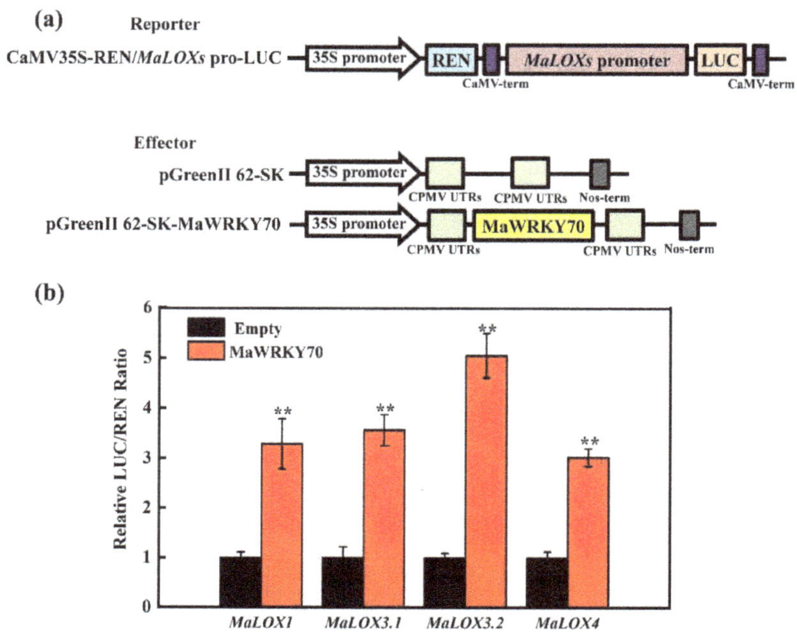

Figure 6. Dual luciferase reporter assay. (**a**) Diagrams of the reporter and effector vectors. (**b**) MaWRKY70 activated transcription of *MaLOX1*, *MaLOX3.1*, *MaLOX3.2*, and *MaLOX4*. The ratio of LUC/REN of the empty vector (62-SK) plus promoter was used as a calibrator (set as 1). Each datum represents the mean ± standard error of three replicates, and the asterisks indicate significant differences by Student's *t*-test (** p-value < 0.01).

4. Discussion

Cold storage is commercially applied in fruit preservation and transport to suppress respiration and metabolism so as to maintain the quality and prolong the shelf-life of fresh horticultural crops [26]. However, the banana fruit is susceptible to CI under low-temperature storage; the development of CI limits the transport and preservation of banana fruit during cold storage [28]. Therefore, a better understanding of CI development is important to improve the quality of postharvest banana fruit. A previous study reported that increased cell membrane permeability was consistent with the development of CI in pineapple fruit during cold storage [29]. Moreover, cold-stored mangoes showed higher levels of cell membrane permeability and more severe CI symptoms compared with control storage mangoes [30]. Here, the CI index and cell membrane permeability exhibited higher levels in cold-stored bananas as compared with control fruits (Figure 1).

The structural integrity of the cell membrane is related to the composition of the membrane lipids. Our previous work showed that membrane lipid degradation and peroxidation enzymes, such as PLD, PLC, DGK, PLA, lipase, and LOX, were increased in line with the CI development process, and the gene expression of *MaLOXs* from the transcriptome was significantly induced by cold [27]. Similarity, the increased activity and expression of PLD, LOX, and lipase in 'Nanguo' pear accelerated membrane lipid degradation and peroxidation, resulting in severe browning and CI [31]. In cucumber fruit, cold stress also induced the expression levels of the *CsPLD* and *CsLOX* genes [10]. CI development was in parallel with the increased PLD and LOX activity and the upregulated expression of *CmPLD-β* and *CmLOX* in cold-stored Hami melon fruits [32]. The upregulated expression of *LOX*, *PLD* and lipase and their enhanced activity were found in cold-stored peaches [33]. *CaPLDα4* was associated with CI development in sweet pepper during cold preservation [34]. Here, RT-qPCR assays were conducted to evaluate the expression levels of *MaLOXs*. Compared to control storage, chilling storage obviously promoted the expression of *MaLOX1*, *MaLOX3.1*, *MaLOX3.2*, and *MaLOX4* (Figure 4), implying the occurrence of membrane lipid injury and CI in cold storage.

The WRKY family members are widely involved in regulating various biological processes in plants, including fruit ripening, leaf senescence, plant hormone signal transduction, and the biological and environmental stress responses [35]. Fifty-eight WRKY proteins in eggplants were identified and the transcriptome showed that the *WRKY* genes were differentially expressed in response to a cold environment [36]. CsWRKY46 confers cold tolerance by modulating cold-stress-responsive genes in cucumber [37]. A cold-induced WRKY gene in grape, *VvWRKY28*, regulated the expression of cold-stress-related genes and played a role in cold stress tolerance [38]. In this study, a CI-inducible WRKY TF, MaWRKY70, was identified. Subcellular localization and transactivation analyses showed that MaWRKY70 was nuclear-localized and displayed a transactivation ability. Moreover, the multiple sequence alignment and phylogenetic tree analyses of MaWRKY70 revealed that it was a homologue of AtWRKY46, AtWRKY70, and AtWRKY54 (Figure 2). In Arabidopsis, AtWRKY46 was found to be a MAPK phosphorylation target involved in plant defense responses [39]. AtWRKY70 could negatively regulate root elongation by inhibiting ROS production [40]. AtWRKY54 interacted with both AvrRps4 and PopP2 and was involved in cell death [41]. These findings suggest the similar stress response function of MaWRKY70.

WRKY TFs could activate or inhibit the transcription of target genes by specifically recognizing the W-box binding elements in the downstream target gene promoter region, and they then participate in plant development and the stress response [42]. For example, FvWRKY50 promoted the expression of *FvCHI* and *FvDFR* by binding to their promoters in strawberry fruit [22]. In banana fruit, *MaWRKY31*, *MaWRKY33*, *MaWRKY60*, and *MaWRKY71* were involved in abscisic acid (ABA)-induced cold tolerance via binding to the W-box motifs in ABA biosynthesis genes' (*MaNCED1* and *MaNCED2*) promoters and promoted their expression [43]. In wheat, TaWRKY19 activated the transcription of *DREB2A*, *RD29A*, *RD29B*, and *Cor6.6* and bound to their promoters to enhance their abiotic stress tolerance [44]. In apple fruit, MdWRKY40 was directly bound to the promoter of *MdDFR* to promote anthocyanin accumulation, and it also interacted with MdMYB15L to modulate its inhibitory effect on MdCBF2, indirectly increasing the expression of *MdCBF2* and enhancing the plant's cold resistance [45]. Similarly, W-box elements were found in the promoter regions of *MaLOX1*, *MaLOX3.1*, *MaLOX3.2*, and *MaLOX4* (Text S1). Therefore, an EMSA and DLR were performed, and the results demonstrated that MaWRKY71 directly bound to the four *LOX* promoters (*MaLOX1*, *MaLOX3.1*, *MaLOX3.2*, and *MaLOX4*) and activated their expression (Figures 5 and 6). Our findings suggest that a banana fruit WRKY TF acts as a positive modulator of CI development, which is associated with their involvement in regulating membrane lipid peroxidation by activating *LOX* expression. A working model of the possible role of MaWRKY70 in CI development in banana fruit is proposed (Figure 7). During cold storage, the expression level of *MaWRKY70* is increased, and it directly activates the transcription of *LOXs* (*MaLOX1*, MaLOX3.1, *MaLOX3.2*, and *MaLOX4*). Enhanced LOX activity leads to membrane lipid degradation and to an apparent CI symptom. Collectively, our findings facilitate advances in understanding the molecular events governing CI development in cold-sensitive fresh fruit.

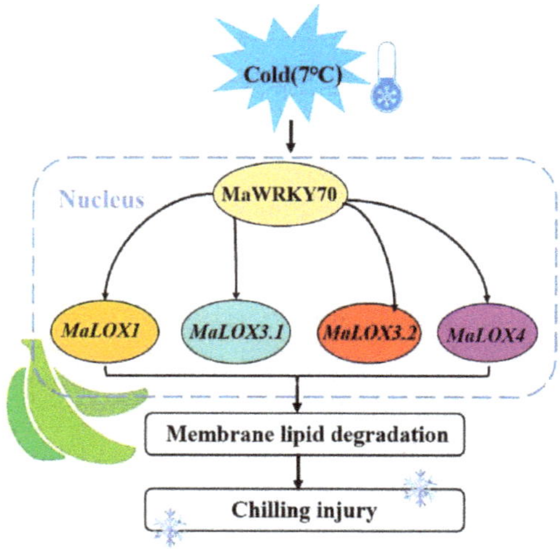

Figure 7. Working model illustrating the regulation of CI development by MaWRKY70. MaWRKY70, which was induced by cold, directly promoted the expression of *LOXs* (*MaLOX1*, *MaLOX3.1*, *MaLOX3.2*, and *MaLOX4*) to accelerate membrane lipid degradation, ultimately leading to CI development.

5. Conclusions

In summary, the results show that cold stress can cause CI in banana fruit during the preservation period, as evidenced by the increasing CI index and cell membrane permeability. More importantly, a postharvest CI-inducible transcriptional activator, MaWRKY70, from banana fruit was identified. MaWRKY70 binds to the promoters of four *LOX* genes, namely *MaLOX1*, *MaLOX3.1*, *MaLOX3.2*, and *MaLOX4*, and activates their transcription. This ultimately accelerates the peroxidation of membrane lipids, resulting in membrane structure damage and peel browning. Although our findings illustrate the transcription-regulatory role of WRKYs in CI development in postharvest banana fruit, the post-translational regulatory module that causes CI in banana needs further investigation.

Supplementary Materials: The following supporting information can be downloaded at: https://www.mdpi.com/article/10.3390/foods13060854/s1, Table S1: Summary of primers used in this study; Text S1: The *cis*-elements of WRKY TFs in the nucleotide sequences of *MaLOX1*, *MaLOX1-2*, *MaLOX3.1*, *MaLOX3.2* and *MaLOX4* promoters. *cis*-elements of WRKY TFs were underlined. The probe nucleotide sequence used in EMSA assay is indicated in bold. Translation start site (ATG) was shown in yellow.

Author Contributions: Conceptualization, H.W.; Methodology, L.B. and Y.W.; Validation, H.L. (Han Lin); Formal analysis, W.S.; Investigation, H.L. (Han Lin) and L.B.; Data curation, W.W., W.S., Y.W. and R.W.; Writing—original draft, H.L. (Han Lin), L.B. and Z.F.; Writing—review & editing, J.C. and H.L. (Hetong Lin); Visualization, W.W.; Project administration, Z.F.; Funding acquisition, Z.F. All authors have read and agreed to the published version of the manuscript.

Funding: This work was supported by the Natural Science Foundation of Fujian Province, China (Grant No. 2021J01110) and the National Natural Science Foundation of China (Grant Nos. 32001747 and 32372776).

Data Availability Statement: The original contributions presented in the study are included in the article/Supplementary Materials, further inquiries can be directed to the corresponding author.

Conflicts of Interest: The authors declare no conflict of interest.

References

1. Qamar, S.; Shaikh, A. Therapeutic potentials and compositional changes of valuable compounds from banana—A review. *Trends Food Sci. Technol.* **2018**, *79*, 1–9. [CrossRef]
2. Asif, M.H.; Lakhwani, D.; Pathak, S.; Gupta, P.; Bag, S.K.; Nath, P.; Trivedi, P.K. Transcriptome analysis of ripe and unripe fruit tissue of banana identifies major metabolic networks involved in fruit ripening process. *BMC Plant Biol.* **2014**, *14*, 316. [CrossRef] [PubMed]
3. Li, D.; Bodjrenou, D.M.; Zhang, S.T.; Wang, B.; Pan, H.; Yeh, K.W.; Lai, Z.X.; Cheng, C.Z. The Endophytic Fungus *Piriformospora indica* Reprograms Banana to Cold Resistance. *Int. J. Mol. Sci.* **2021**, *22*, 4973. [CrossRef] [PubMed]
4. Li, Q.; Lin, H.; Lin, H.T.; Lin, M.S.; Wang, H.; Wei, W.; Chen, J.Y.; Lu, W.J.; Shao, X.F.; Fan, Z.Q. The metabolism of membrane lipid participates in the occurrence of chilling injury in cold-stored banana fruit. *Food Res. Int.* **2023**, *173*, 113415. [CrossRef] [PubMed]
5. Zhou, P.; Enders, T.A.; Myers, Z.A.; Magnusson, E.; Crisp, P.A.; Noshay, J.M.; Gomez-Cano, F.; Liang, Z.; Grotewold, E.; Greenham, K.; et al. Prediction of conserved and variable heat and cold stress response in maize using cis-regulatory information. *Plant Cell* **2022**, *34*, 514–534. [CrossRef] [PubMed]
6. Madani, N.; Kimball, J.S.; Ballantyne, A.P.; Affleck, D.L.R.; van Bodegom, P.M.; Reich, P.B.; Kattge, J.; Sala, A.; Nazeri, M.; Jones, M.O.; et al. Future global productivity will be affected by plant trait response to climate. *Sci. Rep.* **2018**, *8*, 2870. [CrossRef] [PubMed]
7. Eremina, M.; Rozhon, W.; Poppenberger, B. Hormonal control of cold stress responses in plants. *Cell. Mol. Life Sci.* **2015**, *73*, 797–810. [CrossRef]
8. Zhang, X.D.; Liu, T.; Zhu, S.H.; Wang, D.; Sun, S.; Xin, L. Short-term hypobaric treatment alleviates chilling injury by regulating membrane fatty acids metabolism in peach fruit. *J. Food Biochem.* **2022**, *46*, e14113. [CrossRef]
9. Song, C.B.; Wu, M.B.; Zhou, Y.; Gong, Z.H.; Yu, W.W.; Zhang, Y.; Yang, Z.F. NAC-mediated membrane lipid remodeling negatively regulates fruit cold tolerance. *Hortic. Res.* **2022**, *9*, uhac039. [CrossRef]
10. Zhang, Y.P.; Lin, D.; Yan, R.Y.; Xu, Y.H.; Xing, M.Y.; Liao, S.Y.; Wan, C.P.; Chen, C.Y.; Zhu, L.Q.; Kai, W.B.; et al. Amelioration of Chilling Injury by Fucoidan in Cold-Stored Cucumber via Membrane Lipid Metabolism Regulation. *Foods* **2023**, *12*, 301. [CrossRef]
11. Liu, X.H.; Xiao, K.; Zhang, A.D.; Zhu, W.M.; Zhang, H.; Tan, F.; Huang, Q.R.; Wu, X.X.; Zha, D.S. Metabolomic Analysis, Combined with Enzymatic and Transcriptome Assays, to Reveal the Browning Resistance Mechanism of Fresh-Cut Eggplant. *Foods* **2022**, *11*, 1174. [CrossRef] [PubMed]
12. Kong, X.M.; Ge, W.Y.; Wei, B.D.; Zhou, Q.; Zhou, X.; Zhao, Y.B.; Ji, S.J. Melatonin ameliorates chilling injury in green bell peppers during storage by regulating membrane lipid metabolism and antioxidant capacity. *Postharvest Biol. Technol.* **2020**, *170*, 111315. [CrossRef]
13. Yao, W.S.; Xu, T.T.; Farooq, S.U.; Jin, P.; Zheng, Y.H. Glycine betaine treatment alleviates chilling injury in zucchini fruit (*Cucurbita pepo* L.) by modulating antioxidant enzymes and membrane fatty acid metabolism. *Postharvest Biol. Technol.* **2018**, *144*, 20–28. [CrossRef]
14. Meng, D.; Li, Y.Y.; Bai, Y.; Li, M.J.; Cheng, L.L. Genome-wide identification and characterization of WRKY transcriptional factor family in apple and analysis of their responses to waterlogging and drought stress. *Plant Physiol. Biochem.* **2016**, *106*, 71–83. [CrossRef]
15. Wei, H.W.; Chen, S.Y.; Niyitanga, S.; Liu, T.; Qi, J.M.; Zhang, L.W. Genome-wide identification and expression analysis response to GA3 stresses of WRKY gene family in seed hemp (*Cannabis sativa* L.). *Gene* **2022**, *822*, 146290. [CrossRef]
16. Zhang, M.X.; Zhao, R.R.; Huang, K.; Huang, S.Z.; Wang, H.T.; Wei, Z.Q.; Li, Z.; Bian, M.D.; Jiang, W.Z.; Wu, T.; et al. The OsWRKY63-OsWRKY76-OSDREB1B module regulates chilling tolerance in rice. *Plant J.* **2022**, *112*, 383–398. [CrossRef] [PubMed]
17. Kim, C.Y.; Vo, K.T.X.; Nguyen, C.D.; Jeong, D.H.; Lee, S.K.; Kumar, M.; Kim, S.R.; Park, S.H.; Kim, J.K.; Jeon, J.S. Functional analysis of a cold-responsive rice WRKY gene, OsWRKY71. *Plant Biotechnol. Rep.* **2016**, *10*, 13–23. [CrossRef]
18. Zhang, L.L.; Zhao, T.T.; Sun, X.M.; Wang, Y.; Du, C.; Zhu, Z.F.; Gichuki, D.K.; Wang, Q.F.; Li, S.H.; Xin, H.P. Overexpression of VaWRKY12, a transcription factor from Vitis amurensis with increased nuclear localization under low temperature, enhances cold tolerance of plants. *Plant Mol. Biol.* **2019**, *100*, 95–110. [CrossRef]
19. Fei, J.; Wang, Y.S.; Cheng, H.; Su, Y.B.; Zhong, Y.J.; Zheng, L. The *Kandelia obovata* transcription factor KoWRKY40 enhances cold tolerance in transgenic *Arabidopsis*. *BMC Plant Biol.* **2022**, *22*, 274. [CrossRef]
20. Wang, Y.G.; Dong, B.; Wang, N.N.; Zheng, Z.F.; Yang, L.Y.; Zhong, S.W.; Fang, Q.; Xiao, Z.; Zhao, H.B. A WRKY Transcription Factor PmWRKY57 from Prunus mume Improves Cold Tolerance in *Arabidopsis thaliana*. *Mol. Biotechnol.* **2022**, *65*, 1359–1368. [CrossRef]
21. Liu, J.; Li, F.J.; Li, T.T.; Yun, Z.; Duan, X.W.; Jiang, Y.M. Fibroin treatment inhibits chilling injury of banana fruit via energy regulation. *Sci. Hortic.* **2019**, *248*, 8–13. [CrossRef]
22. Chen, Y.T.; Liu, L.P.; Feng, Q.Q.; Liu, C.; Bao, Y.J.; Zhang, N.; Sun, R.H.; Yin, Z.N.; Zhong, C.F.; Wang, Y.H.; et al. FvWRKY50 is an important gene that regulates both vegetative growth and reproductive growth in strawberry. *Hortic Res.* **2023**, *10*, uhad115. [CrossRef] [PubMed]
23. Song, C.B.; Yang, Y.Y.; Yang, T.W.; Ba, L.J.; Zhang, H.; Han, Y.C.; Xiao, Y.Y.; Shan, W.; Kuang, J.F.; Chen, J.Y.; et al. MaMYB4 recruits histone deacetylase MaHDA2 and modulates the expression of ω-3 fatty acid desaturase genes during cold stress response in banana fruit. *Plant Cell Physiol.* **2019**, *60*, 2410–2422. [CrossRef] [PubMed]

24. Sainsbury, F.; Thuenemann, E.C.; Lomonossoff, G.P. pEAQ: Versatile expression vectors for easy and quick transient expression of heterologous proteins in plants. *Plant Biotechnol. J.* **2009**, *7*, 682–693. [CrossRef] [PubMed]
25. Song, J.; Sun, P.P.; Kong, W.N.; Xie, Z.Z.; Li, C.L.; Liu, J.H. SnRK2.4-mediated phosphorylation of ABF2 regulates ARGININE DECARBOXYLASE expression and putrescine accumulation under drought stress. *New Phytol.* **2023**, *238*, 216–236. [CrossRef]
26. Han, Y.C.; Fu, C.C. Cold-inducible MaC2H2s are associated with cold stress response of banana fruit via regulating MaICE1. *Plant Cell Rep.* **2019**, *38*, 673–680. [CrossRef]
27. Li, Z.W.; Zhou, Y.J.; Liang, H.Z.; Li, Q.; Jiang, Y.M.; Duan, X.W.; Jiang, G.X. MaMYB13 is involved in response to chilling stress via activating expression of VLCFAs and phenylpropanoids biosynthesis-related genes in postharvest banana fruit. *Food Chem.* **2023**, *405*, 134957. [CrossRef]
28. Zhu, X.Y.; Luo, J.; Li, Q.M.; Li, T.X.; Wang, R.; Chen, W.X.; Li, X.P. Low temperature storage reduces aroma-related volatiles production during shelf-life of banana fruit mainly by regulating key genes involved in volatile biosynthetic pathways. *Postharvest Biol. Technol.* **2018**, *146*, 68–78. [CrossRef]
29. Nukuntornprakit, O.A.; Chanjirakul, K.; van Doorn, W.G.; Siriphanich, J. Chilling injury in pineapple fruit: Fatty acid composition and antioxidant metabolism. *Postharvest Biol. Technol.* **2015**, *99*, 20–26. [CrossRef]
30. Purwanto, Y.A.; Okvitasari, H.; Mardjan, S.S.; Ahmad, U.; Makino, Y.; Oshita, S.; Kawagoe, Y. Chilling injury in green mature 'Gedong Gincu' mango fruits based on the changes in ion leakage. *Acta Hortic.* **2013**, *1011*, 219–226. [CrossRef]
31. Sun, H.J.; Zhou, X.; Zhou, Q.; Zhao, Y.B.; Kong, X.M.; Luo, M.L.; Ji, S.J. Disorder of membrane metabolism induced membrane instability plays important role in pericarp browning of refrigerated 'Nanguo' pears. *Food Chem.* **2020**, *320*, 126684. [CrossRef]
32. Huang, S.; Bi, Y.; Li, H.; Liu, C.H.; Wang, X.; Wang, X.Y.; Lei, Y.X.; Zhang, Q.; Wang, J. Reduction of Membrane Lipid Metabolism in Postharvest Hami Melon Fruits by n-Butanol to Mitigate Chilling Injury and the Cloning of Phospholipase D-β Gene. *Foods* **2023**, *12*, 1904. [CrossRef]
33. Wang, L.; Bokhary, S.U.F.; Xie, B.; Hu, S.Q.; Jin, P.; Zheng, Y.H. Biochemical and molecular effects of glycine betaine treatment on membrane fatty acid metabolism in cold stored peaches. *Postharvest Biol. Technol.* **2019**, *154*, 58–69. [CrossRef]
34. Kong, X.M.; Zhou, Q.; Zhou, X.; Wei, B.D.; Ji, S.J. Transcription factor CaNAC1 regulates low-temperature-induced phospholipid degradation in green bell pepper. *J. Exp. Bot.* **2020**, *71*, 1078–1091. [CrossRef] [PubMed]
35. Kidokoro, S.; Shinozaki, K.; Yamaguchi-Shinozaki, K. Transcriptional regulatory network of plant cold-stress responses. *Trends Plant Sci.* **2022**, *27*, 922–935. [CrossRef] [PubMed]
36. Yang, Y.; Liu, J.; Zhou, X.H.; Liu, S.Y.; Zhuang, Y. Identification of WRKY gene family and characterization of cold stress-responsive WRKY genes in eggplant. *Plant Biol.* **2020**, *8*, e8777. [CrossRef] [PubMed]
37. Zhang, Y.; Yu, H.J.; Yang, X.Y.; Li, Q.; Ling, J.; Wang, H.; Gu, X.F.; Huang, S.W.; Jiang, W.J. CsWRKY46, a WRKY transcription factor from cucumber, confers cold resistance in transgenic-plant by regulating a set of cold-stress responsive genes in an ABA-dependent manner. *Plant Physiol. Biochem.* **2016**, *108*, 478–487. [CrossRef]
38. Liu, W.; Liang, X.Q.; Cai, W.J.; Wang, H.; Liu, X.; Cheng, L.F.; Song, P.H.; Luo, G.J.; Han, D.G. Isolation and Functional Analysis of VvWRKY28, a Vitis vinifera WRKY Transcription Factor Gene, with Functions in Tolerance to Cold and Salt Stress in Transgenic *Arabidopsis thaliana*. *Int. J. Mol. Sci.* **2022**, *23*, 13418. [CrossRef]
39. Sheikh, A.H.; Eschen-Lippold, L.; Pecher, P.; Hoehenwarter, W.; Sinha, A.K.; Scheel, D.; Lee, J. Regulation of WRKY46 Transcription Factor Function by Mitogen-Activated Protein Kinases in *Arabidopsis thaliana*. *Front. Plant Sci.* **2016**, *7*, 61. [CrossRef]
40. Li, S.; Nayar, S.; Jia, H.; Kapoor, S.; Wu, J.; Yukawa, Y. The *Arabidopsis hypoxia* inducible AtR8 long non-coding RNA also contributes to plant defense and root elongation coordinating with WRKY genes under low levels of salicylic acid. *Non-Coding RNA* **2020**, *6*, 8. [CrossRef]
41. Kim, H.; Kim, J.; Choi, D.S.; Kim, M.S.; Deslandes, L.; Jayaraman, J.; Sohn, K.H. Molecular basis for the interference of the Arabidopsis WRKY54-mediated immune response by two sequence-unrelated bacterial effectors. *Plant J.* 2024; *online ahead of print*. [CrossRef]
42. Zhu, W.J.; Li, H.; Dong, P.F.; Ni, X.T.; Fan, M.L.; Yang, Y.J.; Xu, S.Y.; Xu, Y.B.; Qian, Y.W.; Chen, Z.; et al. Low temperature-induced regulatory network rewiring via WRKY regulators during banana peel browning. *Plant Physiol.* **2023**, *193*, 855–873. [CrossRef]
43. Luo, D.L.; Ba, L.J.; Shan, W.; Kuang, J.F.; Lu, W.J.; Chen, J.Y. Involvement of WRKY Transcription Factors in Abscisic-Acid-Induced Cold Tolerance of Banana Fruit. *J. Agric. Food Chem.* **2017**, *65*, 3627–3635. [CrossRef]
44. Niu, C.F.; Wei, W.; Zhou, Q.Y.; Tian, A.G.; Hao, Y.J.; Zhang, W.K.; Ma, B.; Lin, Q.; Zhang, Z.B.; Zhang, J.S.; et al. Wheat *WRKY* genes *TaWRKY2* and *TaWRKY19* regulate abiotic stress tolerance in transgenic *Arabidopsis* plants. *Plant Cell Environ.* **2012**, *35*, 1156–1170. [CrossRef]
45. Xu, P.Y.; Xu, L.; Xu, H.F.; He, X.W.; He, P.; Chang, Y.S.; Wang, S.; Zheng, W.Y.; Wang, C.Z.; Chen, X.; et al. MdWRKY40 is directly promotes anthocyanin accumulation and blocks MdMYB15L, the repressor of *MdCBF2*, which improves cold tolerance in apple. *J. Integr. Agric.* **2023**, *22*, 1704–1719. [CrossRef]

foods

MDPI

Article

Investigating the Respiratory and Energy Metabolism Mechanisms behind ε-Poly-L-lysine Chitosan Coating's Improved Preservation Effectiveness on *Tremella fuciformis*

Junzheng Sun [1,2,3,†], Yingying Wei [1,2,3,4,†], Longxiang Li [1,2,3,5], Baosha Tang [1,2,3], Yanrong Yang [1,2,3], Zheng Xiao [1,2,3], Junchen Chen [1,2,3] and Pufu Lai [1,2,3,*]

1 Institute of Food Science and Technology, Fujian Academy of Agricultural Sciences, Fuzhou 350003, China; sunjzll@163.com (J.S.); wyy787459024@163.com (Y.W.); 13616901013@163.com (L.L.); tbsty@126.com (B.T.); 18960973035@163.com (Y.Y.); ethyxwat@163.com (Z.X.); junchenccc@163.com (J.C.)
2 National R&D Center for Edible Fungi Processing, Fuzhou 350003, China
3 Key Laboratory of Subtropical Characteristic Fruits, Vegetables and Edible Fungi Processing (Co-Construction by Ministry and Province), Ministry of Agriculture and Rural Affairs, Fuzhou 350003, China
4 Institute of Postharvest Technology of Agricultural Products, College of Food Science, Fujian Agriculture and Forestry University, Fuzhou 350002, China
5 College of Life Science, Fujian Agriculture and Forestry University, Fuzhou 350002, China
* Correspondence: laipufu@163.com
† These authors contributed equally to this work.

Abstract: Freshly harvested *Tremella fuciformis* contains high water content with an unprotected outer surface and exhibits high respiration rates, which renders it prone to moisture and nutrient loss, leading to decay during storage. Our research utilized ε-poly-L-lysine (ε-PL) and chitosan as a composite coating preservative on fresh *T. fuciformis*. The findings revealed that the ε-PL + chitosan composite coating preservative effectively delayed the development of diseases and reduced weight loss during storage compared to the control group. Furthermore, this treatment significantly decreased the respiration rate of *T. fuciformis* and the activity of respiratory metabolism-related enzymes, such as alternative oxidase (AOX), cytochrome c oxidase (CCO), succinic dehydrogenase (SDH), 6-phosphogluconate dehydrogenase, and glucose-6-phosphate dehydrogenase (6-PGDH and G-6-PDH). Additionally, the composite coating preservative also delayed the depletion of ATP and ADP and maintained higher levels of the energy charge while preserving low levels of AMP. It also sustained heightened activities of Mg^{2+}-ATPase, Ca^{2+}-ATPase, and H^{+}-ATPase enzymes. These results demonstrate that utilizing the ε-PL + chitosan composite coating preservative can serve as a sufficiently safe and efficient method for prolonging the shelf life of post-harvest fresh *T. fuciformis*.

Keywords: ε-poly-L-lysine; chitosan; *Tremella fuciformis*; respiratory metabolism; energy metabolism

1. Introduction

Tremella fuciformis (also known as snow fungus) is globally acknowledged as a highly nutritious edible fungus, often lauded as the "crown of fungi" [1,2]. Rich in gelatinous substances, amino acids, minerals, and other nutrients, *T. fuciformis* possesses medicinal and health benefits, including blood sugar reduction, lipid-lowering, immune modulation, anti-aging effects, and antiviral effects [3]. China, a major producer and consumer of edible fungi, with a diverse variety and a long history of mushroom cultivation, leads the world in terms of both cultivation area and production of *T. fuciformis* [4,5]. Owing to its high water content, lack of protective cuticle, and active respiration during storage, fresh *T. fuciformis* is particularly prone to moisture loss, nutrient decomposition, spoilage, and the production of toxic substances, thus losing its edible and economic value [6]. The storage of fresh *T. fuciformis* presents significant challenges, yet there is a growing consumer demand for fresh produce. Consequently, the investigation and improvement of preservation methods for *T. fuciformis* is essential to extend its shelf life, reduce wastage, and maintain its nutritional value.

In the preservation of fruits, vegetables, and edible fungi, respiratory metabolism is one of the critical factors influencing the effectiveness of preservation. Respiration, a fundamental biochemical process, is involved not only in energy production but also in a series of complex metabolic pathways, including the metabolism of carbohydrates, fats, and proteins [7,8]. During the preservation of *T. fuciformis*, an increase in the respiratory metabolism rate is often associated with quality degradation and reduced lifespan. Therefore, controlling the respiratory intensity of *T. fuciformis* is crucial for maintaining its freshness and nutritional value. Moreover, energy metabolism plays an essential role in preserving the cellular vitality and freshness of *T. fuciformis*. Adenosine triphosphate (ATP), along with main metabolites, such as adenosine diphosphate (ADP) and adenosine monophosphate (AMP), is vital in sustaining cellular physiological functions [9,10]. Studies have shown that the levels of ATP and its metabolites vary under different storage conditions, reflecting the energy status and metabolic activity of shiitake mushroom (*Lentinus edodes*) cells [11]. Hence, a thorough investigation of the energy metabolism changes in *T. fuciformis* during storage is important for understanding and improving its preservation outcomes.

ε-Poly-L-Lysine (ε-PL) is a homopolymeric substance comprising lysine residues. ε-PL is characterized by favorable antibacterial properties, biodegradability, thermal stability, and solubility in water. It is non-toxic and harmless to the human body, displays efficient endotoxin removal capabilities, and can prevent the production of toxins by oral bacteria [12,13]. Beyond its applications in the food sector, ε-PL has been extensively researched and applied in fields such as agriculture, pharmaceuticals, and cosmetics [14]. Chitosan, a high molecular weight cationic polysaccharide, is safe, non-toxic, soluble in various organic acids, and exhibits a degree of antimicrobial, antioxidant, and film-forming adsorption properties [15,16]. When used as a food preservative, it can inhibit the growth of microbes, reducing spoilage due to bacterial contamination, and can also form a film on the surface of foods, minimizing moisture loss and delaying direct oxygen contact—thereby suppressing the respiration of fruits and vegetables and reducing nutrient depletion. Additionally, chitosan has antioxidant properties; it can scavenge free radicals such as superoxide anions and inhibit ethylene production, thereby delaying the aging of fruits and vegetables [17]. Edible fungi treated with chitosan retain their original taste and appearance and possess considerable advantages in post-harvest freshness preservation [18–20].

Currently, due to ε-PL's excellent broad-spectrum antimicrobial properties and chitosan's favorable adsorptive film-forming characteristics—and considering that both substances are approved food additives by the market—they present promising application prospects in the post-harvest preservation of fruits and vegetables. However, the effects of a composite coating preservative made from ε-PL and chitosan on the preservation of fresh *T. fuciformis*, as well as its relationship with energy and respiratory metabolism during storage, remain unclear. Therefore, this study aims to build upon the understanding of the composite coating preservative's impact on the storage quality of fresh *T. fuciformis*. It intends to delve into the mechanism by which this preservative enhances the shelf life of the fresh *T. fuciformis* through an analysis of changes in energy and respiratory metabolic pathways.

2. Materials and Methods

2.1. Materials and Treatments

For this study, high-quality *T. fuciformis* cultivated in Gutian County, Fujian Province, was selected as the experimental material. Harvesting was carried out once the *T. fuciformis* reached marketable maturity (cultivated for 90–110 days). The components ε-PL and chitosan implemented in this research were sourced from Zhejiang Yinuo Biotechnology Co., Ltd. (Lanxi, China). The biochemical reagents utilized throughout the assay procedure were uniformly procured from Shanghai Macklin Biochemical Technology Co., Ltd. (Shanghai, China). Furthermore, all enzyme-linked immunosorbent assay (ELISA) kits selected for analyses were acquired from Shanghai Kehansheng Biotechnology Co., Ltd. (Shanghai, China).

Fresh *T. fuciformis* specimens with full caps and uniform size that were free of diseases, pests, and mechanical damage were chosen. In our previous work, it was discovered that a composite coating preservative composed of 150 mg·L^{-1} ε-PL and 5 g·L^{-1} chitosan provided an optimal concentration for *T. fuciformis* preservation. Consequently, this study employed the preservative at this concentration. First, 5 g of chitosan were dissolved in 1 L of deionized water and sonicated for 20 min until completely dissolved. Subsequently, 150 mg of ε-PL was added to the chitosan solution, which was then sonicated for 30 min and left to stand for degassing. This process yielded 150 mg·L^{-1} ε-PL and 5 g·L^{-1} chitosan composite coating preservative.

The fresh *T. fuciformis* specimens were evenly divided into two groups (each group contained 120 specimens of *T. fuciformis*), a control group and a composite coating preservative treatment group (ε-C). Each *T. fuciformis* was uniformly sprayed 15 times with either deionized water, for the control, or the composite coating preservative, for the treatment group, with each spray delivering a dose of 0.5 mL. After the liquid on the surface of each *T. fuciformis* specimen had naturally absorbed and drained, each was placed in a polyethylene film bag, with one *T. fuciformis* specimen per bag, and then deposited in an environment of 25 °C and 80% relative humidity for 5 days. During the period it was deposited, *T. fuciformis* from both the control and ε-C treated groups was sampled daily to obtain the following measurements; each *T. fuciformis* specimen was measured three times.

2.2. Determination of Disease Index

The disease index was calculated as the following formula:

$$\text{Disease Index} = 100 \times \frac{\sum(\text{number of } T.\,fuciformis \text{ at each level} \times \text{representative value of each level})}{\text{total number of } T.\,fuciformis \times \text{maximum representative value}}$$

The method for determining the level of disease index is shown in Table 1.

Table 1. The evaluation method of disease index.

Level	Characteristics
0	disease-free
1	less than 25% of the *T. fuciformis* lobes have few disease spots
2	less than 50% of the *T. fuciformis* lobes have few disease spots or less than 25% of the lobes have several disease spots
3	less than 75% of the *T. fuciformis* lobes are diseased, or less than 25% of the lobes are completely rotted
4	over 75% of the *T. fuciformis* lobes are diseased, less than 50% of the lobes are completely rotten, or the entire *T. fuciformis* is rotten

2.3. Determination of Weight Loss Rate

Referencing the process introduced by Sun et al. [21], weight loss rate was calculated with the gravimetric method as follows:

$$\text{Weight Loss (\%)} = \frac{m - m_1}{m} \times 100\%$$

where m is the initial fresh weight of the *T. fuciformis* (in grams), and m_1 is the fresh weight of the *T. fuciformis* at the time of sampling (in grams).

2.4. Determination of Respiration Rate

Following the procedure introduced by Cliffe-Byrnes and Beirne [22], the respiration rate of the fresh *T. fuciformis* during storage was measured using a fruit and vegetable respirometer (HM-GX, Shandong Hengmei Technology Electronics Co., Ltd., Weifang, China). Before measuring the respiration rate, the test container was sealed tightly, and the respirometer was initiated to undergo a 10-min gas equilibration stabilization procedure. A single specimen of *T. fuciformis* was placed inside the test container, ensuring a tight seal, before commencing the respirometer's respiration rate measurement program. The program was conducted over a period of 20 min, following which, the results were exported.

2.5. Assay of Enzymes Activities Related to Respiratory Metabolism

The detection of alternative oxidase (AOX) enzyme activity was conducted using an ELISA kit. The assay for cytochrome c oxidase (CCO) enzyme activity was performed in accordance with the method developed by Li et al. [23], utilizing an extraction mixture containing PBS buffer, EDTA, and sucrose solution, followed by a thorough reaction with cytochrome c solution, H_2O_2 solution, and dimethyl-para-phenylenediamine dihydrochloride. The absorbance was then measured at a wavelength of 510 nm. The assays for succinic dehydrogenase (SDH) activity in this study was conducted by the methodology described by Zhang et al. [24], utilizing an extraction mixture containing PBS buffer, EDTA, sucrose solution, and resorcinol solution, followed by a thorough reaction with phenazine methyl sulfate solution, 2,6-dichlorophenol solution, and gelatin, and then measured at a wavelength of 600 nm. The 6-phosphogluconate dehydrogenase and glucose-6-phosphate dehydrogenase (6-PGDH and G-6-PDH) activity reflected the method of Li et al. [25], the combination solution of PBS buffer, bovine serum albumin, EDTA, and sucrose was employed for enzyme extraction and subjected to thorough reaction in a solution composed of Tris-HCl, $MgCl_2$, and glucose-6-phosphate. The absorbance values were measured at 340 nm. Each enzyme extract was prepared from 1 g of *T. fuciformis* tissue and the absorbance was recorded at 600 nm and 340 nm wavelengths for the respective enzymes. The enzyme activities were evinced by $U{\cdot}mg^{-1}$ protein.

2.6. Determination of Energy Level

The quantification of ATP, ADP, AMP, and energy charge levels was carried out following the methods described by Xue et al. [26]. A total 5 g of *T. fuciformis* tissue was taken and mixed with 25 mL of 0.6 $mol{\cdot}L^{-1}$ perchlorate acid solution. Following centrifugation, 10 mL of the supernatant was collected. The pH was adjusted to a range from 6.5 to 6.8 using 1 $mol{\cdot}L^{-1}$ KOH solution; the mixture was then subjected to an ice bath for 30 min before another centrifugation was performed to collect the resulting supernatant. A HPLC (LC-2030C, Shimadzu Corporation, Tokyo, Japan) was utilized for detection. The standard curves were created using a mixture of ATP, ADP, and AMP standards for comparison and quantification. The concentrations of ATP, ADP, and AMP were expressed in $mg{\cdot}kg^{-1}$ and the energy charge was calculated according to the formula:

$$\text{Energy charge} = \frac{ATP + ADP \div 2}{ATP + ADP + AMP}$$

2.7. Determination of ATPase Activities

Assays for H^+-ATPase, Ca^{2+}-ATPase, and Mg^{2+}-ATPase activities were conducted using 5g of randomly sampled *T. fuciformis* tissue, as per the methodologies detailed by Sun et al. and Li et al. [12,27]. First, 8 mL of extraction buffer solution (containing sucrose, ascorbic acid, glycerol, EDTA, DTT, and PMSF) was added to the *T. fuciformis* tissue and thoroughly mixed. After centrifugation, 0.2 mL of the supernatant was collected. Based on the determined ATPase activity, 0.5 mL of the corresponding reaction solution (H^+-ATPase: $MgSO_4$, Tris-HCl and KCl; Ca^{2+}-ATPase: $CaCl_2$, EDTA, DTT, Tris-HCl and NaCl; Mg^{2+}-ATPase: $MgCl_2$, EDTA, DTT, Tris-HCl and NaCl) was added and thoroughly mixed. Next, 0.1 mL of 1 $mmol{\cdot}L^{-1}$ ammonium molybdate solution and 0.2 mL of 5 $mmol{\cdot}L^{-1}$ ATP solution was introduced to the compound, which was then incubated in a 36 °C water bath for 20 min. Following that, 0.2 mL of 20% (v/v) trichloroacetic acid was immediately added after incubation, followed by centrifugation. Subsequently, 2 mL of ferrous sulfate-ammonium molybdate reagent was combined with the supernatant and vigorously shaken.

The absorbance of the resulting ATPase enzyme extracts was measured at 660 nm. U·mg^{-1} protein is used to indicate the activities of the ATPases.

2.8. Statistical Analyses

All experiments in this study were performed in triplicate, and the resulting data were imported into SPSS Statistics 22.0 software (IBM Corp., Armonk, NY, USA) for analysis. Differences between the treatment groups and control groups were evaluated using the T-test to assess the statistical significance. Data errors were represented by the STDEV.

3. Results

3.1. Changes in Disease Index

As illustrated in Figure 1, the postharvest disease index of fresh *T. fuciformis* demonstrated an increasing trend with the extension of storage time. The disease index of the control group increased sharply, reaching 0.79 on the fifth day of storage. In contrast, the treatment group's disease index rose more rapidly during the early storage period (0–3 days) and then increased at a slower pace during the later period (3–5 days), consistently remaining lower than that of the control group. Further analysis revealed that the disease index of the treatment group was significantly lower than that of the control group on the second day of storage ($p < 0.05$) and remained significantly lower during the fourth and fifth days of storage ($p < 0.01$). These findings suggest that, compared to the control group, treatment with ε-C effectively delayed the increase of the disease index in postharvest fresh *T. fuciformis*, substantially reducing the occurrence of diseases during the storage period.

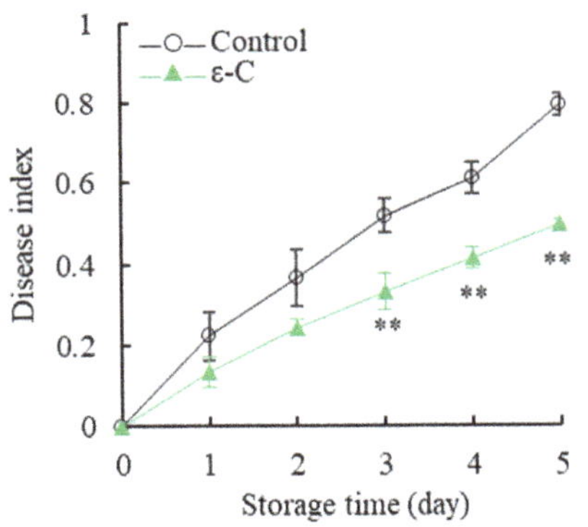

Figure 1. Changes in disease index of fresh *T. fuciformis* during the storage period. ** represents a level of significant difference of $p < 0.01$.

3.2. Changes in Weight Loss Rate

The weight loss rate of postharvest *T. fuciformis* exhibited a continuous upward trend with increasing storage time, as depicted in Figure 2. During the five-day storage period, the control group's weight loss rate accelerated rapidly, while the rate of increase in the treatment group was comparatively more gradual. Upon further comparison, it was found that, after five days of storage, the control group's weight loss rate had increased by 11.48% compared to the day of harvest, while the treatment group's weight loss rate only saw an 8.35% increase. Significance analysis indicated that the weight loss rate of the treatment group was significantly lower than that of the control group after two days of storage ($p < 0.05$) and remained very significantly lower during days 3–5 of the storage period ($p < 0.01$). The results demonstrated that treatment with ε-C effectively reduced the rate of weight loss in postharvest *T. fuciformis*, thereby effectively mitigating water loss and enhancing its shelf life.

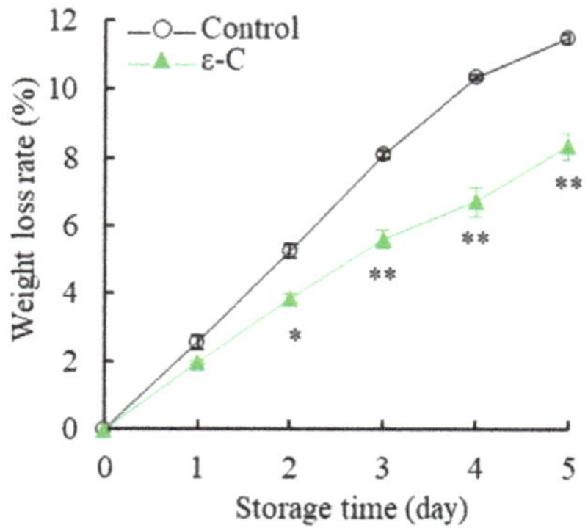

Figure 2. Changes in weight loss rate of fresh *T. fuciformis* during the storage period. * represents a level of significant difference of $p < 0.05$, ** represents a level of significant difference of $p < 0.01$.

3.3. Changes in Respiration Rate

Figure 3 illustrates a trend in the postharvest respiration rate of fresh *T. fuciformis* that initially decreases, then increases, then is followed by a subsequent decrease. The respiration rate of the control group exhibited a rapid increase between the first and fourth days of storage, reaching a peak of 132.31 mg CO_2 $kg^{-1} \cdot h^{-1}$ on the fourth day of storage, while the treatment group recorded a lower peak at 107.78 mg CO_2 $kg^{-1} \cdot h^{-1}$. Statistical analysis indicated that the respiration rate of the treatment group was consistently significantly lower than that of the control group during days 3–4 of storage ($p < 0.05$), and significantly lower on the fifth day of storage ($p < 0.01$). The findings indicate that ε-C treatment contributes to the reduction of respiration intensity in postharvest fresh *T. fuciformis*, thus attenuating its physiological metabolic activity during storage and consequently extending its storage life.

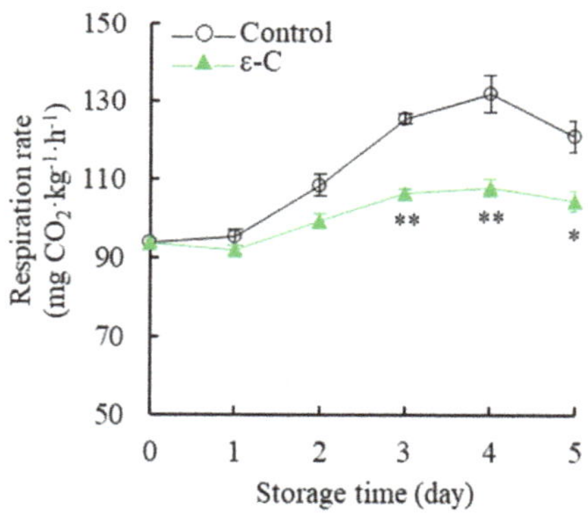

Figure 3. Changes in respiration rate of fresh *T. fuciformis* during the storage period. * represents a level of significant difference of $p < 0.05$, ** represents a level of significant difference of $p < 0.01$.

3.4. Changes in Enzyme Activities of Respiratory Metabolism

As revealed in Figure 4A, the AOX activity in the control group showed a continuously rising trend with the extension of storage time. The AOX activity in the treated group increased slowly during days 0–1, declined slightly during days 1–2, and then surged during days 2–5. Throughout the entire storage period, the AOX activity in the treated group remained consistently lower than that in the control group, being 1.6 times higher in the control group on the fourth day of storage. Significance analysis demonstrated that the treated group's AOX activity was significantly lower than the control group's on the second and fifth days of storage ($p < 0.05$), and very significantly lower during days 3–4 of storage ($p < 0.01$).

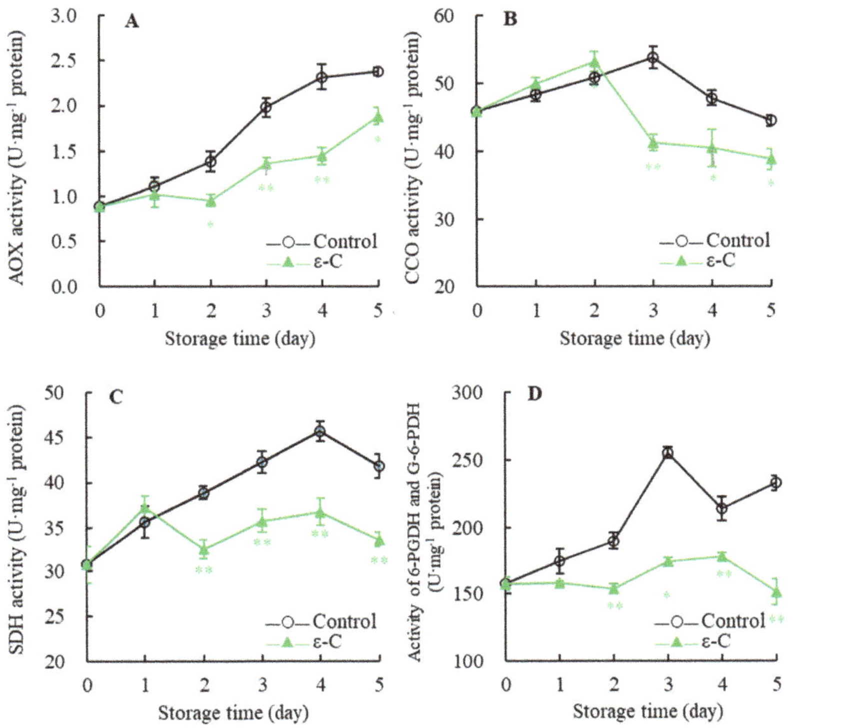

Figure 4. Changes in AOX (**A**), CCO (**B**), SDH (**C**), 6-PGDH and G-6-PDH (**D**) activities of fresh *T. fuciformis* during the storage period. * represents a level of significant difference of $p < 0.05$, ** represents a level of significant difference of $p < 0.01$.

Figure 4B indicates that CCO activity in the control group initially increased and then decreased during storage, peaking at 53.8 U·mg^{-1} protein on the third day. In contrast, the CCO activity in the treated group rose rapidly during days 0–2, dropped sharply during days 2–3, and then decreased slowly during days 3–5. The peak activity for the treated group occurred one day earlier than the control and remained lower than the control in the later stages of storage. Further analysis showed that the CCO activity in the treated group was significantly lower on the third day ($p < 0.01$) and it remained significantly lower during days 4–5 ($p < 0.05$) than it did in the control group.

From Figure 4C, it can be discerned that SDH activity in the control group rapidly increased during days 0–4, reaching the highest value of 45.7 U·mg^{-1} protein on the fourth day, and then swiftly declined. The SDH activity in the treated group surged during days 0–1, peaking at 37.3 U·mg^{-1} protein on the first day, followed by a steep decline during days 1–2, a gradual increase during days 2–4, and a rapid decrease during days 4–5. Statistical analysis revealed that the SDH activity in the treated group was consistently very significantly lower than that in the control group during days 2–5 of storage ($p < 0.01$).

As observed in Figure 4D, the activities of 6-PGDH and G-6-PDH in the control group increased modestly during days 0–2, surged during days 2–3, dramatically decreased during days 3–4, and then steeply increased during days 4–5. In the treated group, the activities of 6-PGDH and G-6-PDH slowly declined during days 0–2, slowly rose during days 2–4, and sharply decreased during days 4–5. Further comparison revealed that the activities of 6-PGDH and G-6-PDH were consistently higher in the control group during the entire storage period. Significance analysis showed that the activities of 6-PGDH and G-6-PDH in the treated group were very significantly lower than those in the control during days 2–5 of storage ($p < 0.01$). These results suggest that ε-C treatment can effectively slow the increase in AOX, CCO, and SDH activities, as well as 6-PGDH and G-6-PDH activities, in postharvest fresh *T. fuciformis*, maintaining lower enzymatic activities.

3.5. Changes in Energy Level

As depicted in Figure 5A, the ATP content trend in both the control and the treated groups during storage was essentially the same, showing a gradual decrease. Compared to the control group, the ATP content in the ε-C group was consistently higher throughout the storage period, with the treated group's ATP content being 1.16 times that of the control group on day five of storage. Statistical analysis indicated that the ATP content in the treated group was significantly higher than that in the control group on day two of storage ($p < 0.05$), and highly significantly higher during days 3–5 ($p < 0.01$).

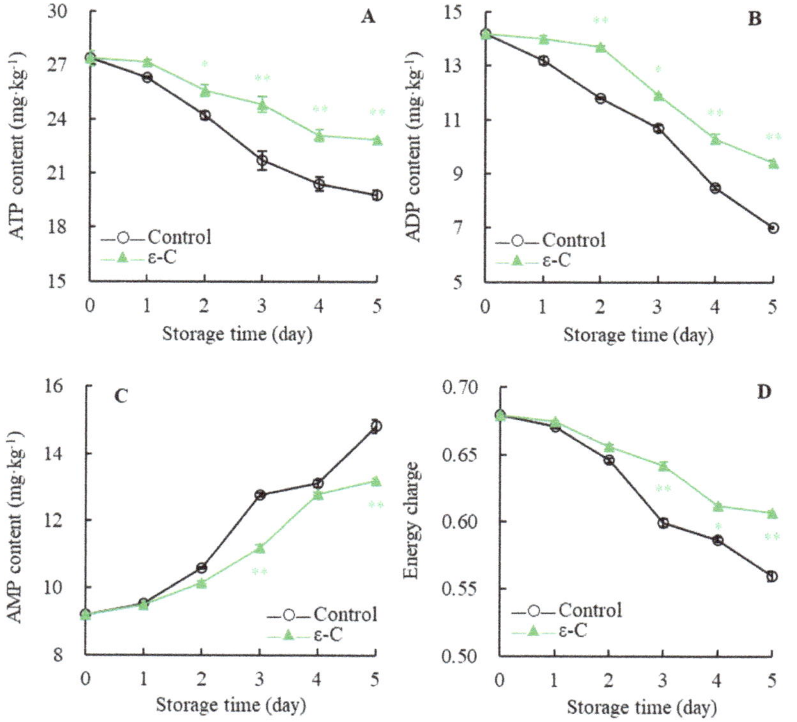

Figure 5. Changes in contents of ATP (**A**), ADP (**B**) and AMP (**C**), energy charge (**D**) of fresh *T. fuciformis* during the storage period. * represents a level of significant difference of $p < 0.05$, ** represents a level of significant difference of $p < 0.01$.

Figure 5B shows a continuous decline in the ADP content of the postharvest fruit in both the control and the treated groups. The rate of decline in the treated group's ADP content was slow during days 0–2 and then accelerated during days 2–5. The control group exhibited a constant sharp decline, with the ADP content dropping to only 7 mg·kg^{-1} by day five. Further analysis revealed that the treated group's ADP content was significantly higher than that of the control group on day three ($p < 0.05$), and very significantly higher on day two and during days 4–5 ($p < 0.01$).

As indicated in Figure 5C, AMP content of control group gradually increased during storage; it rose sharply during days 0–3, increased slowly during days 3–4, and then spiked again during days 4–5, resulting in an AMP content that was 1.61 times higher than it was on the postharvest day. In contrast, the AMP content in the treated group maintained a rapid increase during days 0–4 and then increased slowly during days 4–5. Significance analysis demonstrated that the treated group's AMP content was very significantly lower than that of the control group on day three and day five ($p < 0.01$).

Figure 5D illustrates that the energy charge of both the control and the treated groups displayed a declining trend throughout the storage period. The energy charge in the treated group was consistently higher than in the control group, with the treated group's energy charge decreasing to 0.607 on day five, while the control group's fell to only 0.559. Statistical analysis showed that the energy charge in the treated group was significantly higher than that of the control group on day four ($p < 0.05$), and very significantly higher on day three and day five ($p < 0.01$). The results above indicate that, during the storage period, ε-C treatment of postharvest fresh *T. fuciformis* effectively inhibited the decrease in ATP and AMP contents, delayed the increase in AMP content, and maintained a higher level of energy charge.

3.6. Changes in ATPase Activities

Figure 6A depicts that the Mg^{2+}-ATPase activity in the control group incrementally increased during days 0–2 of storage, declined slowly during days 2–4, and then dramatically decreased during days 4–5. In contrast, the Mg^{2+}-ATPase activity in the treated group gradually increased until day three, when it reached maximum activity; it then sharply declined during days 3–4, and slowly declined during days 4–5. Statistical analysis indicated that the Mg^{2+}-ATPase activity in the treated group was significantly higher than that in the control group on day four of storage ($p < 0.05$), and very significantly higher on days three and five ($p < 0.01$).

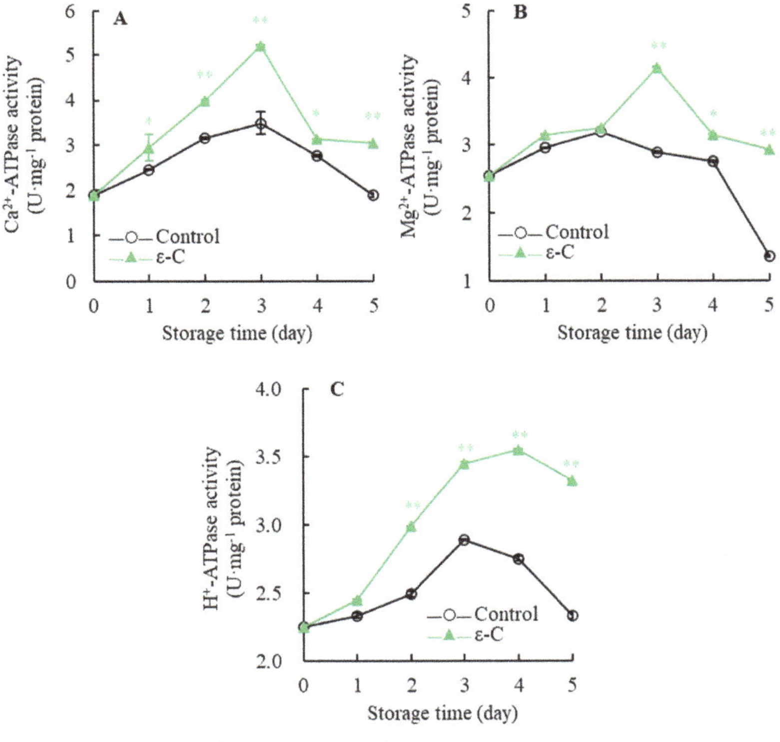

Figure 6. Changes in Ca^{2+}-ATPase (**A**), Mg^{2+}-ATPase (**B**) and H^+-ATPase (**C**) activities of fresh *T. fuciformis* during the storage period. * represents a level of significant difference of $p < 0.05$, ** represents a level of significant difference of $p < 0.01$.

As shown in Figure 6B, Ca^{2+}-ATPase activity of control group slowly increased during days 0–3 and then rapidly decreased during days 3–5. Conversely, the Ca^{2+}-ATPase activity in the treated group displayed a steep increase during days 0–3, then a sudden decline during days 3–4, and a gradual decrease during days 4–5. On day three, both groups' Ca^{2+}-ATPase activity reached their maximum values, with the treated group's activity being 1.49 times that of the control group. Significance analysis revealed that the treated group's Ca^{2+}-ATPase activity was significantly higher than the control group on day one and day four ($p < 0.05$), and very significantly higher during days 2–3 and on day five ($p < 0.01$).

Figure 6C illustrates that the control group's H^+-ATPase activity exhibited a trend of initially increasing and then decreasing with the extension of storage time. The Mg^{2+}-ATPase activity in the fruit flesh plasma membrane of the treated group increased dramatically during days 0–4, with the maximum value reaching 3.55 $U \cdot mg^{-1}$ protein on day four, which was a 1.58-fold increase from the postharvest day. Further analysis showed that the treated group's Mg^{2+}-ATPase activity was consistently very significantly higher than the control group's during days 2–5 of storage ($p < 0.01$). These results suggest that ε-C treatment of postharvest fresh *T. fuciformis* can maintain higher activities of Mg^{2+}-ATPase, Ca^{2+}-ATPase, and H^+-ATPase.

4. Discussion

After harvest, fruits, vegetables, and edible fungi must engage in respiration to obtain energy and maintain their metabolic balance. Intense respiratory activity, however, depletes energy and nutrients, subsequently leading to senescence and a decline in quality [28]. AOX and CCO are typical terminal oxidases in the respiratory chain and influence the level of respiration and the efficiency of electron transport within an organism [29,30]. In our study, following the ε-PL + chitosan coating (ε-C) treatment, the activities of AOX and CCO in *T. fuciformis* were lower than those in the control group, demonstrating a similar pattern to changes in disease index, weight loss, and respiration rate. These data substantiate that ε-C treatment can suppress the activity of terminal respiratory oxidases and reduce the rate of electron transfer and respiratory metabolism, thereby slowing down the consumption of energy and nutrients; and, consequently delaying the senescence and disease progression in *T. fuciformis*. Furthermore, the study of Li et al. showed that the application of high carbon dioxide and low oxygen treatment to *Pleurotus eryngii* slowed down the aging process, which may be attributed to the inhibited activity of terminal respiratory oxidases [23].

The changes in the activity of key respiratory metabolic enzymes can affect respiratory metabolic pathways such as the tricarboxylic acid (TCA) cycle and the pentose phosphate pathway (PPP), subsequently influencing the postharvest aging process [10]. SDH, a TCA cycle enzyme, catalyzes the cleavage of succinate's carboxyl group while concurrently transferring two hydrogen atoms and two electrons to the cofactor flavin adenine dinucleotide (FAD); this reaction is an integral part of the TCA cycle, closely associated with oxidation and electron transfer processes [31]. 6-PGDH and G-6-PDH are two critical enzymes in the PPP. G-6-PDH catalyzes the transformation of glucose-6-phosphate into 6-phosphogluconolactone while generating the reduced form of nicotinamide adenine dinucleotide phosphate (NADPH), and 6-PGDH converts 6-phosphogluconolactone into ribulose 5-phosphate, sustaining the function of the PPP [31,32]. In our study, the SDH, 6-PGDH and G-6-PDH activities of *T. fuciformis* treated with ε-C were lower than those of the control group. We hypothesize that ε-C treatment suppresses respiratory metabolism by reducing the proportion of the TCA cycle and the portion of the PPP, ultimately delaying the senescence of *T. fuciformis*. Moreover, reports by Li et al. have also indicated that aging progression in straw mushroom (*Volvariella volvacea*) can be suppressed by reducing the activities of 6-PGDH and G-6-PDH, thereby decreasing the proportion of the PPP pathway [25].

The energy state of postharvest fruits, vegetables, and edible fungi has a profound impact on their preservation. Once harvested, they transition from a state of growth to one of senescence and decomposition, commencing the continuous consumption of energy substrates [26]. A higher energy state indicates the presence of sufficient substrates within the cells to maintain a relatively slow respiration rate, which can effectively delay the depletion of energy and tissue senescence. Within this context, ATP provides the energy required for various physiological processes, including active transport, macromolecule synthesis, cell division, and various metabolic pathways. Moreover, the levels of ADP, AMP, and the energy charge are critical markers for evaluating the energy state [12,33,34]. In this experiment, *T. fuciformis* treated with ε-C had higher contents of ATP and ADP and an elevated energy charge, as well as lower levels of AMP and a lower disease index and weight loss rate, compared to the control group. These results suggest that ε-C treatment can slow down the consumption of energy, maintain the integrity of the cellular structure of *T. fuciformis*, and delay its senescence and disease progression. Similarly, Yao et al. have proposed that Nanocomposite-based packaging treatment can slow down the decline in ATP content and energy charge in white *Hypsizygus marmoreus*, thereby inhibiting the quality loss of white *H. marmoreus* [35].

Ca^{2+}-ATPase, Mg^{2+}-ATPase, and H^{+}-ATPase are crucial enzymes in the process of energy metabolism, playing roles in the hydrolysis of energy from ATP molecules to facilitate the transport of various molecules across cell membranes, intracellular signal transduction, and other biological processes [36]. Moreover, these ATPases are significant for the formation and maintenance of the electrochemical gradient across cell membranes, as well as for sustaining cellular pH and osmotic pressure. Thus, maintaining high ATPase activity is essential for the preservation of intracellular energy [37]. In this study, the activities of Ca^{2+}-ATPase, Mg^{2+}-ATPase, and H^{+}-ATPase in *T. fuciformis* treated with ε-C were observed to be higher than those in the control group. When coupled with previous experimental results, these data indicate that novel water-based phase change coolant (PCC) treatment can enhance ATPase activities and maintain an optimal level of energy supply, thereby delaying the senescence of shiitake mushrooms (*Lentinula edodes*) [38].

Moreover, it has been found that the application of a chitosan-based composite coating effectively prolongs the shelf life of and combats bacterial contamination in shiitake mushrooms (*Lentinus edodes*) and enoki mushrooms (*Flammulina velutipes*) [18,39]. These findings indicate that composite coatings possess considerable research merit and application potential within the realm of edible fungi preservation.

5. Conclusions

Studies indicate that ε-PL and chitosan composite coating preservatives can effectively delay senescence and quality deterioration of fresh *T. fuciformis* during storage. Further analysis reveals that these preservatives can suppress the electron transport chain and the TCA cycle and PPP in the respiratory metabolism, thereby moderating the respiration rate through the alteration of respiratory metabolic pathways. Additionally, the treatments are observed to enhance the activities of Ca^{2+}-ATPase, Mg^{2+}-ATPase, and H^{+}-ATPase, thus elevating the energy levels of *T. fuciformis* during storage. Consequently, it is believed that ε-PL and chitosan composite coatings hold significant potential for the preservation of fresh edible fungi.

Author Contributions: Conceptualization, J.S. and Y.W.; methodology, J.S., Y.W. and L.L.; data curation, Y.W. and L.L.; writing—original draft preparation, J.S. and Y.W.; writing—review and editing, B.T., Y.Y., Z.X., J.C. and P.L. All authors have read and agreed to the published version of the manuscript.

Funding: This work was supported by the Fujian Provincial People's Government-China Academy of Agricultural Sciences High-quality Development of Agriculture Beyond the "5511" Collaborative Innovation Engineering Project (XTCXGC2021014), the Fujian provincial department of science and technology, China (2023J01377, 2023R1099 and 2022R1032005), and the Finance Special Project of the Province of Fujian-Technology and Innovation Team (CXTD2021018-2).

Institutional Review Board Statement: Not applicable.

Informed Consent Statement: Not applicable.

Data Availability Statement: The original contributions presented in the study are included in the article, further inquiries can be directed to the corresponding author.

Conflicts of Interest: The authors declare no conflicts of interest.

References

1. Wang, L.; Wu, Q.; Zhao, J.; Lan, X.; Yao, K.; Jia, D. Physicochemical and rheological properties of crude polysaccharides extracted from *Tremella fuciformis* with different methods. *CYTA J. Food* **2021**, *19*, 247–256. [CrossRef]
2. Zheng, Z.; Wu, L.; Li, Y.; Deng, W.; Chen, S.; Song, H. Effects of different blanching methods on the quality of *Tremella fuciformis* and its moisture migration characteristics. *Foods* **2023**, *12*, 1669. [CrossRef] [PubMed]
3. Wu, Q.; Zheng, C.; Ning, Z.; Yang, B. Modification of low molecular weight polysaccharides from *Tremella fuciformis* and their antioxidant activity in vitro. *Int. J. Mol. Sci.* **2007**, *8*, 670–679. [CrossRef]
4. Huang, S.; Zou, Y.; Ye, Z.; Chen, J.; Luo, J.; Lan, Y.; Guo, L.; Lin, J.; Zheng, Q. A comparative study on the physio-chemical properties, antioxidant and immuno-stimulating activities of two national geographical indication products of *Tremella fuciformis* in china. *Int. J. Food Sci. Technol.* **2021**, *56*, 2904–2914. [CrossRef]
5. Wang, X.; Zhang, Z.; Zhao, M. Carboxymethylation of polysaccharides from *Tremella fuciformis* for antioxidant and moisture-preserving activities. *Int. J. Biol. Macromol.* **2015**, *72*, 526–530. [CrossRef]
6. Lin, Y.; Lai, D.; Wang, D.; Zhou, F.; Tan, B.K.; Zhang, Z.; Hu, J.; Lin, S. Application of curcumin-mediated antibacterial photodynamic technology for preservation of fresh *Tremella fuciformis*. *LWT* **2021**, *147*, 111657. [CrossRef]
7. Li, C.; Sun, L.; Zhu, J.; Cheng, Y.; Huang, R.; Fan, Y.; Guo, M.; Ge, Y. Trehalose maintains the quality of *Malus domestica* by mediating sucrose and respiratory metabolism. *Sci. Hortic.* **2022**, *295*, 110857. [CrossRef]
8. Li, L.; Lv, F.; Guo, Y.; Wang, Z. Respiratory pathway metabolism and energy metabolism associated with senescence in postharvest broccoli (*Brassica oleracea l. Var. Italica*) florets in response to O_2/CO_2 controlled atmospheres. *Postharvest Biol. Technol.* **2016**, *111*, 330–336. [CrossRef]
9. Li, L.; Sun, H.; Kitazawa, H.; Wang, X. Effects of a high O_2 dynamic-controlled atmosphere technology on the browning of postharvest white mushroom (*Agaricus bisporus*) in relation to energy metabolism. *Food Sci. Technol. Int.* **2017**, *23*, 385–395. [CrossRef]
10. Tang, J.; Ren, H.; Chen, X.; Ma, F.; Jiang, F.; Sun, B. Effects of short-term n2 anaerobic treatment on respiratory metabolism and oxidation status of *Agaricus bisporus*. *Postharvest Biol. Technol.* **2021**, *181*, 111692. [CrossRef]
11. Zhang, S.; Fang, X.; Wu, W.; Tong, C.; Chen, H.; Yang, H.; Gao, H. Effects of negative air ions treatment on the quality of fresh shiitake mushroom (*Lentinus edodes*) during storage. *Food Chem.* **2022**, *371*, 131200. [CrossRef]
12. Wei, M.; Ge, Y.; Li, C.; Chen, Y.; Wang, W.; Duan, B.; Li, X. Antifungal activity of ε-poly-L-lysine on *Trichothecium roseum* in vitro and its mechanisms. *Physiol. Mol. Plant P* **2018**, *103*, 23–27. [CrossRef]
13. Dou, Y.; Dhanasekaran, S.; Ngea, G.L.N.; Yang, Q.; Zhang, X.; Zhao, L.; Wang, K.; Zhang, H. Transcriptome analysis provides insights into potential mechanisms of epsilon-poly-l-lysine inhibiting *penicillium expansum* invading apples. *Postharvest Biol. Technol.* **2024**, *207*, 112622. [CrossRef]
14. Wang, D.; Xu, P.; Sun, J.; Yuan, J.; Zhao, J. Effects of ethanol stress on epsilon-poly-l-lysine (ε-PL) biosynthesis in *Streptomyces albulus* x-18. *Enzym. Microb. Technol.* **2022**, *153*, 109907. [CrossRef]
15. Adiletta, G.; Di Matteo, M.; Petriccione, M. Multifunctional role of chitosan edible coatings on antioxidant systems in fruit crops: A review. *Int. J. Mol. Sci.* **2021**, *22*, 2633. [CrossRef]
16. Wang, J.; Wang, B.; Jiang, W.; Zhao, Y. Quality and shelf life of mango (*Mangifera Indica* L. cv. 'Tainong') coated by using chitosan and polyphenols. *Food Sci. Technol. Int.* **2007**, *13*, 317. [CrossRef]
17. Ackah, S.; Bi, Y.; Xue, S.; Yakubu, S.; Han, Y.; Zong, Y.; Atuna, R.A.; Prusky, D. Post-harvest chitosan treatment suppresses oxidative stress by regulating reactive oxygen species metabolism in wounded apples. *Front. Plant Sci.* **2022**, *13*, 959762. [CrossRef]
18. Li, F.; Han, Q.; Wang, W.; Wu, S. Carboxymethyl chitosan-based coatings loaded with glutathione extend the shelf-life of harvested enoki mushrooms (*Flammulina velutipes*). *LWT* **2022**, *166*, 113807. [CrossRef]
19. Eissa, H.A.A. Effect of chitosan coating on shelf life and quality of fresh-cut mushroom. *J. Food Qual.* **2007**, *30*, 623–645. [CrossRef]
20. Jiang, T.; Feng, L.; Zheng, X. Effect of chitosan coating enriched with thyme oil on postharvest quality and shelf life of shiitake mushroom (*Lentinus edodes*). *J. Agric. Food Chem.* **2012**, *60*, 188–196. [CrossRef]
21. Sun, T.; Bian, J.; Wang, Y.; Hu, J.; Yun, X.; Chen, E.; Dong, T. One-step synthesis of poly(l-lactic acid)-based soft films with gas permselectivity for white mushrooms (*Agaricus bisporus*) preservation. *Foods* **2023**, *12*, 586. [CrossRef]
22. Cliffe Byrnes, V.; O' Beirne, D. Effects of gas atmosphere and temperature on the respiration rates of whole and sliced mushrooms (*Agaricus bisporus*)— implications for film permeability in modified atmosphere packages. *J. Food Sci.* **2007**, *72*, E197–E204. [CrossRef]
23. Li, P.; Zhang, X.; Hu, H.; Sun, Y.; Wang, Y.; Zhao, Y. High carbon dioxide and low oxygen storage effects on reactive oxygen species metabolism in *Pleurotus eryngii*. *Postharvest Biol. Technol.* **2013**, *85*, 141–146. [CrossRef]
24. Zhang, Z.; Zhang, X.; Xin, G.; Gong, X.; Wang, Y.; Wang, L.; Sun, B. Umami taste and its association with energy status in harvested *Pleurotus geesteranus* stored at different temperatures. *Food Chem.* **2019**, *279*, 179–186. [CrossRef]

25. Li, N.; Chen, F.; Cui, F.; Sun, W.; Zhang, J.; Qian, L.; Yang, Y.; Wu, D.; Dong, Y.; Jiang, J.; et al. Improved postharvest quality and respiratory activity of straw mushroom (*Volvariella volvacea*) with ultrasound treatment and controlled relative humidity. *Sci. Hortic.* **2017**, *225*, 56–64. [CrossRef]
26. Xue, X.; Wang, F.; Zhou, J.; Chen, F.; Li, Y.; Zhao, J. Online cleanup of accelerated solvent extractions for determination of adenosine 5′-triphosphate (ATP), adenosine 5′-diphosphate (ADP), and adenosine 5′-monophosphate (AMP) in royal jelly using high-performance liquid chromatography. *J. Agric. Food Chem.* **2009**, *57*, 4500–4505. [CrossRef]
27. Li, D.; Wang, D.; Fang, Y.; Belwal, T.; Li, L.; Lin, X.; Xu, Y.; Chen, H.; Zhu, M.; Luo, Z. Involvement of energy metabolism and amino acid metabolism in quality attributes of postharvest *Pleurotus eryngii* treated with a novel phase change material. *Postharvest Biol. Technol.* **2021**, *173*, 111427. [CrossRef]
28. Wu, Y.; Hu, Q.; Li, Z.; Pei, F.; Mugambi Mariga, A.; Yang, W. Effect of nanocomposite-based packaging on microstructure and energy metabolism of *Agaricus bisporus*. *Food Chem.* **2019**, *276*, 790–796. [CrossRef]
29. Hu, Y.; Li, J.; Lin, H.; Liu, P.; Zhang, F.; Lin, X.; Liang, J.; Tao, Y.; Jiang, Y.; Chen, B. Ultrasonic treatment decreases *Lyophyllum* decastes fruiting body browning and affects energy metabolism. *Ultrason. Sonochem.* **2022**, *89*, 106111. [CrossRef]
30. Gualanduzzi, S.; Baraldi, E.; Braschi, I.; Carnevali, F.; Gessa, C.E.; De Santis, A. Respiration, hydrogen peroxide levels and antioxidant enzyme activities during cold storage of zucchini squash fruit. *Postharvest Biol. Technol.* **2009**, *52*, 16–23. [CrossRef]
31. Zhou, Y.; Liu, X.; Liang, X.; Li, H.; Lai, J.; Liao, Y.; Liu, K. Biochemical and metabolomics analyses reveal the mechanisms underlying ascorbic acid and chitosan coating mediated energy homeostasis in postharvest papaya fruit. *Food Chem.* **2024**, *439*, 138168. [CrossRef]
32. Lin, Y.; Lin, Y.; Lin, M.; Chen, L.; Li, H.; Lin, H. Propyl gallate postharvest treatment improves the storability of longans by regulating the metabolisms of respiratory and disease-resistance substances. *Postharvest Biol. Technol.* **2023**, *206*, 112556. [CrossRef]
33. Lin, Y.; Zhan, L.; Shao, P.; Sun, P. Phase-change materials and exogenous melatonin treatment alleviated postharvest senescence of *Agaricus bisporus* by inhibiting browning and maintaining cell membrane integrity. *Postharvest Biol. Technol.* **2022**, *192*, 112009. [CrossRef]
34. Li, L.; Kitazawa, H.; Wang, X.; Sun, H. Regulation of respiratory pathway and electron transport chain in relation to senescence of postharvest white mushroom (*Agaricus bisporus*) under high O_2/CO_2 controlled atmospheres. *J. Agric. Food Chem.* **2017**, *65*, 3351–3359. [CrossRef]
35. Yan, M.; Yuan, B.; Cheng, S.; Huang, H.; Huang, D.; Chen, J.; Cao, C. Nanocomposite-based packaging affected the taste components of white *Hypsizygus marmoreus* by regulating energy status. *Food Chem.* **2020**, *311*, 125939. [CrossRef]
36. Li, M.; Lin, H.; Chen, Y.; Chen, Y.; Lin, Y.; Fan, Z.; Wang, H. The role of energy transport system in pericarp browning of harvested longan fruit. *Postharvest Biol. Technol.* **2024**, *207*, 112619. [CrossRef]
37. Yang, B.; Han, Y.; Gao, H.; Liu, R.; Xu, F.; Liu, R.; Xiao, S.; Li, B.; Chen, H. Application of melatonin delays lignification in postharvest water bamboo shoots in association with energy metabolism. *Postharvest Biol. Technol.* **2023**, *196*, 112149. [CrossRef]
38. Li, D.; Wang, D.; Fang, Y.; Li, L.; Lin, X.; Xu, Y.; Chen, H.; Zhu, M.; Luo, Z. A novel phase change coolant promoted quality attributes and glutamate accumulation in postharvest shiitake mushrooms involved in energy metabolism. *Food Chem.* **2021**, *351*, 129227. [CrossRef] [PubMed]
39. Guo, Y.; Chen, X.; Gong, P.; Guo, J.; Deng, D.; He, G.; Ji, C.; Wang, R.; Long, H.; Wang, J.; et al. Effect of shiitake mushrooms polysaccharide and chitosan coating on softening and browning of shiitake mushrooms (*Lentinus edodes*) during postharvest storage. *Int. J. Biol. Macromol.* **2022**, *218*, 816–827. [CrossRef] [PubMed]

foods MDPI

Article

Investigation of Melatonin Incorporated CMC-Gelatin Based Edible Coating on the Alleviation of Chilling Injury Induced Pericarp Browning in Longkong

Karthikeyan Venkatachalam [1], Narin Charoenphun [2], Somwang Lekjing [1] and Paramee Noonim [1,*]

[1] Faculty of Innovative Agriculture and Fishery Establishment Project, Prince of Songkla University, Surat Thani Campus, Makham Tia, Mueang, Surat Thani 84000, Thailand; karthikeyan.v@psu.ac.th (K.V.); somwang.s@psu.ac.th (S.L.)

[2] Faculty of Science and Arts, Burapha University Chanthaburi Campus, Chanthaburi 22170, Thailand; narinch@buu.ac.th

* Correspondence: paramee.n@psu.ac.th

Abstract: Longkong (*Aglaia dookkoo* Griff.) fruit is prone to rapid pericarp browning and shortened shelf life (<7 days) under prolonged low-temperature storage. This study investigates the effect of an edible coating, comprising carboxymethyl cellulose (CMC) and gelatin in a fixed 3:1 ratio, integrated with various concentrations of melatonin (MT) (0.4, 0.8, and 1.2 mM/L) to mitigate chilling injury in longkong fruit. Coated longkong fruits were stored at 13 °C with 90% relative humidity for 18 days and underwent physicochemical evaluations every three days. Samples coated with CMC-Gel without MT and uncoated fruits were served as controls. The findings indicated that the CMC-Gel-MT coating significantly mitigated pericarp browning, chilling injury, weight loss, and respiration rate increase under extended cold storage conditions. High concentrations of MT ($\geq$0.8 mM/L) in the coating notably inhibited the activities of cellular degrading enzymes such as lipoxygenase and phospholipase D. This inhibition contributed to reduced membrane permeability, lower reactive oxygen species accumulation (H_2O_2, OH^-, $O_2{}^-$), and decreased malondialdehyde levels in the longkong pericarp. Furthermore, the CMC-Gel-MT coating increased the activity of phenylalanine ammonia lyase, leading to an enhancement in phenolic content. Consequently, it improved the fruit's ability to scavenge DPPH (2,2-diphenyl-1-picrylhydrazyl) and ABTS (2,20-azino-di-3-ethylbenzthiazoline sulfonic acid) radicals. Control samples exhibited high levels of pericarp browning-related enzymes (polyphenol oxidase, peroxidase), whereas CMC-Gel-MT-coated fruits, particularly at higher MT concentrations, showed significant reductions in those enzyme activities. In conclusion, incorporating high concentrations of MT in a CMC-Gel-based edible coating is a promising alternative for mitigating chilling injury in longkong fruit.

Keywords: longkong pericarp; chilling injury; melatonin; CMC-gelatin; storage; enzyme activity; fruit quality

1. Introduction

Longkong (*Aglaia dookkoo* Griff.) is a non-climacteric tropical fruit, abundant in health-beneficial phytonutrients, that demonstrates potent antioxidant and antimicrobial properties [1]. Additionally, the fruit is characterized by a unique flavor profile, primarily recognized for its pleasant taste, which starts sweet and then shifts to a mild sourness [2]. Although the fruit offers significant health benefits, only the flesh is edible, while other parts, including the seed and pericarp, are nonedible. The fruit typically contains 3–5 green seeds enveloped in white, pale flesh. It is recommended to be cautious while consuming longkong fruit, as inadvertent ingestion of the seeds may lead to an adverse sensory experience attributable to their pronounced bitterness [3]. The pericarp, in contrast, is tasteless. However, both pericarp and seeds have been utilized in various traditional medicines, and recent studies suggest they may contain key components useful in modern drug production [4]. Longkong fruit is rich in essential nutrients, but its short shelf life limits its market reach, consequently diminishing its economic value [5]. Predominantly grown in Thailand and other tropical countries in Southeast Asia, the fruit suffers from rapid deterioration due to browning and microbial spoilage, which has led to limited recognition and restricted market coverage. Several studies have been conducted to investigate the causes of spoilage, with rapid pericarp deterioration, mainly browning and consequently a short shelf life, typically under ambient storage conditions (less than 5–7 days) or chilling temperatures (less than 5 days), being the primary issues [6–8]. To expand its market reach, longkong fruit must be chilled for transportation over long distances, yet this low-temperature storage can exacerbate chilling injuries [9]. Additionally, environmental variables may contribute to the extrinsic aging of longkong fruits. This causes the production of reactive oxygen species (ROS), which is thought to be an essential indicator of longkong pericarp resiliency to solar ultraviolet radiation and chilling injury. In terms of chilling injury, the ROS are produced and accumulate when cold stress occurs to longkong fruits after harvest. This initiates an oxidative chain reaction that leads to oxidative damage and loss of function in the cell membrane [4,6]. Numerous postharvest studies have explored ways to extend longkong fruit's shelf life, employing various active and passive modified atmospheric packaging techniques as well as phytohormone treatments to control chilling injury [6,10,11]. While some of these methods are easy to use, they may struggle to maintain quality, and vice versa. This situation presents an ongoing opportunity for research into longkong fruit, aiming to find a feasible method to achieve extended shelf life with technology that is both easy to use and effective for commercial purposes.

Edible coatings, specifically those involving polymers, are a prevalent postharvest technique used to extend the shelf life of a broad array of plant produce, especially fruits [12]. Carboxymethyl cellulase (CMC), a polysaccharide-based edible polymer, boasts commendable coating properties due to its pronounced hydrophilicity. Notably, CMC can be integrated with various functional ingredients, which enhances its efficiency in preserving fruit quality and extending shelf life. Additionally, this polymer can form a colorless, odorless, nontoxic, stable, and uniform coating, although it sometimes lacks sufficient mechanical strength [13]. To address this, recent studies have highlighted the potential of blending polymers to bolster the coating's strength and prevent fruit quality loss [14]. Gelatin (Gel), a protein-based edible polymer, has gained popularity due to its biodegradable attributes and superior film and coating formation capabilities [15]. Combining CMC and Gel has demonstrated significant potential in prolonging the shelf life of agricultural products, with an optimal concentration blend reported at a 3:1 ratio [16]. The combined use of CMC and Gel offers an ecofriendly, biodegradable alternative for edible food coatings [17,18]. This blend exhibits increased viscoelasticity, resistance to oil/fat, enhanced coating properties, and a proven capability to prolong fruit shelf life [17]. Furthermore, the CMC–gelatin biopolymeric matrix can produce transparent edible coatings and serve as a carrier for active ingredients, including phytohormones, antioxidants, and antimicrobial agents [19]. Recent studies have shown that the addition of phytohormones such as melatonin (MT) into the coating emulsion can offer immense advantages, including enhanced disease resistance, antioxidant activity, resistance to pathogens, decay inhibition, and extended fruit shelf life [20]. MT is renowned for its protective effects against cellular and DNA damage, and its role in inhibiting stress-induced peroxidation enhances oxidation resistance in plants [21]. It mitigates pericarp browning, onset of free radicals, reactive oxygen species, and phenolic oxidations. Moreover, MT has been shown to control damage from extended chilling and suppress lipoxygenase activity, which in turn manages alterations in cell membrane fatty acids [22]. The efficacy of MT in postharvest fruit treatments varies with the concentration used and its effectiveness also varies among fruit types [23]. When MT was applied exogenously, it dramatically improved disease resistance and decreased decay in bananas, grapes, kiwis, strawberries, and plums [20].

The present study utilized the opportunity to use of this unique combination including CMC-Gel coating that incorporated various concentrations of MT to investigate the effects of controlling pericarp browning in longkong fruit stored under prolonged chilling temperatures by regulating ROS and the membrane metabolism.

2. Materials and Methods

2.1. Raw Materials, Chemicals, and Reagents

Longkong fruit samples were collected from a designated garden in the Nasan District, Surat Thani Province, Thailand. The fruit samples were gathered 14 weeks post-anthesis, after which they were assessed based on the maturity index (color, total soluble solids, and titratable acidity) recommended by Venkatachalam and Meenune [7]. The collected fruits were transported to the laboratory within 4–6 h for further inspection to identify any visible damage caused by abrasion, impact, or microbial attack. Once the fruits were selected for treatment, they were individually detached from their bunches with the raceme end still attached and washed with distilled water to remove any dust or dirt on the surface and submerged in sodium hypochlorite solution (200 μL/L) for 2 min, followed by washing again with distilled water. Subsequently, excess moisture on the fruits was eliminated with paper towels, followed by exposure to an electric fan to ensure the complete removal of any residual moisture on their surface. The fruits were then prepared for the coating application, as detailed in Section 2.3.

For coating, food-grade chemicals and reagents such as CMC, gelatin, melatonin, and glycerol were used, while all other chemicals utilized for analyses in this study were of analytical grade. All the chemicals and reagents used for this study were procured from Sigma Aldrich, Bangkok, Thailand.

2.2. Formulation of Coating and Treatment

The CMC-Gel coating emulsion was prepared following the method described by Vargas-Torrico et al. [24], with certain modifications. The emulsion, consisting of 1.5% CMC, 0.5% Gel, and 0.5% glycerol, was mixed in 500 mL of deionized water. This mixture was continuously stirred at 600 rpm and heated on a magnetic stirrer hot plate at 90 ± 2 °C for 30 min. After this, the emulsion was cooled to a temperature range of 45–50 °C. MT was then introduced separately in four different concentrations: 0.4, 0.8, and 1.2 mM/L, resulting in treatments CMC-Gel-MT1, MT2, MT3, respectively. The CMC-Gel coating emulsion without MT was served as treatment–control and fruit with no coating was served as control (C). Each emulsion was continuously mixed for 15 min at the aforementioned rpm, then allowed to cool to room temperature. Upon reaching approximately 25 °C, the emulsions were applied to the fruits using an immersion method. The prepared longkong fruits were divided into five groups, with 30 fruits each, for every replication per treatment group. They were meticulously immersed in their respective emulsions, ensuring a uniform coating. The fruits were then subjected to airflow at a low speed (approximately 300–400 rpm) for 20 min to dry the surface. Once completely dried, the fruits were placed in perforated punnets and stored under chilling conditions (13 °C with 90% relative humidity) for 18 days. Every three days, the fruits were assessed for various quality attributes, as outlined in Section 2.4. Figure 1 shows the infographic representation of raw material preparation, CMC-Gel-MT coating formation and application on longkong fruit.

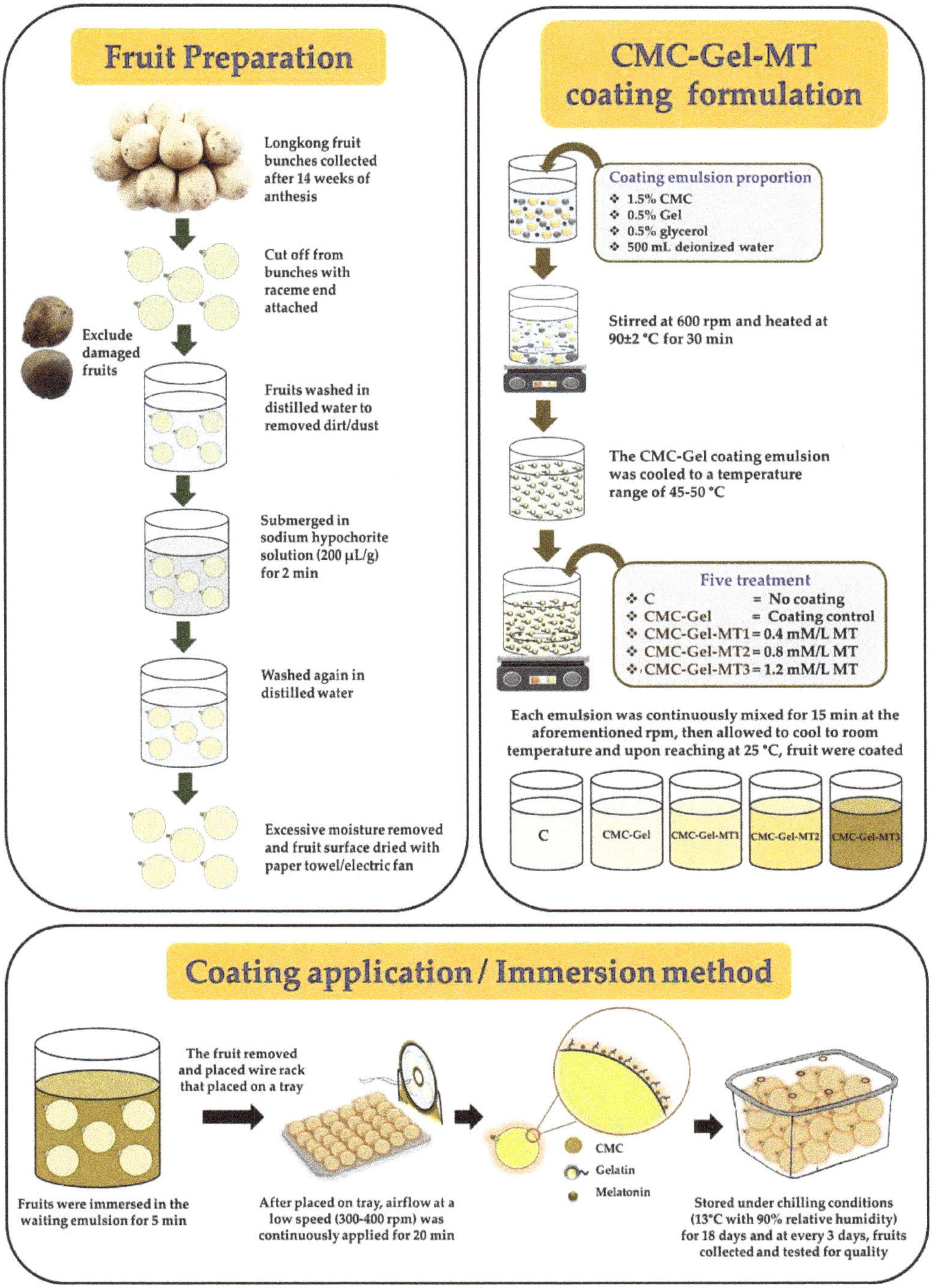

Figure 1. Infographic presentation of raw material preparation, coating formulation and application on longkong fruit surface.

2.3. *Quality Analysis*

2.3.1. Pericarp Color and Observation of Fruit Surface Appearance

To quantify the fruit's color properties, this study measured 15 individual fruits from each replicate. A colorimeter (HunterLab, Reston, VA, USA) was employed to determine the lightness (L*), redness (a*), and yellowness (b*) color parameters of the fruit's pericarp (outer skin) by following the procedure proposed by Charoenphun et al. [6]; four random surfaces on each fruit were selected for analysis. For the observation of fruit surface appearance, the fruits from initial and final day of storage were carefully collected—any surface coating and moisture removed—and digitally photographed using a handheld Sony digital camera (Model ZV-1F, Tokyo, Japan), following the method of Venkatachalam [25].

2.3.2. Chilling Injury Index (CI)

To assess CI in longkong fruit pericarp (outer skin), this study employed visual inspection by following the method of Nguyen et al. [26]. Longkong fruit develops a distinctive brown discoloration on its pericarp when chilled, which we quantified using a 5-point scale ranging from 1 (no CI) to 5 (severe injury).

2.3.3. Electrolytic Leakage (EL)

EL measurement in longkong fruit pericarp was adapted from the method of Wang et al. [27]. Twenty longkong pericarp discs from each replication were cut out using a metal borer. The discs were then rinsed with distilled water and air-dried on absorbent paper. Each disc was placed in an individual test tube holding 20 mL of distilled water. These tubes were then shaken in a water bath at 25 °C for half an hour. Then, they were measured initial conductivity (L0) with a conductivity meter and then the tubes were subjected to a temperature of 100 °C for 20 min, followed by rapid cooling. Post-cooling, the pericarp discs were measured for final electrical conductivity (L1). The rate of electrolyte leakage in longkong fruit was calculated as the percentage ratio of L0 to L1 ((L0/L1) × 100%).

2.3.4. Malondialdehyde (MDA) Content

MDA content in longkong fruit pericarp was assessed using the procedure proposed by Noonim and Venkatachalam [28]. Two grams of pericarp was blended with 8 mL of sodium phosphate buffer (50 mmol/L, pH 7.8) and then the mixture was centrifuged at 12,000× *g* and 4 °C for 15 min. Two milliliters of the resulting supernatant was collected and combined with 3 mL of thiobarbituric acid solution (5 g/L) in 0.1% trichloroacetic acid (TCA). Then, this mixture was heated in a boiling water bath for 15 min, quickly cooled, and centrifuged under the same conditions as above. The absorbance of the final supernatant was recorded at wavelengths of 450, 532, and 600 nm. The MDA levels were determined, and the findings of the samples were reported as nmol/g of fresh weight (FW).

2.3.5. Fruit Weight Loss

The weight loss of longkong fruit was determined by following the method of Chen et al. [29]. This involved daily weighing of the fruit samples and calculating the weight loss over time by comparing the weights at different storage intervals. The data were then presented as percentage of weight loss.

2.3.6. Respiration Rate

The respiration rate of longkong fruit was assessed using the method of Caleb et al. [30]. For analysis, 30 fruits for each replication were collected and placed in individual 3 L sealed glass chambers. These chambers were then kept at room temperature for 3 h. Subsequently, a 1 mL gas sample was extracted with a syringe and analyzed using a gas chromatograph (GC; Auto systems XL, Perkin Elmer, Waltham, MA, USA), which was equipped with a Porapack Q column (80/100 mesh) and a thermal conductivity detector. The respiration rates were quantified as mL CO_2 per mg kg^{-1} h^{-1}.

2.3.7. Determination of Hydrogen Peroxide (H_2O_2)

H_2O_2 content in longkong fruit pericarp was assessed by using the method of Patterson et al. [31]. A 2 g pericarp sample was homogenized in 15 mL of acetone under cold conditions and then the mixture was centrifuged at 6000× g for 15 min at 4 °C. After that, 1 mL of the supernatant was added along with 0.1 mL of 5% $TiOSO_4$ and 0.2 mL of NH_3 to the centrifuge tube and then centrifuged again under similar conditions for 10 min. The resultant pellet was then dissolved in 3 mL of 10% H_2SO_4 and centrifuged at 5000× g for 10 min at 4 °C. Then, the final supernatant was used to measure the H_2O_2 content. The measurement was based on the absorbance at 410 nm, using a UV-Vis spectrophotometer (model: Mini UV 1240, Shimadzu, Kyoto, Japan). The results were then expressed as nmol g^{-1} FW.

2.3.8. Determination of Hydroxyl Radical (OH^-)

The OH^- content in longkong fruit pericarp was assessed using the 2-deoxyribose oxidation method by Chung et al. [32]. A 10 g pericarp sample was homogenized in 50 mL of methanol, followed by filtration and centrifugation at 2000× g for 10 min at 4 °C. The supernatant obtained was then used for analysis. In a test tube, 0.2 mL of a reagent mixture containing 10 mmol of iron (II) sulfate heptahydrate and 10 mmol of ethylenediaminetetraacetic acid (EDTA) was combined with 0.2 mL of 2-deoxyribose. After thorough mixing, 0.8 mL of the sample was added to this mixture, along with 1 mL of 0.1 mol sodium phosphate buffer at pH 7.4, followed by the addition of 200 µL of 10 mmol hydrogen peroxide to the mixture, which was then incubated at 37 °C for 4 h. Subsequently, 1 mL of 2.8% trichloroacetic acid (TCA) solution was added, and the mixture was heated in a boiling water bath for 10 min before cooling to room temperature. The absorbance of the mixture was measured at 520 nm using a UV-Vis spectrophotometer (model: Mini UV 1240, Shimadzu, Kyoto, Japan). The results were then expressed as nmol g^{-1} FW.

2.3.9. Determination of Superoxide Anion (O_2^-) Radical

To measure the O_2^- content [33] in longkong fruit pericarp, a 4 g pericarp sample was homogenized under cold conditions (4–8 °C) in 12 mL of 50 mmol/L potassium phosphate buffer (pH 7.8) with 1% polyvinylpyrrolidone. The homogenate was then centrifuged at 5000× g for 15 min at 4 °C. Subsequently, 1 mL of the resulting supernatant was combined with 0.9 mL of the same potassium phosphate buffer and 0.1 mL of 10 mmol/L hydroxylamine hydrochloride and this mixture was thoroughly mixed and incubated at 25 °C for 30 min. Following this, 1 mL of the incubated mixture was added to 1 mL of 17 mmol/L metanilic acid and 1 mL of 7 mmol/L 1-naphthylamine. After thorough mixing, this final mixture was incubated again at 25 °C for 30 min. The absorbance of the mixture was then measured at 530 nm using a UV-Vis spectrophotometer (model Mini UV 1240, Shimadzu, Kyoto, Japan). The results were then expressed as nmol g^{-1} FW.

2.3.10. Total Phenolic Content (TPC)

The determination of TPC in longkong fruit pericarp was performed using a colorimetric method [34]. Five grams of the pericarp was homogenized with 25 mL of 95% ethanol, the mixture was centrifuged at 12,000× g and 4 °C for 15 min, and the supernatant was collected and proceeded for the TPC measurement. The procedure involved mixing 8.4 mL of distilled water with 100 µL of the fruit sample and 500 µL of Folin–Cioclateu reagent in a test tube. This mixture was left to stand for 3 min, after which 1.0 mL of 20% sodium carbonate was added. The mixture was then thoroughly vortexed (Vision Scientific Co., Ltd., Daegu, Republic of Korea). After an hour, the absorbance at 720 nm was measured using a spectrophotometer (model: Mini UV 1240, Shimadzu, Kyoto, Japan). The phenolic content was quantified by comparing the absorbance to a standard curve created with chlorogenic acid, and the results were expressed as µg gallic equivalent (GAE) g^{-1} FW.

2.3.11. Antioxidant Activities

For determining the antioxidant activities, a 5 g sample of pericarp was homogenized with 25 mL of 95% ethanol, the mixture was centrifuged at 12,000× g and 4 °C for 15 min, and the supernatant was collected and processed for antioxidant measurement. For DPPH radical scavenging activity [35] of longkong fruit, a 100 µL sample of the extract was thoroughly mixed with 3.9 mL of 60 µmol/L DPPH solution in a test tube. This mixture was then left to incubate for 30 min in darkness at room temperature. The absorbance was measured at 515 nm using a spectrophotometer (model: Mini UV 1240, Shimadzu, Kyoto, Japan). The results were expressed as percentages. For ABTS radical scavenging activity [36], a 100 µL sample of extract was mixed with 100 µL of ABTS reagent, and then this mixture was incubated for 6 min at ambient temperature. After that, the absorbance readings were taken at 734 nm using a spectrophotometer (model: Mini UV 1240, Shimadzu, Kyoto, Japan). The results were expressed as percentages.

2.3.12. Pericarp Enzyme Activities

For measuring phenylalanine ammonia lyase (PAL) activity [29], 4 g of pericarp sample was homogenized with 20 mL of borate buffer (50 mM, pH 8.5), which included 5 mM β-mercaptoethanol and polyvinylpyrrolidone (PVP) (25 g/L) and after that, the mixture was centrifuged at 15,000× g at a temperature of 4 °C for 20 min. The resulting supernatant served as the crude enzyme extract. To measure PAL activity, 1 mL of this extract was mixed with 1 mL of 20 mM l-phenylalanine and 2 mL of distilled water. This mixture was then incubated at 40 °C for 2 h. The reaction was halted by adding 0.1 mL of 6 M HCl, and the absorbance was recorded at 290 nm using a UV-Visible spectrophotometer (model: Mini UV 1240, Shimadzu, Kyoto, Japan). PAL activity was quantified based on the production of 1 µmol of cinnamic acid per hour, and the specific activity was reported as unit g^{-1} FW.

To measure phospholipase D (PLD) activity [37], 5 g of pericarp sample was homogenized in 20 mL of 100 mM acetate buffer (pH 5.5), and then the mixture was centrifuged at 14,000× g for 10 min at 4 °C. Separately, 0.4 g of phosphatidylcholine (lecithin) was dissolved in 50 mL of ether. The solution was evaporated using a rotary evaporator at 35 °C until it was completely dry. The dried residue was reconstituted in 1 L of 100 mM acetate buffer (pH 5.5) containing 5 mM dithiothreitol and 25 mM $CaCl_2$ to form the reaction substrate. For the PLD assay, 1 mL of the crude enzyme extract was mixed with 3 mL of the prepared reaction substrate. This mixture was then vigorously agitated for 1 h at 28 °C. Afterwards, the mixture was washed with petroleum ether. The aqueous phase was separated and mixed with 2 g of Reinecke salt (ammonium tetrarhodanatodiammonchromate) dissolved in 100 mL of methanol, resulting in the formation of a precipitate. The precipitate was collected by centrifugation at 16,000× g and redissolved in 3 mL of acetone. The absorbance of the clear supernatant was measured at 520 nm using a UV-Visible spectrophotometer (model: Mini UV 1240, Shimadzu, Kyoto, Japan), and the specific activity was reported as unit g^{-1} FW.

For measuring the activity of lipoxygenase (LOX) [38], a 5 g pericarp sample was homogenized in a chilled environment at 4 °C with 10 mL of potassium phosphate buffer (50 mM, pH 7). Then, the homogenate was centrifuged at 15,000× g for 15 min at 4 °C, and the supernatant was collected and analyzed for LOX activity. In this assay, 0.2 mL of supernatant was combined with 2.8 mL of same potassium phosphate extraction buffer that contained 25 mM of sodium linoleate. Then, the reaction mixture was measured at 234 nm using a UV-Visible spectrophotometer (model: Mini UV 1240, Shimadzu, Kyoto, Japan). LOX activity was quantified as the enzyme quantity required to alter the absorbance by 0.01 per minute, with specific activity reported as unit g^{-1} FW.

For measuring the polyphenol oxidase (PPO) activity [39], a 5 g pericarp sample was homogenized in 20 mL of 0.05 M phosphate buffer (pH 7.0) with 0.5 g of insoluble PVP at a cold temperature of 4 °C. The homogenate was then filtered through a muslin cloth and centrifuged at 19,000× g for 20 min at 4 °C. Then, the resulting supernatant was collected and used for PPO activity analysis. To measure PPO activity, 0.1 mL of supernatant was mixed with 2.9 mL of 10 mM 4-methyl catechol. The change in the reaction mixture was measured at 420 nm using a UV-Vis spectrophotometer (model: Mini UV 1240, Shimadzu, Kyoto, Japan), recorded every minute. Enzymatic activity was defined as the amount causing a 0.001 change in absorbance per minute, with specific activity reported as unit g^{-1} FW.

For measuring POD activity [40], a 2 g pericarp sample was homogenized in a mortar and pestle at 4 °C with 20 mL of 0.05 M phosphate buffer (pH 7) and 0.2 g of PVP. Then, the homogenate was strained through cheesecloth and centrifuged at 19,000× *g* for 20 min at 4 °C. The resulting supernatant was used for POD activity analysis, which involved a reaction mixture of 3 mL, comprising 25 µL of the supernatant, 2.78 mL phosphate buffer, 0.1 mL of 20 mM H_2O_2, and 0.1 mL of 20 mM guaiacol. The increase in absorbance was measured at 470 nm using a UV-Visible spectrophotometer (model: Mini UV 1240, Shimadzu, Kyoto, Japan) for over 2 min. One unit of enzyme activity was defined as a 0.01 absorbance change per minute, with specific activity reported as unit g^{-1} FW.

2.4. Statistical Analysis

All experiments were conducted in triplicate, and the results were expressed as the mean ± standard deviation. A one-way analysis of variance (ANOVA) and Duncan's multiple range test were employed to analyze the statistical significance of the mean values. SPSS software for Windows (v12) was used for the statistical analyses in this study.

3. Results and Discussion

3.1. Pericarp Color and Appearance

Generally, color is the primary characteristic used to determine harvest maturity and purchasing decisions for longkong pericarp [7]. Changes in longkong pericarp color properties including lightness (L*), redness (a*), and yellowness (b*) of longkong fruit pericarp coated with or without CMC-Gel-MT are shown in Figure 2A–C. The overall trend of color changes in longkong fruit showed that the L* and b* values of the pericarp continuously decreased throughout storage in all samples, whereas the a* values increased in the pericarp samples during storage. Among the coated and uncoated samples, the fruits coated with CMC-Gel and CMC-Gel-MT showed better retainment of color characteristics, and furthermore, the application of MT in the coating materials provided additional protection over the control of pericarp color loss. MT performance on preventing pericarp color degradation was significantly increased when the concentration of MT increased >0.4 mM/L. Overall, significant color changes in the pericarp samples were observed after the third day of storage, continuing until storage's end. The tropical longkong fruit and its pericarp are highly susceptible to color degradation under both warm and cold storage conditions, and particularly under cold conditions, the severity and damage are more pronounced [41]. Several studies have established that the darkening effect on the longkong pericarp under various storage conditions is mainly attributed to cellular degradation and the activation of enzymatic browning, especially by oxidoreductase enzymes including PPO, POD, and LOX [6,28]. LOX and PAL enzymes do not directly induce browning on the fruit; however, the PAL enzyme produces significant levels of polyphenolics under chilling stress, which are major substrates for PPO and POD [42]. Similarly, LOX activity increases rapidly under extended chilling conditions, using cellular materials as substrates [43]. This leads to the breakdown of the cell wall and cell membrane of the longkong pericarp, making it more accessible to PPO and POD. This aligns with findings by Lin et al. [44]. The application of CMC-based coating mitigated these changes; likely, the CMC coating acts as a barrier, preventing external environmental interactions with the fruit. Ali et al. [45] reported that application of CMC-based coating to mango exhibited significantly fewer color changes compared those untreated. Saowakon et al. [45] stated that CMC coating slightly controlled rambutan fruit browning compared to controls. Zheng et al. [46] found that the addition of melatonin boosts the fruit's antioxidant activities, controlling lipid peroxidation and reducing cellular degradation, resulting in less pericarp browning. This study is in concordance with their findings, as the application of CMC-Gel-

MT coating indeed controlled the ROS, LOX, and PPO activities in the longkong pericarp under prolonged chilling conditions. Similarly, the appearance of the CMC-Gel-MT-coated longkong pericarp on the initial day and at the end of storage days is shown in Figure 3. Photographs of the longkong pericarp were taken after the fruits were washed of the applied coating and our observation showed no visible decay. However, prolonged storage affected almost 90% of the surface with browning in the control fruit, followed by 60% of the surface showing browning in the CMC-Gel-coated fruits; on the other hand, <40% of surface browning was observed in CMC-Gel-MT-coated fruits.

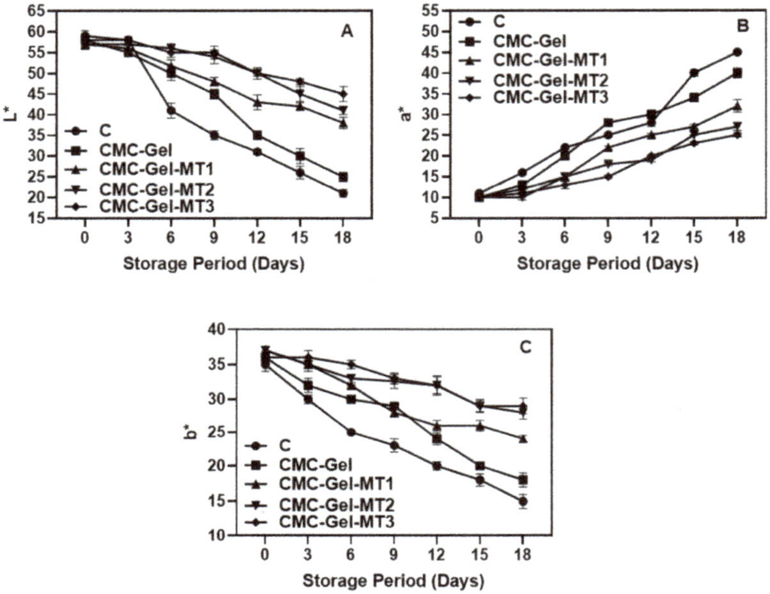

Figure 2. Changes in lightness (L*) (**A**), redness (a*) (**B**), and yellowness (b*) (**C**) of longkong fruit coated with or without CMC-Gel-MT at varying concentrations and stored under prolonged low temperature conditions.

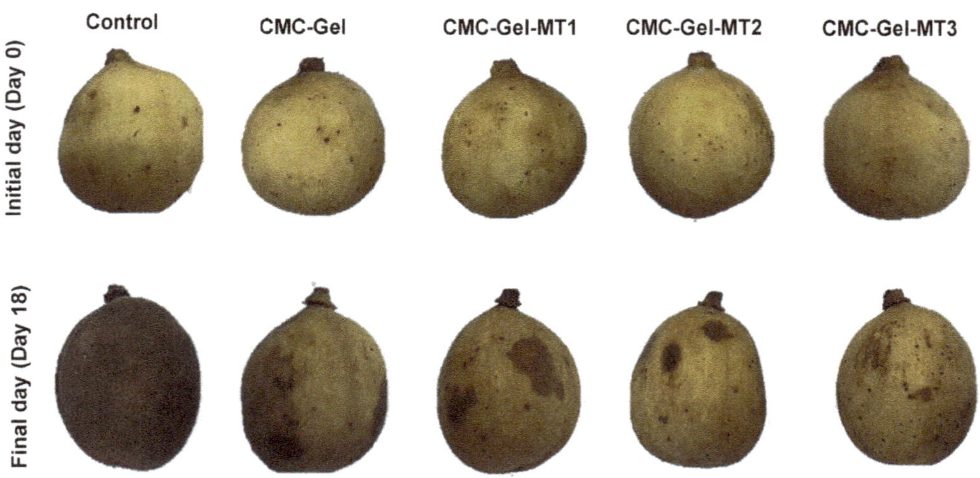

Figure 3. Surface observation of longkong fruit coated with or without CMC-Gel-MT at varying concentrations and stored under prolonged low temperature conditions.

3.2. CI Index, EL, and MDA

The CI index measures the degree of cold-induced damage in plants, assisting in the refinement of storage and handling practices to improve the quality and lifespan of produce [47]. The considerable postharvest loss of longkong fruit can primarily be attributed to the damaging effects of CI. Figure 4A displays the CI index of longkong fruit pericarp coated with and without CMC-Gel-MT. The data reveal that the onset of the CI in the examined fruit pericarp varied significantly across the different sample types. CI began to appear in control samples at the beginning of the third day of storage, while CMC-Gel-MT-treated samples showed slightly visible symptoms after the ninth day of storage, and they tended to further intensify with extended storage days. The control of CI incidence using CMC-Gel-MT was MT dose-dependent, with higher concentrations providing better control against CI incidence compared to lower concentrations and control. The CI in control samples increased intensively as the storage period increased. Generally, the longkong fruit pericarp is highly susceptible to CI. Its trichomes on the pericarp surface can act as a signaling agent, inducing an adverse response (such as the production of oxidoreductase enzymes and reactive oxygen species) against stressful environmental conditions, particularly chilling temperatures. The elevated production of reactive oxygen species results in oxidative damage to cellular membranes, ultimately compromising membrane stability [48]. This deterioration often presents as an increased level of electrolyte leakage compared to normal conditions (see Figure 4B). The utilization of edible coatings, especially CMC-Gel and CMC-Gel-MT, can enhance preservation by stabilizing the membrane structure and stability of horticultural produce. This, in turn, reduces electrolyte leakage and extends the shelf life of coated fruits [19,49].

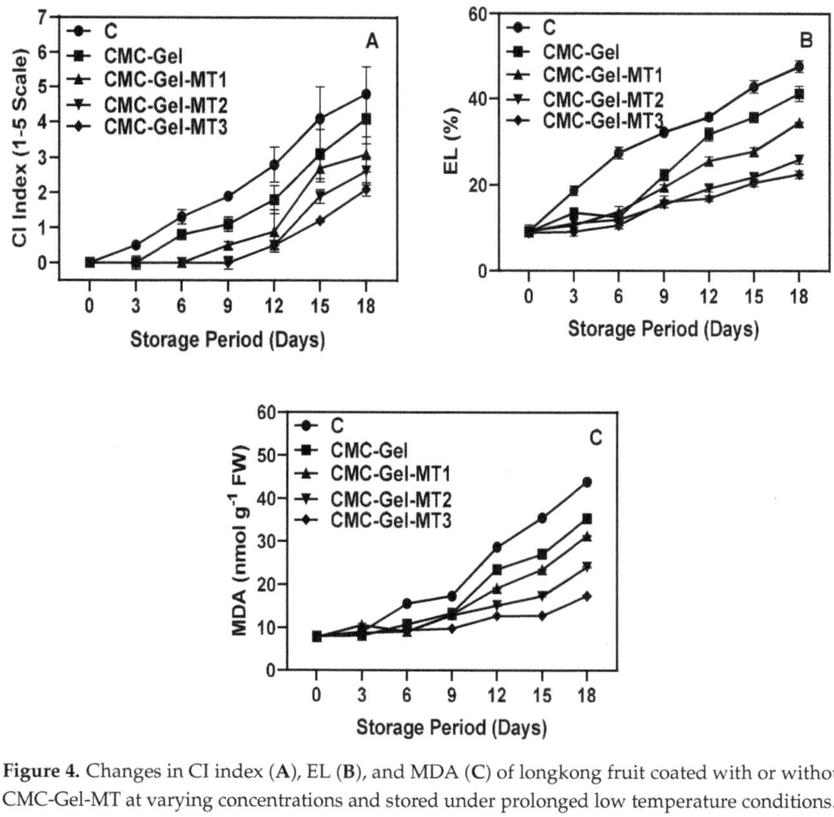

Figure 4. Changes in CI index (**A**), EL (**B**), and MDA (**C**) of longkong fruit coated with or without CMC-Gel-MT at varying concentrations and stored under prolonged low temperature conditions.

The changes in EL level in the pericarp of longkong fruit coated with Gel-CMC-MT at different concentrations are shown in Figure 4B. The results suggest that the extended storage at low temperatures significantly increased the EL level in all sample types. Furthermore, the severity of the leakage seems to augment with the duration of storage. Comparatively, longkong pericarps coated with CMC-Gel and CMC-Gel-MT exhibited a substantial reduction in leakage compared to the noncoated control samples. In comparison with CMC-Gel and its MT variation, the CMC-Gel-MT showed better control of EL and, furthermore, MT2 and MT3 samples exhibited better EL control in the fruits as compared to their low-concentration variation. EL is regarded as a key identifier of chilling damage in fruits [50]. Electrolytes in living cells reside within compartments bound by membranes, and when these cells encounter stress factors like chilling temperatures that can cause cold damage, or during natural aging, the proteins and lipids within the membranes undergo degradation and oxidation [51]. This leads to modifications in their structure, which in turn weakens their integrity and increases membrane permeability. The present study demonstrates that an escalation in EL is an indicator of increased cold injury in longkong pericarp. Low-temperature-induced membrane damage was noticeably significant after just 6 days of storage, possibly due to biochemical changes occurring within the bilayers following the initial chilling temperature storage. However, the addition of MT in the CMC-Gel coating significantly controlled those changes and reduced the chilling injury impact on the fruit's pericarp.

MDA is an important indicator of membrane dysfunction, and its increasing level corresponds to an increase in oxidative damage and membrane degradation [52]. The MDA content in the CMC-Gel-MT-coated and uncoated longkong pericarp continuously increased in trend (Figure 4C). Control fruits had the predominant level of MDA accumulation in their pericarp, followed by the treated samples. However, among the treated ones, the higher concentration of CMC-Gel-MT in the treated samples had the lowest MDA level in their pericarp ($p > 0.05$). However, when compared with other samples in this study, MT2 and MT3 were highly significant in controlling the MDA content ($p < 0.05$). CMC-Gel-MT coating may be able to suppress the membrane damage by altering the fatty acid arrangements and limiting the ROS production in the cell membrane. Composite coating made of protein and polysaccharide in the presence of crosslinkers creates strong interactions via covalent bonding, which helps reduce structural degradation, water loss, and maintain the firmness of fruits [53,54]. Saowakon et al. [45] found that the CMC-based composite coating increased the vitamin C content in rambutan fruit, which is one of the strong antioxidant vitamins that can alleviate the severity of lipid oxidation plants.

3.3. Fruit Weight Loss and Respiration Rate

The weight loss of the longkong fruit pericarp increased steadily over the storage period in all samples, as depicted in Figure 5A. Notably, the control fruits, which were not coated, exhibited particularly severe weight loss. On the other hand, the longkong fruits coated with CMC-Gel-MT displayed significantly less weight loss. The extended low-temperature storage period induced physiologically based weight loss via endogenous biochemical activities that occurred due to CI severity, whereas the CMC-Gel and CMC-Gel-MT coatings had a significant impact on the adverse changes and controlled the sample's weight loss. The results show that a sharp increase in weight loss of longkong fruit, particularly in the control group, was found on the third day and the trend continued until the end of the storage. While the CMC-Gel-MT-coated samples did exhibit some weight loss, it was significantly less than that observed in the control group. Among the different CMC-Gel-MT-coated samples, the differences in weight loss were not significantly higher, especially in the MT2 and MT3 samples. Samples that were coated with CMC-Gel and CMC-Gel with low MT concentrations showed slightly lower performance in controlling the weight loss. Generally, the loss of fresh mass in longkong fruit pericarp could lead to visible symptoms, particularly dryness and browning on the pericarp surface, which could result in significant economic loss [6,55]. In fresh produce, moisture loss is critical to maintaining the weight of the product [56]. Stress-induced respiration might induce a higher transpiration rate in the fruits, which also might contribute to weight loss [57]. Continuous moisture loss in food may result in the desiccation of the product, rendering it unsuitable. Coating creates barriers in fresh produce and could control the impediment of water vapor transmission by covering the surface and any minor cracks, ultimately reducing moisture loss [19,58]. Although the CMC-Gel-MT-based coating could control weight loss, a slight increase in weight loss was still observed during storage. This could be due to the formation of a semipermeable layer between the coated fruit and storage conditions.

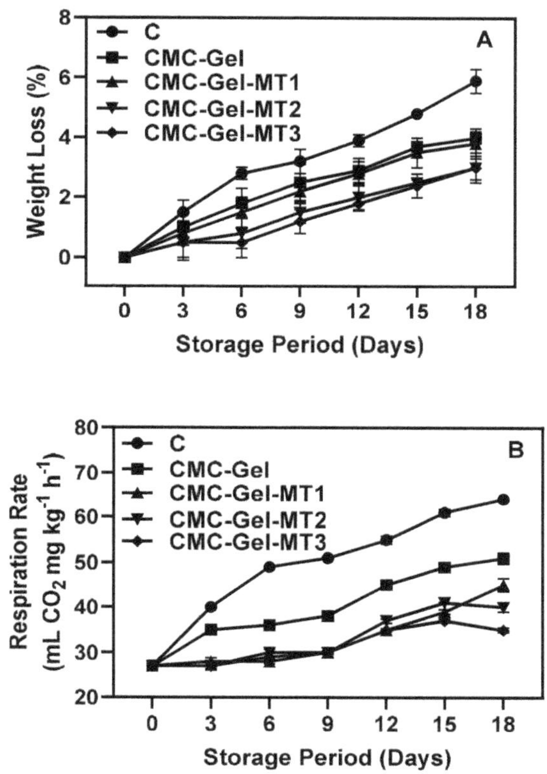

Figure 5. Changes in weight loss (**A**) and respiration rate (**B**) of longkong fruit coated with or without CMC-Gel-MT at varying concentrations and stored under prolonged low temperature conditions.

The respiration rate is a dependable measure for tracking the quality of fruit during storage, and alterations in this rate can substantially reflect in the fruit's quality and shelf life [59]. The respiration rate of the longkong fruit coated with or without CMC-Gel-MT were tested, and the results are shown in Figure 5B. The present study found that the fruit's overall respiration rate tended to increase in the samples all over the storage. The respiration rate in the CMC-Gel-coated samples was controlled slightly better than in the control. Furthermore, incorporating a higher concentration of MT into the coating showed better overall performance in controlling the respiration rate. Generally, changes in the respiration rate of fruit reflect the changes in the physiological state of the fruit and indicate the advancement of the senescence process [60]. Normally, the chilling injury in tropical fruits leads to a sharp increase in respiration rate, and this increase is often a stress response from the fruit to chilling conditions; however, when the chilling stress is prolonged for a long period, the cellular functions could be impaired in the fruit and eventually decrease the respiration rate [61]. This is in accordance with the present study. Our study results showed that longkong pericarp treated with CMC-Gel-MT significantly controlled the rate of respiration, exhibited minimal changes in respiration rate, and maintained its physiological state better than in the control. Ozden and Bayindirli [62] reported that edible coatings on fruits could lead to accumulation of high CO_2 and low O_2 concentrations in the interstitial space between the coating and the fruit surface, potentially decreasing the respiration rate.

3.4. Production of Reactive Oxygen Species

Changes in reactive oxygen species (ROS), including hydrogen peroxide (H_2O_2), hydroxyl radical (OH^-), and superoxide radical anion (O_2^-), in CMC-Gel-MT-coated and uncoated longkong fruit stored under prolonged chilling stress are depicted in Figure 6A–C. Generally, CI induces strong oxidative stress in horticultural produce, leading to increased ROS production, which in turn causes irreversible damage to macromolecules [61]. H_2O_2 is a potent free radical that inflicts oxidative damage on fresh produce, thereby accelerating aging and reducing shelf life [63]. H_2O_2 levels in the pericarp of CMC-Gel-MT-coated and uncoated longkong fruits were observed to continuously increase under prolonged chilling stress. However, these levels were significantly lower compared to those in fruits with only the CMC-Gel coating and the control group. Among the MT samples, those with a higher concentration were more effective in scavenging H_2O_2. The overproduction of H_2O_2 under low-temperature conditions can be detrimental due to the resultant excessive free radicals that damage cell membranes, typically caused by an imbalance between free radical production and antioxidant levels. Similarly, the levels of OH^- and O_2^- radicals also tended to rise in longkong fruit during extended storage under chilling conditions, with O_2^- radical production surpassing that of the other two ROS. CMC-Gel-MT at various concentrations succeeded in minimizing the accumulation of OH^- and O_2^- radicals. This finding aligns with the study by Mirshekari et al. [64], who discovered that MT treatment on sapota fruit effectively regulated ROS production compared to controlling fruits. The application of CMC-Gel-based coating to the fruit surface strengthens the membrane by deeply penetrating the pores and cracks, thereby preserving cell membrane integrity and consequently reducing H_2O_2 production. Besides its ROS scavenging capacity, MT also bolsters the activity of the ROS scavenging system, not only by stimulating the expression of ROS scavenging genes and enhancing the enzyme activity but also by reducing LOX activity and generating additional ascorbic acid [65]. These actions contribute to maintaining membrane integrity, as evidenced by the decreased accumulation of MDA [66]. In addition, MT can regulate ROS signaling pathways, such as the MAPK and Nrf2 pathways, and these pathways play a role in regulating the production and detoxification of ROS [67,68]. MT has also been shown to induce the expression of heat shock proteins (HSPs) in fruits. HSPs are a group of proteins that are expressed in response to stress, and they can help to protect cells from damage caused by ROS and other stressors [69]. Furthermore, the alleviation of ROS in longkong pericarp can also be achieved by their secondary metabolic product, particularly phenolic compounds, which is rich in longkong [28,70]. Phenolic compounds exhibit antioxidant properties by directly scavenging the ROS by themselves and/or chelating transition metals, predominantly iron and copper, through their hydroxyl and carboxyl functional groups, thereby mitigating oxidative stress [71]. Recent studies have reported that phenolic acids are able to upregulate the expression of antioxidant enzymes, such as superoxide dismutase, catalase, and peroxidase activities, thus controlling ROS production and accumulation in the chilling-injured plant crops [72].

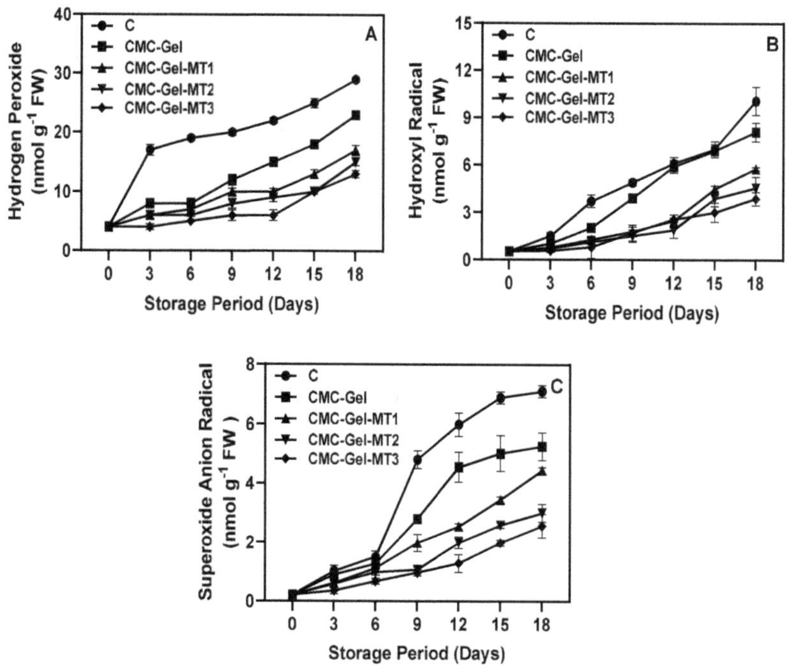

Figure 6. Changes in hydrogen peroxide (**A**), hydroxyl radical (**B**), and superoxide anion radical (**C**) of longkong fruit coated with or without CMC-Gel-MT at varying concentrations and stored under prolonged low temperature conditions.

3.5. Pericarp PAL Enzyme Activity, Phenolic Content, and Antioxidant Activities

The biosynthetic phenylpropanoid pathway can be readily triggered in plants under abiotic stress, especially under extreme temperature conditions, both high and low, and this activation leads to the buildup of phenolic compounds. PAL is the pioneer enzyme in the phenolic metabolism of plants, which generates phenolics in the plants via the phenylpropanoid pathway [73]. PAL catalyzes the conversion of phenylalanine to trans-cinnamic acid as well as other varieties of phenolic compounds including flavonoids, coumarins, hydrolysable tannins, monolignols, lignans, and lignin [74]. This study showed that CMC-Gel-MT-coated and uncoated longkong fruit pericarp samples showed an increase in PAL activity with fluctuations (Figure 7A). In comparison with control and CMC-Gel-coated samples, MT-incorporated (MT1–4) samples showed increased PAL activity. The higher MT concentration in the coating emulsion significantly increased longkong pericarp PAL activity. This is in accordance with the study by Esmaeili et al. [75], who reported that an increase in MT concentration positively increased the PAL enzyme and, consequently, abundance of phenolic acids produced. Studies have reported that increased PAL activity by MT is mainly due to enhanced antioxidant activity as well as an upregulation of the genes that encode PAL [76]. Similarly, the phenolic contents in the plants tend to increase as the abiotic stress level increases, which is an indicator for stimulus response to scavenge or prevent stress-induced damage in plants, particularly by reactive oxygen species [77]. The present study showed that the phenolic content in longkong pericarp tended to decrease gradually throughout storage, regardless of sample type. However, CMC-Gel-MT-coated pericarp samples had significantly higher phenolics, in accordance with the PAL activity in the samples (Figure 7B). On the other hand, the phenolic contents of the CMC-Gel and control samples were lower than in the CMC-Gel-MT samples, but the CMC-Gel samples were slightly better than the control. The changes in phenolic level in pericarp could be attributed to chilling stress, oxidation, and antioxidant properties. The phenolic content in the MT-coated samples was directly in concordance with the amounts of MT concentrations. Plant phenolics play a crucial role in controlling abiotic stress, as this stress could disturb the balance between ROS generation and the acceleration of ROS-induced damage in vital macromolecules such as nucleic acids, proteins, carbohydrates, and lipids [78]. Plant phenolics act as physiological chemical messengers that inhibit the catabolic reactions in plants through stress activation of various symbiotic processes [79]. In general, longkong pericarp is very sensitive due to surface trichomes, and always tends to produce more phenolics than usual due to accidental stimuli, which is sometimes a drawback, as it might unwantedly induce browning; however, other than that, it possesses excellent antioxidant properties. Longkong pericarp polyphenolics may be able to scavenge stubborn free radicals, especially when produced under extended chilling stress [28]. The antioxidant activities, mainly DPPH radical and ABTS radical scavenging activities of CMC-Gel-MT-coated and uncoated longkong results, are shown in Figure 7C,D. Both radicals were significantly controlled in longkong pericarp using the tested coating; however, in comparison with ABTS radials, the coating material exhibited better scavenging efficiency against DPPH radicals. The antioxidant potential of uncoated longkong fruit pericarps initially increases but gradually decreases during storage, leading to browning. This decrease in antioxidants corresponds to elevated ROS levels (O_2 and H_2O_2), causing membrane damage, as evidenced by increased MDA content and charge leakage [80]. Enzymes PPO and POD catalyze the oxidation of phenolic compounds, further promoting browning. The reduced antioxidant efficiency accelerates ROS accumulation, intensifying membrane oxidative damage and resulting in brown longkong fruit pericarps during storage [81]. Furthermore, the increased MT concentration in the coating emulsion significantly increased the scavenging efficiency of the longkong fruit against radicals. The antioxidant potency in longkong pericarp is mainly attributed to its polyphenolic contents [82]. This study proves that the application of MT in coating emulsion increases PAL activities and consequently increases secondary metabolic products, particularly phenolic content and antioxidant potency.

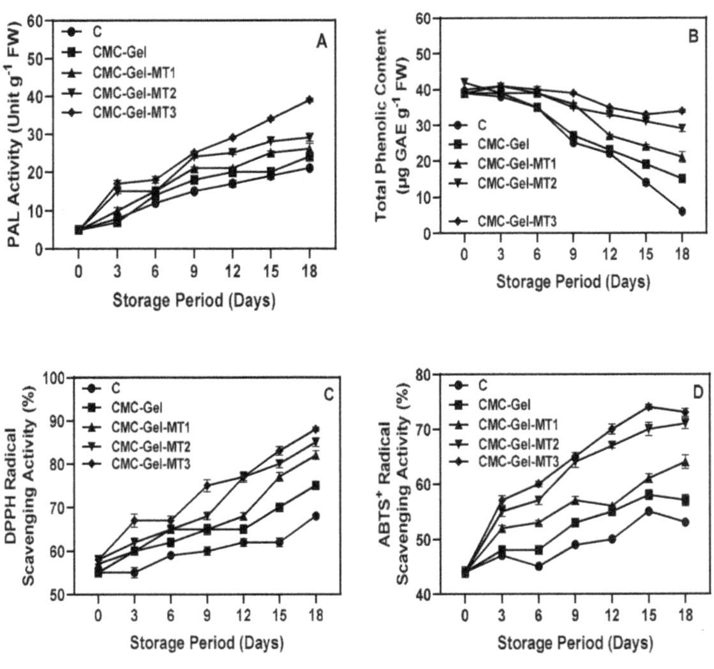

Figure 7. Changes in PAL activity (**A**), total phenolic content (**B**), DPPH radical scavenging activity (**C**), and $ABTS^+$ radical scavenging activity (**D**) of longkong fruit coated with or without CMC-Gel-MT at varying concentrations and stored under prolonged low temperature conditions.

3.6. Lipid Metabolism-Based Enzyme Activities

PLD and LOX are crucial enzymes involved in plant senescence and stress-induced deterioration in fruits; furthermore, their activities notably intensify in fruits and vegetables subjected to prolonged chilling stress [83]. By catalyzing the oxidation of unsaturated fatty acids, both enzymes contribute to membrane lipid peroxidation, significantly impacting maturation, senescence, and stress responses in organisms [84]. Specifically, LOX facilitates the formation of oxylipins from polyunsaturated fatty acids, such as linoleic and linolenic acid, which are vital for plant development and resilience against various stresses [85]. This study tested the LOX and PLD activities in coated and uncoated longkong fruit pericarp during storage under extended chilling stress (Figure 8A,B). The results showed that despite the samples tested, LOX and PLD levels continuously increased in the fruit pericarp throughout storage. The severity of the increment in the tested activities were noticed after the fourth day of storage in control samples for PLD activity and sixth day of storage for LOX activities. For the CMC-Gel and CMC-Gel-MT samples, the onset of mild severeness in the enzyme activities was noticed after 8 days of storage. Among the samples, the CMC-Gel-MT-coated longkong pericarp was able to control the intensity of enzyme activities, followed by the CMC-Gel and control samples. The control samples showed elevated activity of these enzymes, indicating extreme cell membrane phospholipid degradation. On the other hand, the CMC-Gel-MT-coated samples efficiently controlled lipid metabolism-based enzyme activities, the efficiency of MT in controlling such enzyme activities was dose-dependent, and higher MT concentration effectively controlled those enzyme activities. CMC-Gel samples performed slightly better than the control samples, which might be due to the ability of structural strengthening by the coating polymers. Magri et al. [86] tested the efficiency of CMC-based composite coating, and their study effectively controlled the LOX activity in the minimally processed apple fruit under chilling conditions. MT ability to interact directly with the hydrophobic lipid phase of cell membranes due to its lipid solubility and potential to alter PLD functionality may be key factors in controlling PLD activity and pericarp cellular damage. MT contributes to the reduction in LOX activity mainly through its antioxidant properties, modulation of gene expression, direct regulation of enzyme activities such as PLD, and general alleviation of stress responses in plants [87]. However, MT has a positive effect on controlling PLD and LOX activities, but the exact mechanism can be complex and may involve a combination of various factors, possibly varying among different plant species or under different types of stress.

3.7. Browning Related Enzyme Activities

PPO and POD are the primary enzymes responsible for oxidizing phenolic content in the pericarp of longkong fruit, leading to browning during extended storage and prolonged exposure to stress [88]. The current study demonstrated that extended storage under continuous chilling stress significantly increased PPO and POD activities in the pericarp of longkong fruit, regardless of sampling variability (Figure 9A,B). However, the severity of PPO and POD activities was markedly reduced in fruit pericarps coated with CMC-Gel-MT at various concentrations, followed by those coated with CMC-Gel. The uncoated control samples exhibited the highest PPO and POD activities. Although both PPO and POD activities continually affected the longkong fruit pericarp, PPO activity was more predominant compared to POD. Studies have reported that an increase in PPO induces more browning in longkong pericarp compared to POD activities [41]. Increases in POD enzyme activity is also associated with the accumulation of H_2O_2 in the fruit pericarp, as POD utilizes H_2O_2 as a co-substrate alongside its organic substrate, phenolic acid [89]. Generally, the POD enzyme acts as a secondary stimulator of the browning reaction by converting the phenolic substrate into quinones, since the activation of POD requires two substrates [90]. In contrast, PPO rapidly reacts with phenols in the presence of oxygen. Both PPO and POD produce quinones, which are further polymerized into melanin, the browning agent that adversely affects the fruit's marketability and shelf life [91]. Xiao et al. [92] reported that MT treatment significantly downregulated the expression of PPO

(PPO1, PPO2, PPO3, and PPO4) and POD (POD1, POD4, POD72, and POD-N) genes in fresh-cut taro. Wei et al. [93] found that MT treatment helps protect against subcellular decompartmentalization in rambutan pericarp, thereby reducing enzymatic browning. Magri et al. [86] reported that LOX-driven lipid peroxidation produces MDA and H_2O_2, leading to cellular degradation and fruit discoloration. However, highbush blueberries treated with melatonin exhibit reduced MDA and H_2O_2, suggesting minimized cellular harm and browning from decreased PPO and LOX activities during cold storage. This aligns with the present study, which noted reduced MDA and H_2O_2 levels in longkong samples coated with CMC-Gel-MT.

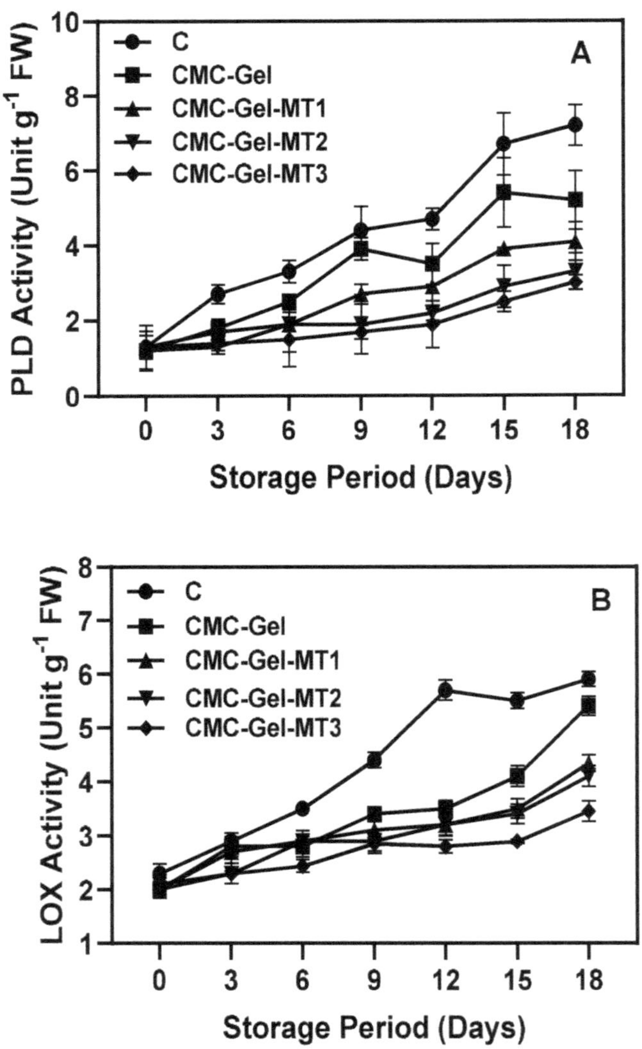

Figure 8. Changes in PLD activity (**A**) and LOX activity (**B**) in longkong fruit coated with or without CMC-Gel-MT at varying concentrations and stored under prolonged low temperature conditions.

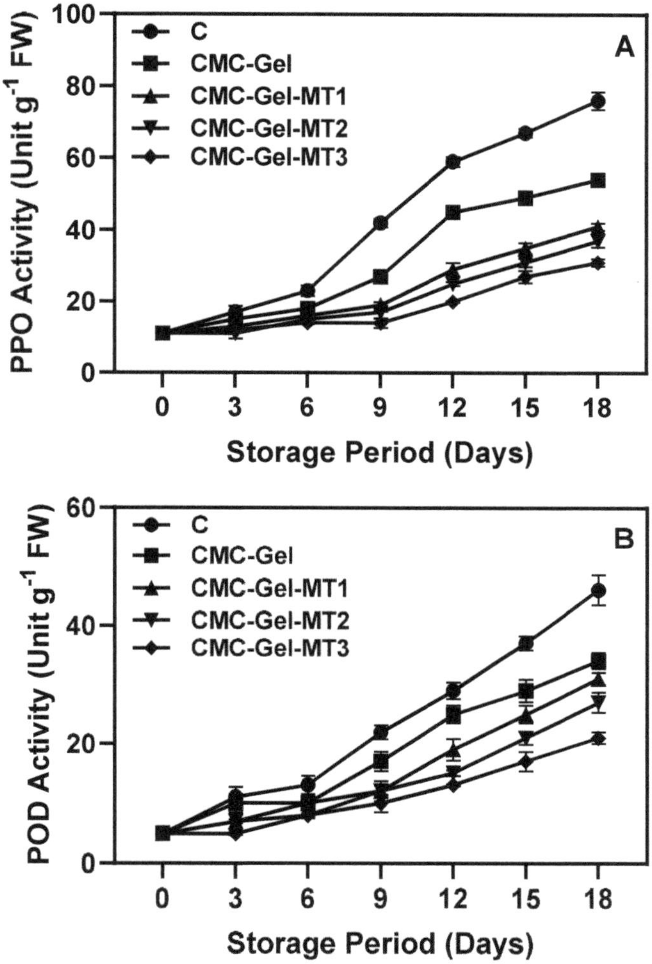

Figure 9. Changes in PPO activity (**A**) and POD activity (**B**) in longkong fruit coated with or without CMC-Gel-MT at varying concentrations and stored under prolonged low temperature conditions.

4. Conclusions

The application of CMC-Gel and CMC-Gel-MT coatings on longkong fruit stored under extended chilling conditions had significantly enhanced postharvest preservation through a series of adaptable biochemical and physiological mechanisms. These coatings effectively inhibited the activity of key enzymes responsible for browning, namely, polyphenol oxidase and peroxidase. This enzymatic suppression played a pivotal role in mitigating color deterioration and chilling injury symptoms. Additionally, the formation of a semipermeable barrier by these coatings was instrumental in reducing moisture loss from the fruit. CMC-Gel-MT coating regulated the respiration rate of longkong fruit, effectively decelerated the senescence process, and prolonged shelf life. Furthermore, the CMC-Gel-MT coating effectively controlled the extensive accumulation of reactive oxygen species in longkong pericarp. The CMC-Gel and CMC-Gel-MT coatings contributed to maintaining the structural and functional stability of the fruit by modulating the activities of phospholipase D and lipoxygenase. Overall, this study showed that addition of melatonin in the coating significantly improved the longkong fruit's quality; furthermore, the action of MT on the longkong fruit was dose-dependent, and higher concentrations performed better than the lower concentrations. Concentrations of 0.8 and 1.2 mM/L MT showed similar effects on the longkong fruits and therefore this study recommends the use of a 0.8 mM/L concentration for optimal performance regarding shelf life enhancement and quality retention of longkong fruit. In summary, CMC-Gel-MT coatings present a promising approach in the realm of postharvest fruit treatment technologies, offering practical and effective means to address the challenges of fruit preservation during storage. The application of CMC-Gel-MT coating to fresh longkong fruit shows promise for commercial production processes. This method effectively maintains quality and extends the shelf life of longkong fruits. Additionally, it is a cost-effective and environmentally friendly treatment.

Author Contributions: Conceptualization, K.V. and P.N.; methodology, K.V., S.L. and P.N; software, N.C.; validation, K.V., N.C., S.L. and P.N.; formal analysis, K.V. and P.N.; investigation, K.V. and P.N.; resources, K.V., N.C. and S.L.; data curation, K.V. and P.N.; writing—original draft preparation, K.V., N.C., S.L. and P.N.; writing—review and editing, K.V., N.C., S.L. and P.N.; visualization, K.V.; supervision, P.N.; project administration, P.N.; funding acquisition, P.N. All authors have read and agreed to the published version of the manuscript.

Funding: This research received no external funding.

Institutional Review Board Statement: Not applicable.

Informed Consent Statement: Not applicable.

Data Availability Statement: Data is contained within the article.

Acknowledgments: The authors would like to greatly acknowledge the Prince of Songkla University and Food Innovation and Product Development Laboratory (FIPD) for their provided supplies, equipment, and laboratory spaces to complete this research. The authors would also like to express their gratitude to Burapha University Chanthaburi Campus for the additional support to finish this research.

Conflicts of Interest: The authors declare no conflicts of interest.

References

1. Keawpeng, I.; Paulraj, B.; Venkatachalam, K. Antioxidant and Antimicrobial Properties of Mung Bean Phyto-Film Combined with Longkong Pericarp Extract and Sonication. *Membranes* **2022**, *12*, 379. [CrossRef] [PubMed]
2. Sangkasanya, S. Flavor and Its Related Quality in Longkong (*Aglaia dookkoo* Griff.) during On-Tree Maturation and Storage. Ph.D. Thesis, Prince of Songkla University, Songkhla, Thailand, 2014.
3. Venkatachalam, K. Postharvest Physiology and Handling of Longkong Fruit: A Review. *Fruits* **2016**, *71*, 289–298. [CrossRef]
4. Abdallah, H.M.; Mohamed, G.A.; Ibrahim, S.R.M. Lansium Domesticum—A Fruit with Multi-Benefits: Traditional Uses, Phytochemicals, Nutritional Value, and Bioactivities. *Nutrients* **2022**, *14*, 1531. [CrossRef] [PubMed]
5. Sathya, R.; Rasane, P.; Singh, J.; Kaur, S.; Bakshi, M.; Gunjal, M.; Kaur, J.; Sharma, K.; Sachan, S.; Singh, A.; et al. Strategic Advances in the Management of Browning in Fruits and Vegetables. *Food Bioprocess Technol.* **2023**. [CrossRef]
6. Charoenphun, N.; Ali, A.M.M.; Paulraj, B.; Venkatachalam, K. Effect of Aqueous N-Butanol Treatments on Shelf-Life Extension of Longkong Fruit during Ambient Storage. *Horticulturae* **2023**, *9*, 938. [CrossRef]
7. Venkatachalam, K.; Meenune, M. Physiological Changes of Longkong Fruit during Different Storage Conditions. *Adv. Environ. Biol.* **2014**, *8*, 362–369.
8. Techavuthiporn, C.; Kanlayanarat, S. Quality changes of different form of minimally processed longkong at low temperature storage. *Acta Hortic.* **2010**, *877*, 635–640. [CrossRef]
9. Wongs-Aree, C.; Noichinda, S. Chapter 10—Postharvest Physiology and Quality Maintenance of Tropical Fruits. In *Postharvest Handling*, 3rd ed.; Florkowski, W.J., Shewfelt, R.L., Brueckner, B., Prussia, S.E., Eds.; Academic Press: San Diego, CA, USA, 2014; pp. 275–312. ISBN 978-0-12-408137-6.
10. Noichinda, C.W.-A. Sompoch Rambutan and Longkong. In *Postharvest Physiological Disorders in Fruits and Vegetables*; CRC Press: Boca Raton, FL, USA, 2019; ISBN 978-1-315-26747-0.
11. Venkatachalam, K. Changes in Quality and Enzymes of Longkong (*Aglaia dookkoo* Griff.) Fruit during Storages as Affected by Maturation, Package and Methyl Jasmonate Treatment. Ph.D. Thesis, Prince of Songkla University, Yai, Thailand, 2013.
12. Al-Tayyar, N.A.; Youssef, A.M.; Al-Hindi, R.R. Edible Coatings and Antimicrobial Nanoemulsions for Enhancing Shelf Life and Reducing Foodborne Pathogens of Fruits and Vegetables: A Review. *Sustain. Mater. Technol.* **2020**, *26*, e00215. [CrossRef]
13. Ballesteros, L.F.; Teixeira, J.A.; Cerqueira, M.A. Active Carboxymethyl Cellulose-Based Edible Coatings for the Extension of Fresh Goldenberries Shelf-Life. *Horticulturae* **2022**, *8*, 936. [CrossRef]

14. Liyanapathiranage, A.; Dassanayake, R.S.; Gamage, A.; Karri, R.R.; Manamperi, A.; Evon, P.; Jayakodi, Y.; Madhujith, T.; Merah, O. Recent Developments in Edible Films and Coatings for Fruits and Vegetables. *Coatings* **2023**, *13*, 1177. [CrossRef]
15. Lu, Y.; Luo, Q.; Chu, Y.; Tao, N.; Deng, S.; Wang, L.; Li, L. Application of Gelatin in Food Packaging: A Review. *Polymers* **2022**, *14*, 436. [CrossRef] [PubMed]
16. Samsi, M.S.; Kamari, A.; Din, S.M.; Lazar, G. Synthesis, Characterization and Application of Gelatin–Carboxymethyl Cellulose Blend Films for Preservation of Cherry Tomatoes and Grapes. *J. Food Sci. Technol.* **2019**, *56*, 3099–3108. [CrossRef] [PubMed]
17. Esteghlal, S.; Niakousari, M.; Hosseini, S.M.H. Physical and Mechanical Properties of Gelatin-CMC Composite Films under the Influence of Electrostatic Interactions. *Int. J. Biol. Macromol.* **2018**, *114*, 1–9. [CrossRef] [PubMed]
18. Yildirim-Yalcin, M.; Tornuk, F.; Toker, O.S. Recent Advances in the Improvement of Carboxymethyl Cellulose-Based Edible Films. *Trends Food Sci. Technol.* **2022**, *129*, 179–193. [CrossRef]
19. Kocira, A.; Kozłowicz, K.; Panasiewicz, K.; Staniak, M.; Szpunar-Krok, E.; Hortyńska, P. Polysaccharides as Edible Films and Coatings: Characteristics and Influence on Fruit and Vegetable Quality—A Review. *Agronomy* **2021**, *11*, 813. [CrossRef]
20. Moradinezhad, F.; Ranjbar, A. Advances in Postharvest Diseases Management of Fruits and Vegetables: A Review. *Horticulturae* **2023**, *9*, 1099. [CrossRef]
21. Galano, A.; Tan, D.-X.; Reiter, R.J. Melatonin: A Versatile Protector against Oxidative DNA Damage. *Molecules* **2018**, *23*, 530. [CrossRef] [PubMed]
22. Vazquez-Hernandez, M.; Blanch, M.; Sanchez-Ballesta, M.T.; Merodio, C.; Escribano, M.I. High CO_2 Alleviates Cell Ultrastructure Damage in Autumn Royal Table Grapes by Modulating Fatty Acid Composition and Membrane and Cell Oxidative Status during Long-Term Cold Storage. *Postharvest Biol. Technol.* **2020**, *160*, 111037. [CrossRef]
23. Njie, A.; Zhang, W.; Dong, X.; Lu, C.; Pan, X.; Liu, Q. Effect of Melatonin on Fruit Quality via Decay Inhibition and Enhancement of Antioxidative Enzyme Activities and Genes Expression of Two Mango Cultivars during Cold Storage. *Foods* **2022**, *11*, 3209. [CrossRef]
24. Vargas-Torrico, M.F.; Von Borries-Medrano, E.; Valle-Guadarrama, S.; Aguilar-Méndez, M.A. Development of Gelatin-Carboxymethylcellulose Coatings Incorporated with Avocado Epicarp and Coconut Endocarp Extracts to Control Fungal Growth in Strawberries for Shelf-Life Extension. *CyTA J. Food* **2022**, *20*, 27–38. [CrossRef]
25. Venkatachalam, K. Exogenous Nitric Oxide Treatment Impacts Antioxidant Response and Alleviates Chilling Injuries in Longkong Pericarp. *Sci. Hortic.* **2018**, *237*, 311–317. [CrossRef]
26. Nguyen, T.B.T.; Ketsa, S.; van Doorn, W.G. Relationship between Browning and the Activities of Polyphenoloxidase and Phenylalanine Ammonia Lyase in Banana Peel during Low Temperature Storage. *Postharvest Biol. Technol.* **2003**, *30*, 187–193. [CrossRef]
27. Wang, Y.; Luo, Z.; Du, R.; Liu, Y.; Ying, T.; Mao, L. Effect of Nitric Oxide on Antioxidative Response and Proline Metabolism in Banana during Cold Storage. *J. Agric. Food Chem.* **2013**, *61*, 8880–8887. [CrossRef] [PubMed]
28. Noonim, P.; Venkatachalam, K. Combination of Salicylic Acid and Ultrasonication for Alleviating Chilling Injury Symptoms of Longkong. *Food Qual. Saf.* **2022**, *6*, fyab032. [CrossRef]
29. Chen, X.; Ren, L.; Li, M.; Qian, J.; Fan, J.; Du, B. Effects of Clove Essential Oil and Eugenol on Quality and Browning Control of Fresh-Cut Lettuce. *Food Chem.* **2017**, *214*, 432–439. [CrossRef]
30. Caleb, O.; Mahajan, P.; Opara, U.; Witthuhn, C. Modeling the Respiration Rates of Pomegranate Fruit and Arils. *Postharvest Biol. Technol.* **2012**, *64*, 49–54. [CrossRef]
31. Patterson, B.D.; MacRae, E.A.; Ferguson, I.B. Estimation of Hydrogen Peroxide in Plant Extracts Using Titanium(IV). *Anal. Biochem.* **1984**, *139*, 487–492. [CrossRef]
32. Chung, S.-K.; Osawa, T.; Kawakishi, S. Hydroxyl Radical-Scavenging Effects of Spices and Scavengers from Brown Mustard (*Brassica nigra*). *Biosci. Biotechnol. Biochem.* **1997**, *61*, 118–123. [CrossRef]
33. Wang, A.G.; Luo, G.H. Quantitative Relation between the Reaction of Hydroxylamine and Superoxide Anion Radicals in Plants. *Zhi Wu Sheng Li Xue Tong Xun* **1990**, *26*, 55–57.
34. Alberti, A.; Zielinski, A.A.F.; Zardo, D.M.; Demiate, I.M.; Nogueira, A.; Mafra, L.I. Optimisation of the Extraction of Phenolic Compounds from Apples Using Response Surface Methodology. *Food Chem.* **2014**, *149*, 151–158. [CrossRef]
35. Brand-Williams, W.; Cuvelier, M.E.; Berset, C. Use of a Free Radical Method to Evaluate Antioxidant Activity. *LWT Food Sci. Technol.* **1995**, *28*, 25–30. [CrossRef]
36. Lo'ay, A.A.; Dawood, H.D. Chilling injury, fruit color maturity stages, and antioxidant enzyme activities of lemon baladi CV fruits under cold storage stress. *Sci. Hortic.* **2019**, *257*, 108676. [CrossRef]
37. Yi, C.; Qu, H.X.; Jiang, Y.M.; Shi, J.; Duan, X.W.; Joyce, D.C.; Li, Y.B. ATP-Induced Changes in Energy Status and Membrane Integrity of Harvested Litchi Fruit and Its Relation to Pathogen Resistance. *J. Phytopathol.* **2008**, *156*, 365–371. [CrossRef]
38. Lin, Y.; Lin, Y.; Lin, H.; Zhang, S.; Chen, Y.; Shi, J. Inhibitory Effects of Propyl Gallate on Browning and Its Relationship to Active Oxygen Metabolism in Pericarp of Harvested Longan Fruit. *LWT Food Sci. Technol.* **2015**, *60*, 1122–1128. [CrossRef]
39. Jiang, Y. Role of Anthocyanins, Polyphenol Oxidase and Phenols in Lychee Pericarp Browning. *J. Sci. Food Agric.* **2000**, *80*, 305–310. [CrossRef]
40. Zhang, Z.; Pang, X.; Xuewu, D.; Ji, Z.; Jiang, Y. Role of Peroxidase in Anthocyanin Degradation in Litchi Fruit Pericarp. *Food Chem.* **2005**, *90*, 47–52. [CrossRef]

41. Lichanporn, I.; Srilaong, V.; Wongs-Aree, C.; Kanlayanarat, S. Postharvest Physiology and Browning of Longkong (*Aglaia dookkoo* Griff.) Fruit under Ambient Conditions. *Postharvest Biol. Technol.* **2009**, *52*, 294–299. [CrossRef]
42. Yingsanga, P.; Srilaong, V.; Kanlayanarat, S.; Noichinda, S.; McGlasson, W.B. Relationship between Browning and Related Enzymes (PAL, PPO and POD) in Rambutan Fruit (Nephelium Lappaceum Linn.) Cvs. Rongrien and See-Chompoo. *Postharvest Biol. Technol.* **2008**, *50*, 164–168. [CrossRef]
43. Singh, P.; Arif, Y.; Miszczuk, E.; Bajguz, A.; Hayat, S. Specific Roles of Lipoxygenases in Development and Responses to Stress in Plants. *Plants* **2022**, *11*, 979. [CrossRef]
44. Lin, Y.; Lin, H.; Lin, Y.; Zhang, S.; Chen, Y.; Jiang, X. The Roles of Metabolism of Membrane Lipids and Phenolics in Hydrogen Peroxide-Induced Pericarp Browning of Harvested Longan Fruit. *Postharvest Biol. Technol.* **2016**, *111*, 53–61. [CrossRef]
45. Saowakon, K.; Deewatthanawong, R.; Khurnpoon, L. Effect of Carboxymethyl Cellulose as Edible Coating on Postharvest Quality of Rambutan Fruit under Ambient Temperature. *Int. J. Agric. Technol.* **2017**, *13*, 1449–1457.
46. Zheng, H.; Liu, W.; Liu, S.; Liu, C.; Zheng, L. Effects of Melatonin Treatment on the Enzymatic Browning and Nutritional Quality of Fresh-Cut Pear Fruit. *Food Chem.* **2019**, *299*, 125116. [CrossRef]
47. Pham, T.T.; Nguyen, L.L.P.; Dam, M.S.; Baranyai, L. Application of Edible Coating in Extension of Fruit Shelf Life: Review. *AgriEngineering* **2023**, *5*, 520–536. [CrossRef]
48. Juan, C.A.; Pérez de la Lastra, J.M.; Plou, F.J.; Pérez-Lebeña, E. The Chemistry of Reactive Oxygen Species (ROS) Revisited: Outlining Their Role in Biological Macromolecules (DNA, Lipids and Proteins) and Induced Pathologies. *Int. J. Mol. Sci.* **2021**, *22*, 4642. [CrossRef]
49. Matloob, A.; Ayub, H.; Mohsin, M.; Ambreen, S.; Khan, F.A.; Oranab, S.; Rahim, M.A.; Khalid, W.; Nayik, G.A.; Ramniwas, S.; et al. A Review on Edible Coatings and Films: Advances, Composition, Production Methods, and Safety Concerns. *ACS Omega* **2023**, *8*, 28932–28944. [CrossRef]
50. Dolhaji, N.H.; Muhammad, I.D.; Yaakob, H.; Mohd Marsin, A. Chilling Injury in Pineapple Fruits: Physical Quality Attributes and Antioxidant Enzyme Activity. *Food Res.* **2020**, *4*, 86–95. [CrossRef]
51. Marcantonini, G.; Bartolini, D.; Zatini, L.; Costa, S.; Passerini, M.; Rende, M.; Luca, G.; Basta, G.; Murdolo, G.; Calafiore, R.; et al. Natural Cryoprotective and Cytoprotective Agents in Cryopreservation: A Focus on Melatonin. *Molecules* **2022**, *27*, 3254. [CrossRef]
52. Jové, M.; Mota-Martorell, N.; Pradas, I.; Martín-Gari, M.; Ayala, V.; Pamplona, R. The Advanced Lipoxidation End-Product Malondialdehyde-Lysine in Aging and Longevity. *Antioxidants* **2020**, *9*, 1132. [CrossRef]
53. Liu, J.; Zhang, L.; Liu, C.; Zheng, X.; Tang, K. Tuning Structure and Properties of Gelatin Edible Films through Pullulan Dialdehyde Crosslinking. *LWT Food Sci. Technol.* **2021**, *138*, 110607. [CrossRef]
54. Hosseini, S.F.; Gómez-Guillén, M.C. A State-of-the-Art Review on the Elaboration of Fish Gelatin as Bioactive Packaging: Special Emphasis on Nanotechnology-Based Approaches. *Trends Food Sci. Technol.* **2018**, *79*, 125–135. [CrossRef]
55. Venkatachalam, K.; Meenune, M. Physical and Chemical Quality Changes of Longkong (*Aglaia dookkoo* Griff.) during Passive Modified Atmospheric Storage. *Int. Food Res. J.* **2012**, *19*, 795–800.
56. Lufu, R.; Ambaw, A.; Opara, U.L. Water Loss of Fresh Fruit: Influencing Pre-Harvest, Harvest and Postharvest Factors. *Sci. Hortic.* **2020**, *272*, 109519. [CrossRef]
57. Xanthopoulos, G.T.; Templalexis, C.G.; Aleiferis, N.P.; Lentzou, D.I. The Contribution of Transpiration and Respiration in Water Loss of Perishable Agricultural Products: The Case of Pears. *Biosyst. Eng.* **2017**, *158*, 76–85. [CrossRef]
58. Lin, D.; Zhao, Y. Innovations in the Development and Application of Edible Coatings for Fresh and Minimally Processed Fruits and Vegetables. *Compr. Rev. Food Sci. Food Saf.* **2007**, *6*, 60–75. [CrossRef]
59. Nair, S.; Singh, Z. Chilling Injury during Storage Affects Respiration Rate and Fruit Quality in Kensington Pride Mango Fruit. *Acta Hortic.* **2006**, *820*, 737–744. [CrossRef]
60. Biale, J.B.; Young, R.E.; Olmstead, A.J. Fruit Respiration and Ethylene Production. *Plant Physiol.* **1954**, *29*, 168–174. [CrossRef]
61. Valenzuela, J.L.; Manzano, S.; Palma, F.; Carvajal, F.; Garrido, D.; Jamilena, M. Oxidative Stress Associated with Chilling Injury in Immature Fruit: Postharvest Technological and Biotechnological Solutions. *Int. J. Mol. Sci.* **2017**, *18*, 1467. [CrossRef]
62. Özden, Ç.; Bayindirli, L. Effects of Combinational Use of Controlled Atmosphere, Cold Storage and Edible Coating Applications on Shelf Life and Quality Attributes of Green Peppers. *Eur. Food Res. Technol.* **2002**, *214*, 320–326. [CrossRef]
63. Ransy, C.; Vaz, C.; Lombès, A.; Bouillaud, F. Use of H_2O_2 to Cause Oxidative Stress, the Catalase Issue. *Int. J. Mol. Sci.* **2020**, *21*, 9149. [CrossRef]
64. Mirshekari, A.; Madani, B.; Yahia, E.M.; Golding, J.B.; Vand, S.H. Postharvest Melatonin Treatment Reduces Chilling Injury in Sapota Fruit. *J. Sci. Food Agric.* **2020**, *100*, 1897–1903. [CrossRef] [PubMed]
65. Bhattacharyya, A.; Chattopadhyay, R.; Mitra, S.; Crowe, S.E. Oxidative Stress: An Essential Factor in the Pathogenesis of Gastrointestinal Mucosal Diseases. *Physiol Rev.* **2014**, *94*, 329–354. [CrossRef]
66. Jannatizadeh, A. Exogenous Melatonin Applying Confers Chilling Tolerance in Pomegranate Fruit during Cold Storage. *Sci. Hortic.* **2019**, *246*, 544–549. [CrossRef]
67. Schieber, M.; Chandel, N.S. ROS Function in Redox Signaling and Oxidative Stress. *Curr. Biol.* **2014**, *24*, R453–R462. [CrossRef]
68. Saha, M.; Das, S.; Manna, K.; Saha, K.D. Melatonin Targets Ferroptosis through Bimodal Alteration of Redox Environment and Cellular Pathways in NAFLD Model. *Biosci. Rep.* **2023**, *43*, BSR20230128. [CrossRef]

69. Hu, C.; Yang, J.; Qi, Z.; Wu, H.; Wang, B.; Zou, F.; Mei, H.; Liu, J.; Wang, W.; Liu, Q. Heat Shock Proteins: Biological Functions, Pathological Roles, and Therapeutic Opportunities. *MedComm (2020)* **2022**, *3*, e161. [CrossRef]
70. Venkatachalam, K. *Bioactive Compounds of Longkong Fruit (Lansium Domesticum Corr.)*; Springer International Publishing: New York, NY, USA, 2020; pp. 107–122. ISBN 978-3-030-30181-1.
71. Zhu, Y.; Wang, K.; Jia, X.; Fu, C.; Yu, H.; Wang, Y. Antioxidant Peptides, the Guardian of Life from Oxidative Stress. *Med. Res. Rev.* **2023**, *25*, 21986. [CrossRef]
72. Zandi, P.; Schnug, E. Reactive Oxygen Species, Antioxidant Responses and Implications from a Microbial Modulation Perspective. *Biology* **2022**, *11*, 155. [CrossRef]
73. Sharma, A.; Shahzad, B.; Rehman, A.; Bhardwaj, R.; Landi, M.; Zheng, B. Response of Phenylpropanoid Pathway and the Role of Polyphenols in Plants under Abiotic Stress. *Molecules* **2019**, *24*, 2452. [CrossRef]
74. Vogt, T. Phenylpropanoid Biosynthesis. *Mol. Plant* **2010**, *3*, 2–20. [CrossRef]
75. Esmaeili, S.; Sharifi, M.; Ghanati, F.; Soltani, B.M.; Samari, E.; Sagharyan, M. Exogenous Melatonin Induces Phenolic Compounds Production in Linum Album Cells by Altering Nitric Oxide and Salicylic Acid. *Sci. Rep.* **2023**, *13*, 4158. [CrossRef] [PubMed]
76. Sanchez-Ballesta, M.; Zacarías, L.; Granell, A.; Lafuente, M.T. Accumulation of Pal Transcript and Pal Activity as Affected by Heat-Conditioning and Low-Temperature Storage and Its Relation to Chilling Sensitivity in Mandarin Fruits. *J. Agric. Food Chem.* **2000**, *48*, 2726–2731. [CrossRef]
77. Šamec, D.; Karalija, E.; Šola, I.; Vujčić Bok, V.; Salopek-Sondi, B. The Role of Polyphenols in Abiotic Stress Response: The Influence of Molecular Structure. *Plants* **2021**, *10*, 118. [CrossRef]
78. Salam, U.; Ullah, S.; Tang, Z.-H.; Elateeq, A.A.; Khan, Y.; Khan, J.; Khan, A.; Ali, S. Plant Metabolomics: An Overview of the Role of Primary and Secondary Metabolites against Different Environmental Stress Factors. *Life* **2023**, *13*, 706. [CrossRef]
79. Dehghanian, Z.; Habibi, K.; Dehghanian, M.; Aliyar, S.; Asgari Lajayer, B.; Astatkie, T.; Minkina, T.; Keswani, C. Reinforcing the Bulwark: Unravelling the Efficient Applications of Plant Phenolics and Tannins against Environmental Stresses. *Heliyon* **2022**, *8*, e09094. [CrossRef]
80. Toivonen, P.M.A.; Brummell, D.A. Biochemical Bases of Appearance and Texture Changes in Fresh-Cut Fruit and Vegetables. *Postharvest Biol. Technol.* **2008**, *48*, 1–14. [CrossRef]
81. Chomkitichai, W. Oxidative Damage and Total and Antioxidant Capacity Involved in Pericarp Browning during Storage of Longkong Fruit. *Burapha Sci. J.* **2020**, *25*, 102–116.
82. Nagarajan, M.; Rajasekaran, B.; Benjakul, S.; Venkatachalam, K. Influence of Chitosan-Gelatin Edible Coating Incorporated with Longkong Pericarp Extract on Refrigerated Black Tiger Shrimp (*Penaeus monodon*). *Curr. Res. Food Sci.* **2021**, *4*, 345–353. [CrossRef]
83. Li, L.; Yi, P.; Huang, F.; Tang, J.; Sun, J.; Duan, X.; Li, J.; Su, Z.; Ling, D.; Tang, Y.; et al. Effects of Phospholipase D Inhibitors Treatment on Membrane Lipid Metabolism of Postharvest Banana Fruit in Response to Mechanical Wounding Stress. *Horticulturae* **2022**, *8*, 901. [CrossRef]
84. Hamsanathan, S.; Gurkar, A.U. Lipids as Regulators of Cellular Senescence. *Front. Physiol.* **2022**, *13*, 796850. [CrossRef]
85. Upadhyay, R.K.; Handa, A.K.; Mattoo, A.K. Transcript Abundance Patterns of 9- and 13-Lipoxygenase Subfamily Gene Members in Response to Abiotic Stresses (Heat, Cold, Drought or Salt) in Tomato (*Solanum lycopersicum* L.) Highlights Member-Specific Dynamics Relevant to Each Stress. *Genes* **2019**, *10*, 683. [CrossRef] [PubMed]
86. Magri, A.; Petriccione, M. Melatonin Treatment Reduces Qualitative Decay and Improves Antioxidant System in Highbush Blueberry Fruit during Cold Storage. *J. Sci. Food Agric.* **2022**, *102*, 4229–4237. [CrossRef] [PubMed]
87. Wang, T.; Hu, M.; Yuan, D.; Yun, Z.; Gao, Z.; Su, Z.; Zhang, Z. Melatonin Alleviates Pericarp Browning in Litchi Fruit by Regulating Membrane Lipid and Energy Metabolisms. *Postharvest Biol. Technol.* **2020**, *160*, 111066. [CrossRef]
88. Ismail, H.A.; Richard, I.; Ramaiya, S.D.; Zakaria, M.H.; Lee, S.Y. Browning in Relation to Enzymatic Activities and Phytochemical Content in Terap Peel (*Artocarpus odoratissimus* Blanco) during Postharvest Ripening. *Horticulturae* **2023**, *9*, 57. [CrossRef]
89. González-Gordo, S.; Muñoz-Vargas, M.A.; Palma, J.M.; Corpas, F.J. Class III Peroxidases (POD) in Pepper (*Capsicum annuum* L.): Genome-Wide Identification and Regulation during Nitric Oxide (NO)-Influenced Fruit Ripening. *Antioxidants* **2023**, *12*, 1013. [CrossRef] [PubMed]
90. Doğan, S.; Doğan, M.; Arslan, O. Enzymatic Browning in Foods and Its Prevention. 2009. Available online: https://dspace.balikesir.edu.tr/xmlui/handle/20.500.12462/9199 (accessed on 21 December 2023).
91. Zhang, S. Recent Advances of Polyphenol Oxidases in Plants. *Molecules* **2023**, *28*, 2158. [CrossRef]
92. Xiao, Y.; Xie, J.; Wu, C.; He, J.; Wang, B. Effects of Melatonin Treatment on Browning Alleviation of Fresh-Cut Foods. *J. Food Biochem.* **2021**, *45*, e13798. [CrossRef]
93. Wei, D.; Yang, J.; Xiang, Y.; Meng, L.; Pan, Y.; Zhang, Z. Attenuation of Postharvest Browning in Rambutan Fruit by Melatonin Is Associated With Inhibition of Phenolics Oxidation and Reinforcement of Antioxidative Process. *Front. Nutr.* **2022**, *9*, 905006. [CrossRef]

foods

MDPI

Article

Effects of Near-Freezing Temperature Combined with Jujube Polysaccharides Treatment on Proteomic Analysis of 'Diaogan' Apricot (*Prunus armeniaca* L.)

Zhipeng Wang, Wei Wang *, Wei Li, Rui Yang, Yanbo Li, Lusi Zhang, Mengying Zhang and Xuewen Li

College of Food Science and Pharmacy, Xinjiang Agricultural University, Urumqi 830052, China
* Correspondence: ww2shz@163.com

Abstract: This study involved the extraction of polysaccharides from jujube for application in apricot storage. Although near-freezing temperature (NFT) storage is commonly employed for preserving fresh fruit, its effectiveness is somewhat limited. Incorporating jujube polysaccharides was proposed to augment the preservative effect on apricots. Our findings demonstrated that the combined use of NFT and jujube polysaccharides can maintain fruit color, and effectively inhibit decay. Additionally, Tandem Mass Tag (TMT) quantitative proteomic technology was utilized to analyze protein variations in 'Diaogan' apricots during storage. This dual approach not only markedly lowered the activity of polyphenol cell wall-degrading enzymes ($p < 0.05$) but also revealed 1054 differentially expressed proteins (DEPs), which are related to sugar and energy metabolism, stress response and defense, lipid metabolism, and cell wall degradation. The changes in DEPs indicated that the combined use of NFT and jujube polysaccharides could accelerate the conversion of malic acid to oxaloacetic acid and regulate antioxidant ability, potentially extending the storage lifespan of apricot fruit.

Keywords: apricot fruit; proteomics; cell wall degrading enzyme; cold storage

1. Introduction

In recent years, growing concerns about pesticide use and residues on fruits have led to increasing attention to the development of safer fruit preservatives that are effective but do not harm our health. Although reducing the temperature for fruit storage can cut losses caused by metabolism, this approach is usually insufficient to fully protect harvested fruits [1,2]. The 'Diaogan' apricot (*Prunus armeniaca* L.), a typical climacteric fruit, exhibits a concentrated harvesting period and rapid respiratory climacteric post-harvest. Improper storage and preservation can result in swift softening and deterioration of the fruit within a brief timeframe [3,4]. Storing fresh produce at a near-freezing temperature (NFT) is a preservation technique that maintains low respiration rates and other metabolic processes. Prior research has highlighted the advantages of the initial storage of fresh apricots at NFT in enhancing their quality [5].

During prolonged cold storage, the fruit undergoes softening, which is associated with ripening and softening in apricots related to cell wall degradation by cell wall degradation enzymes [6]. Recently, plant polysaccharides have become a new focus in the field of postharvest fruit due to their ability to regulate the metabolism of fruit by controlling respiration rate, extending shelf life, and improving the preservation of rigidity [1]. The application of exogenous plant polysaccharides in postharvest treatments has notably enhanced the preservation of various crops, including potatoes [7], mangoes [8], and strawberries [9]. Jujube polysaccharides (natural plant polysaccharides) are of interest, but their impact on the storage and preservation of apricot fruits, as well as their potential mechanisms, remain unclear.

Proteomics, an increasingly important analytical technique, has been progressively applied in fruit and vegetable preservation to decipher the mechanisms of different preservation methods at the protein level. Considering the complexity of fruit ripening and aging, a comprehensive assessment of the molecular processes in fruits using proteomics is crucial [10]. Recent advancements in this field have led to significant discoveries, such as the role of vacuolar membranes in apple fruit senescence [11], metabolic shifts in kiwifruit revealed through iTRAQ-based quantitative proteomics [12], and the exploration of salicylic acid's antagonistic effects on ethylene, auxin, and glucose during pear fruit senescence [13]. Nonetheless, proteomic studies focusing on the regulatory mechanisms of fruit quality maintenance and the impact of cold storage are limited [14,15].

This study investigated the relationship between changes in cell wall characteristics and differential protein expression levels during storage from the perspective of cell wall metabolism and proteomics, with the aim of exploring the potential mechanism of combined treatment (NFT + jujube polysaccharides) to delay the senescence of apricot fruit.

2. Materials and Methods

2.1. Fruit Materials and Treatment

Apricots were post harvested from local orchards in Yining, Xinjiang, China. The fruit had a freezing point of −2 °C. After removing the field heat for 2 h, the harvested fruits (day 0) were randomly divided into two groups.

The jujube polysaccharides were prepared as described previously [16]. Briefly, the dried jujube powder sample was treated twice with distilled water in the proportion of 1:40 (g/mL) at 90 °C for 4 h. The aqueous extracts were concentrated under vacuum and mixed with five volumes of anhydrous ethanol for 12 h at 4 °C. After centrifugation at 4000× *g* for 10 min, the precipitates were collected and washed successively with ether, anhydrous ethanol, and acetone, affording the crude polysaccharides.

The fruit freezing point was −2 °C and stored for 49 d for weekly sampling. (1) NFT group: apricots were stored at NFT (−1.5 to −1 °C) with 85–90% humidity for 7 weeks and were separately sampled at 0 d and 49 d; (2) NFT + jujube polysaccharide group: fruits were stored at temperatures ranging from −1.5 to −1 °C with 1.5 g/L jujube polysaccharide soak treatment for 10 min and were separately sampled at 0 d and 49 d. The samples collected from different storages were immediately frozen in liquid nitrogen and stored at −80 °C for further analysis.

2.2. Cell Wall Metabolism Enzyme-Related Enzymes

A total of 2 g of frozen apricot tissue was homogenized in an ice bath with 4 mL of pre-cooled 95% ethanol at 4 °C and 12,000 rpm for 10 min. Subsequently, the supernatant was discarded, and another 4 mL of pre-cooled 80% ethanol was added, followed by centrifugation under the same conditions. The tissue was then mixed with 8 mL of an 8.8% NaCl solution containing 2.5% polyvinyl polypyrrolidone (PVPP). After centrifugation at 12,000 rpm for 20 min at 4 °C, the supernatant obtained was further centrifuged [5]. Polygalacturonase (PG) activity was determined using the method described by Yang et al. [17], with results expressed in U/g. Cellulase (CX) activity was detected using the method outlined by Win et al. [18], and the results are presented in units of $mg{\cdot}h^{-1}{\cdot}g^{-1}$. The activities of β-galactosidase (β-GAL) and β-glucosidase (β-Gul) were determined according to Hagerman et al. [19], with the results expressed as $\mu mol{\cdot}h^{-1}{\cdot}g^{-1}$ and $mg{\cdot}h^{-1}{\cdot}g^{-1}$, respectively.

Frozen apricot tissue (2 g), mixed with 1 mL of phosphate-buffered solution (pH = 7.5), was centrifuged at 1200× *g*, and the supernatant was stored at 4 °C. Pectin methyl esterase (PME) activities were determined following the method described by Liu et al. [20], with the results expressed as U/g.

2.3. Proteins Extraction

Apricot fruits were pulverized into powder using liquid nitrogen, and 0.1 g of this sample was combined with 1 mL of phenol extraction solution (0.25 M Tris-HCl, pH 8.0, 24% *w*/*v* sucrose, 0.1 M NaCl, 0.04 M EDTA·2Na, 0.01 M DTT), supplemented with proteinase inhibitors. The mixture was homogenized by pipetting and incubated for 10 min. Subsequently, 1 mL of phenol-Tris-HCl (pH 7.8) was added, and the supernatant was separated by centrifugation at 6000× *g* for 15 min, isolating the phenol layer. The protein in the supernatant was precipitated by adding 5 volumes of 0.1 M ammonium acetate solution and incubating at 20 °C for 24 h, followed by centrifugation at 4 °C and 12,000× *g* for 10 min. The pellet was sequentially washed with methanol and acetone. Next, it was maintained at 20 °C for 12 h and centrifuged at 4 °C and 12,000× *g* for 10 min to collect the sediment.

The sediment was dried at room temperature and redissolved in 10 mL of SDS Cracking solution (50 mmol·L^{-1} Tris-HCl, pH 7.2, 2% *w*/*v* β-mercaptoethanol, and 1 mmol·L^{-1} phenylmethylsulfonyl fluoride) for 3 h. The mixture was centrifuged at 12,000× *g* for 20 min, and the supernatant was collected and stored at −80 °C [21].

2.4. Protein Enzymatic Hydrolysis and TMT Labeling

A total of 100 µg of the protein samples were incubated with 0.5 µL of 10 mM dithiothreitol at 56 °C for 20 min, followed by cooling to 25 °C. Iodoacetamide was added to achieve a final concentration of 10 mM, and the mixture was thoroughly combined and incubated in the dark for 30 min. The proteins were then digested using trypsin (Promega, Madison, WI, USA) at a protein-to-enzyme ratio of 50:1 at 37 °C for 12 h, and the resulting peptides were stored at 4 °C.

For the labeling of peptides, the TMT Labeling Reagent Kit (Thermo Scientific, Rockford, IL, USA) was employed. The samples were mixed with 100 mM TEAB. TMT reagent vials were sequentially arranged from smallest to largest. Seventy µL of the enzymatic hydrolysates was reacted with the contents of one TMT reagent vial. Each TMT reagent was first mixed with 65 µL of acetonitrile, and then 25 µL of the sample was added, and the mixture was incubated with 5% hydroxylamine for 15 min.

2.5. LC-MS/MS Analysis

LC-MS/MS analysis was conducted using an EASY-nLC1200 system (Thermo Fisher Scientific Rockford, IL, USA). One microgram of each sample was loaded onto a reversed-phase C18AQ column (Reprisal Ur 120, 1.9 µm, 100 µm × 15 cm, Thermo Fisher Scientific, Rockford, IL, USA) at a flow rate of 0.3 µL/min. Mobile phase A consisted of 0.1% formic acid (FA) and 98% water (2% acetonitrile), while mobile phase B consisted of 0.1% FA in 80% acetonitrile (with 20% water). The flow rate of the mobile phase was maintained at 300 nL/min with a gradient duration of 90 min. The gradient profile for elution was set as follows: 2–5% B for 2 min, 5–22% B over 68 min, 22–45% B for 16 min, 45–95% B for 2 min, and a final hold at 95% B for 2 min.

MS scans were performed on a Q Exactive HFX Orbitrap mass spectrometer (Thermo Fisher Scientific, Rockford, IL, USA) equipped with a nanoelectrospray Flex-Ion source (Thermo Fisher Scientific, Rockford, IL, USA). The instrument settings were as follows: a full scan range of 350−1600 m/z, a mass resolution of 120,000, and an automatic gain control (AGC) target of 3×10^6. The top 20 most intense ions were selected for fragmentation by higher-energy collisional dissociation (HCD) with a normalized collision energy (NCE) of 32%. MS/MS spectra were acquired with a mass resolution of 4500, an AGC target of 1×10^5, and a maximum injection time (MIT) of 96 ms.

2.6. Data and Bio Informatics Analysis

Data interpretation and quantification were conducted using Marquand (version 1.6.15.0) on a Linux system (Debian 9). The database utilized in this study was "uniprot-Oryza-sativa. Festa". Protein functional annotation was carried out using the Blast2GO program. Protein groups from three sets of 'Diaogan' apricot samples were compared pairwise. To minimize false peptide identification, only proteins with peptide confidence intervals (≥20) exceeding 95% and containing at least one unique peptide were considered significant. Proteins with a p-value < 0.05 and a fold change > 1.2 were identified as significantly different. Those exhibiting significant differences across all three parallel samples were classified as differentially expressed proteins (DEPs). Gene Ontology (GO) enrichment analysis for protein identification was conducted using the website http://geneontology.org/ (accessed on 12 December 2022). Additionally, the Kyoto Encyclopedia of Genes and Genomes (KEGG) pathway enrichment analysis was performed to elucidate functions, accessible at https://www.kegg.jp/ (accessed on 12 December 2022). Hierarchical clustering for visualizing different protein expression levels was performed using the heatmap. Heatmap was plotted by https://www.bioinformatics.com.cn (last accessed on 10 November 2023), an online platform for data analysis and visualization.

2.7. Statistic Analysis

The Tandem Mass Tag (TMT) labeling technique was employed to analyze three biological and technical replicates for each treatment. Sample stability was assessed using IBM SPSS Statistics 20 (IBM Corp., Armonk, NY, USA). Kyoto Encyclopedia of Genes and Genomes (KEGG) pathways were retrieved from the KEGG database. Other charts pertinent to the proteomics analysis were generated using Origin 8.5 (OriginLab, Northampton, MA, USA).

3. Results

3.1. Cell-Wall-Modifying Enzyme Activities

Figure 1 shows that the apricot samples stored in NFT and NFT + jujube polysaccharides can maintain fruit color and effectively inhibit decay. The implementation of NFT coupled with jujube polysaccharide storage emerges as a feasible strategy for extending favorable preservation outcomes. Figure 2A–E show an increase in the activities of PG, CX,β-Gal, and β-Glu throughout the storage period from 0 to 49 days. On day 49, the activities of PG, CX, β-Gal, and β-Glu in the NFT + jujube polysaccharide group were 14%, 22%, 18%, and 38% lower, respectively, than those in the NFT group ($p < 0.05$). Additionally, at the end of storage, the activities of PG, CX, β-Gal, and β-Glu were 1.1, 1.2, 2.27, and 1.09 times higher, respectively, in the NFT + jujube polysaccharide group compared to day 0. However, in the NFT group, these enzyme activities were 1.3, 1.6, 2.7, and 1.77 times higher, respectively, at the end of storage. The activity of PME fluctuated throughout the storage period in both groups. On day 49, the PME activity in the NFT + jujube polysaccharides group was 25% lower than that in the NFT group ($p < 0.05$).

Figure 1. Appearance of apricot fruits stored in NFT and NFT + jujube polysaccharides at 0 d and 49 d.

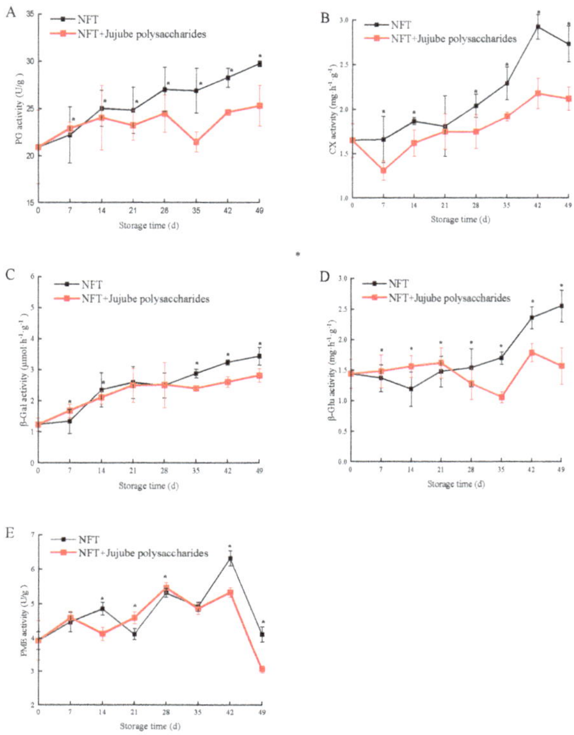

Figure 2. Changes in PG (**A**), CX (**B**), β-GAL (**C**), β-Glu (**D**), and PME (**E**) of NFT and NFT + jujube polysaccharides. Vertical bars represent the standard errors of the mean (n = 3). The symbol (*) within the storage time is significantly different ($p < 0.05$).

3.2. Principal Component Analysis (PCA)

Prior to proteomic analysis, the semi-quantitative SDS-PAGE method was utilized to evaluate the protein samples. The results confirmed that a sufficient quantity of protein was extracted, and the purity of the samples was appropriate for subsequent experiments. Given that the SDS-PAGE method provides only semi-quantitative insights, a TMT-labeled proteomic approach was employed to identify differentially abundant proteins. A total of 7701 proteins were identified, of which 7550 credible proteins were selected for quantitative analysis after excluding data with any missing quantified values. The PCA revealed distinct separations between different storage stages of 'Diaogan' apricot samples based on the identified proteins. The PCA score plot (Figure 3) indicated that the first principal component (PC1) accounted for 68.9% of the total variance, while the second principal component (PC2) contributed 13.7%, collectively explaining 82.6% of the total variance within the 95% confidence interval of the samples, delineated by Hotelling's T-squared ellipse. The groups stored under NFT and NFT combined with jujube polysaccharides were distinctly separated. Each group, represented by three biological replicates, clustered closely, demonstrating the excellent stability and reproducibility of the proteomic dataset.

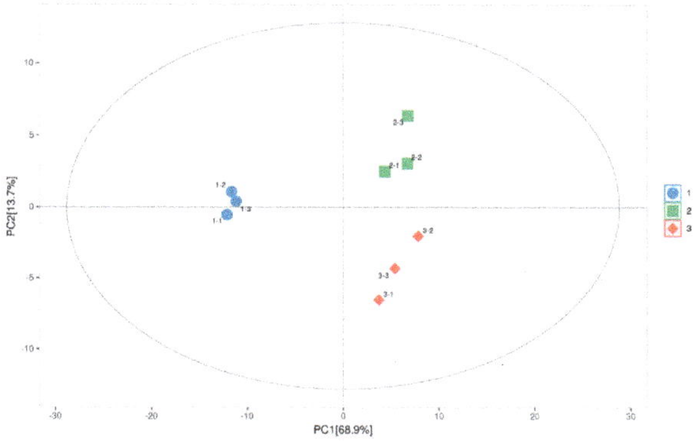

Figure 3. Score scatter plot for PCA model: "1" represents samples of storage at 0 d; "2" represents samples of NFT storage at 49 d; "3" represents samples of NFT + jujube polysaccharides storage at 49 d.

3.3. Identification of Differentially Abundant Proteins

The volcano plot provides a visual representation of the distribution of DEPs among the samples, and in Figure 4, it can be seen that there are 1054 DEPs between groups 2 and 3. Among these, 690 proteins were upregulated, and 364 were downregulation. Notably, as the number of DEPs increased, their significance also intensified. Hierarchical clustering was utilized to visually differentiate protein sub-groups and DEPs within these clusters, facilitating the quantification of proteins. Figure 5 indicates that proteins are color-coded to represent DEPs with distinct expression patterns within the protein sets. The color representation in group 1 is markedly different from groups 2 and 3. The significant color variations between groups 2 and 3 suggest that polysaccharide treatment in jujube influences protein expression during apricot storage differently, leading to varied storage effects. Some color discrepancies observed among the three replicates within each group might be attributed to individual variability among the apricots.

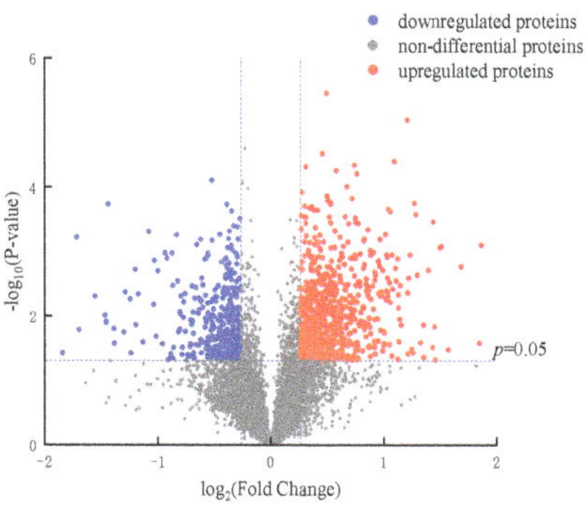

Figure 4. The volcano plot of 2 vs. 3. Proteins exhibiting significant upregulation in expression are represented in red, while proteins exhibiting significant downregulation in expression are represented in blue. Proteins with non-significant differences in expression are represented in gray.

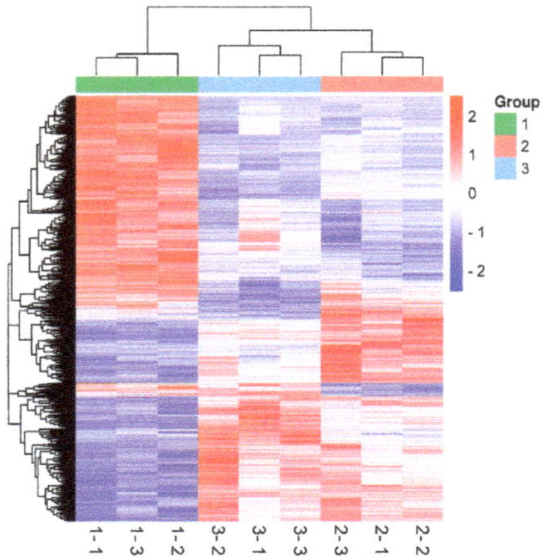

Figure 5. Hierarchical clustering analysis: "1" represents samples of storage at 0 d; "2" represents samples of NFT storage at 49 d; "3" represents samples of NFT + jujube polysaccharides storage at 49 d. The color key values ranged from 2 to −2. The x-axis and y-axis represent the multiples of the difference and significant difference (logarithmic transformation of 2 at the bottom), respectively.

3.4. Bioinformatics Analysis

Gene Ontology (GO), a standardized international gene function classification system, was utilized to categorize the functional classifications of the DEPs in this study. All DEPs were annotated in accordance with the standard GO criteria. In the comparison between groups 2 and 3, the top 10 cellular components associated with the DEPs included parts of the chloroplast and plastid, chloroplast stroma, plastid stroma, ribosomal subunit, ribosome, plastid envelope, and small ribosomal subunit. The most prominent molecular functions comprised rRNA binding, structural constituent of ribosome, structural molecule activity, RNA binding, nucleic acid binding, electron transfer activity, phosphate ion binding, electron carrier activity, electron transfer within the cytochrome b6/f complex, and photosystem II activity, including large and small ribosomal subunit rRNA binding.

KEGG pathway analysis revealed that out of all identified proteins, a total of 4796 proteins were mapped to various KEGG pathways. In the comparison between groups 2 and 3, 100 DEPs were associated with 10 distinct pathways, displaying significant enrichment in 10 metabolic pathways ($p < 0.05$). The top five significantly enriched metabolic pathways for these DEPs were related to the ribosome, photosynthesis, pyruvate metabolism, fatty acid metabolism, and fatty acid biosynthesis.

4. Discussion

Over the years, the preservation of apricot fruit has remained a focal point of scholarly research. However, postharvest storage longevity of apricot fruit is limited. Extensive research has indicated that fruit softening is intricately linked to cell wall-hydrolyzing enzymes, which degrade cell wall polysaccharides and contribute to fruit softening through coordinated actions [22]. In this study, compared to NFT storage, NFT combined with jujube polysaccharides significantly inhibited the activity of PG, PME, β-Gal, CX, and β-Glu, thereby delaying the detrimental effects caused by the depolymerization of cell wall polysaccharides [23,24]. Storage temperature is a crucial factor influencing fruit ripening and senescence. Numerous studies have focused on the physiological and biochemical changes in apricot fruit under NFT storage. However, the use of jujube polysaccharides in fruit preservation has been relatively unexplored. Fruit ripening processes are complex and involve pathways unique to each fruit species. Therefore, in this research, we employed TMT proteomic technology to analyze protein expression during the ripening and senescence of apricot fruit under both NFT and NFT + jujube polysaccharide conditions.

Initially, our research focused on carbohydrate metabolism, a crucial process for energy production and conservation in apricot fruit. In this study, a majority of the identified DEPs were involved in carbohydrate metabolism pathways, including glycolysis, the tricarboxylic acid cycle (TCA), the pentose phosphate pathway, and Glyceraldehyde-3-phosphate dehydrogenase (GAPDH), a key enzyme in the glycolytic pathway [25]. Notably, a downregulation of GAPDH (Table S1) was observed in both the 2 vs. 3 and 2 vs. 1 group comparisons. This suggests that storage under NFT and NFT combined with jujube polysaccharides might similarly influence the apricot glycolytic pathway.

Cytosolic malate dehydrogenase (cy MDH) plays a pivotal role as a precursor in various metabolic pathways of fruit cells. It acts as a respiratory substrate in the mitochondrial tricarboxylic acid (TCA) cycle and can be converted to soluble sugars in the gluconeogenesis pathway [26]. Our findings revealed one upregulated protein (cy MDH) in the 2 vs. 1 group comparison. In contrast, an upregulation of cy MDH was noted in the 2 vs. 3 group, indicating that polysaccharides treatment potentially enhances the TCA cycle, accelerating the conversion of L-malic acid to oxaloacetic acid and facilitating NADH production [27]. Glucose-6-phosphate dehydrogenase (G6PD), involved in the initial step of the sub-pathway synthesizing D-ribulose 5-phosphate from D-glucose 6-phosphate (oxidative phase) [28], showed a general downregulation in our study, suggesting an inhibitory effect of NFT storage on the pentose phosphate pathway of glucose. These results support the hypothesis that NFT storage may preserve function through sustained metabolic activity [29].

Additionally, reactive oxygen species (ROS) are inevitable by-products of fruit ripening and senescence processes. The excessive accumulation of ROS, however, can lead to lipid peroxidation, membrane system damage, and oxidation of proteins and DNA, disrupting normal metabolism and ultimately causing cell death [30]. Plants have developed a comprehensive antioxidant system, including both antioxidant enzymes and metabolites, to mitigate oxidative damage. In this study, significant upregulations in the expressions of five peroxidases (POD), one catalase (CAT), two ascorbate peroxidases (APX), two superoxide dismutases (SOD), three glutathione peroxidases (GSH-PPx), and three glutathione transferases (GST) were observed in the 2 vs. 1 group. In the 2 vs. 3 group, there were noteworthy upregulations of nine POD, one APX, and one GSH-PPx. This upregulation of antioxidant enzymes suggests that jujube polysaccharides treatment can enhance the activity of these enzymes in apricots during cold storage, reducing ROS accumulation and maintaining redox homeostasis. Additionally, peroxidases can remove H_2O_2 and contribute to the cross-linking of cell wall components [31].

Heat shock proteins (HSPs) are crucial for enhancing plant stress tolerance and responding to stress. The accumulation of HSPs has been shown to protect tomato fruits from cold stress [32]. In this study, a small heat shock protein (HSP20) domain-containing protein was identified. In the 2 vs. 3 group, the expression of 6 HSP20 domain-containing proteins was upregulated. In the 2 vs. 1 group, seven proteins from the HSP20 family were upregulated, with one downregulated HSP20 protein. D'Ambrosio suggested that small HSPs were not deregulated, but rather heat shock cognate protein species (HSC70) were induced [33]. This suggests that jujube polysaccharides can induce the expression of HSP20, enhancing the stress resistance of apricot fruits during cold storage.

The softening of apricots post-harvest is primarily due to cell wall degradation, which leads to a reduction in pectin substances [34]. PG, PEM, β-Gal, and PEL are key enzymes involved in this degradation process [35]. Our study found that in the 2 vs. 3 group, there was a downregulation of one PEL, one PEM, one β-Gal, and one α-Gal. The downregulation of these enzyme activities indicates that NFT storage can delay fruit softening by inhibiting the expression of hydrolytic and glycoside hydrolysis-related proteins. Therefore, we infer that the treatment of apricots with jujube polysaccharides significantly inhibits the expression of PEM and β-Gal proteins, thereby delaying the degradation of pectin substances during cold storage [36].

Polygalacturonate-inhibitory proteins (PGIPs) are a type of defense protein associated with plant cell walls. They inhibit polygalacturonate enzymes (PGs) produced by pathogens and pests [37,38]. This study observed an upregulation of PGIPs in both the 2 vs. 1, and 2 vs. 3 groups. PGIPs, characterized by leucine-rich repeats (LRRs), specifically bind to cell walls and inhibit the activity of endogenous PGs. The interaction between PGIP and PGs facilitates the production and accumulation of oligogalacturonides (OGs) in plants, playing a vital role in plant defense signal transduction. This mechanism enhances plant disease resistance, effectively blocks pathogen invasion, and inhibits the onset and progression of related diseases [39].

Lipoxygenases (LOX) play a significant role in cellular processes, catalyzing the production of hydroperoxides from unsaturated fatty acids, and concurrently generating a substantial amount of reactive oxygen species (ROS). LOXs are involved in the peroxidation of cellular membrane lipids, leading to membrane damage and cell apoptosis [40]. At the end of storage, a downregulation of three LOX proteins was noted in the NFT storage group, whereas only one LOX protein was downregulated in the NFT + jujube polysaccharides storage group. This suggests that NFT combined with jujube polysaccharides can inhibit the upregulation of LOX activity during cold storage. Furthermore, the 2 vs. 3 comparison group exhibited a downregulation of one phospholipase C (PI-PLC) protein. Studies have linked increased PLC activity with membrane degradation induced by low-temperature stress [41].

5. Conclusions

During the ripening of 'Diaogan' apricots, there is a rapid softening and breakdown of cell wall components. The application of NFT combined with jujube polysaccharides effectively inhibits this cell wall disassembly. This inhibition is achieved by reducing the degradation of cell wall polysaccharides, which is evident in the decreased activities of enzymes such as PG, β-Glu, CX, β-GAL, and PME. Employing the TMT quantitative proteomic technique revealed 768 DEPs, showing more than 1.2-fold change in abundance, across 25 functional categories or biological processes. Key processes influenced during fruit storage and senescence include carbohydrate and energy metabolism, cell wall and membrane degradation, signal transduction, and stress response and defense mechanisms. Notably, NFT and jujube polysaccharide treatment effectively mitigates oxidative stress, likely due to the increased expression of antioxidant proteins such as POD, APX, and GSH-PPx. Identifying these functionally crucial proteins lays the groundwork for further exploration of the molecular mechanisms underlying apricot fruit storage, using genetic, transcriptional, and additional methodologies.

Supplementary Materials: The following supporting information can be downloaded at: https://www.mdpi.com/article/10.3390/foods12244504/s1, Table S1: List of some important differentially expressed proteins during apricot fruit storage at different manage.

Author Contributions: Z.W.: Writing—review and editing, Data curation. W.W.: Conceptualization, Funding acquisition. W.L.: Investigation. R.Y.: Investigation. Y.L.: Investigation. L.Z.: Investigation. M.Z.: Investigation. X.L.: Supervision, Project administration. All authors have read and agreed to the published version of the manuscript.

Funding: This research received no external funding.

Data Availability Statement: Data is contained within the article or supplementary material.

Acknowledgments: This work was supported by the National Natural Science Foundation of China (32260586), Major Science and Technology Projects of Xinjiang Uygur Autonomous Region (2022A02006-1-4), the Key Research and Development Project of Xinjiang Uygur Autonomous Region (2022B02018-3), the Key Research and Development Project of Xinjiang Uygur Autonomous Region (2022B02005-2-3), and the Key Research and Development Project of Xinjiang Uygur Autonomous Region (2022B02045-3).

Conflicts of Interest: The authors declare no conflict of interest.

References

1. Yuan, Z.X.; Shi, Y.Q.; Cai, F.H.; Zhao, J.R.; Xiong, Q.P.; Wang, Y.P.; Wang, X.L.; Zheng, Y.H. Isolation and identification of polysaccharides from pythium archenemies and application to strawberry fruit (*Fragaria ananassa* Duch.) preservation. *Food Chem.* **2020**, *309*, 125604. [CrossRef] [PubMed]
2. Ghaouth, A.E.; Wilson, C.; Wisniewski, M. Biologically-based alternatives to synthetic fungicides for the control of postharvest diseases of fruit and vegetables. *Dis. Fruits Veg.* **2004**, *2*, 511–535.
3. Li, Y.L.; Zhao, Y.T.; Zhang, Z.C.; He, H.; Shi, L.; Zhu, X.; Cui, K.B. Near-freezing temperature storage improves shelf-life and suppresses chilling injury in postharvest apricot fruit (*Prunus armeniaca* L.) by regulating cell wall metabolism. *Food Chem.* **2022**, *387*, 132921. [CrossRef] [PubMed]
4. Liu, G.S.; Li, H.L.; Grierson, D.; Fu, D.Q. NAC Transcription Factor Family Regulation of Fruit Ripening and Quality: A Review. *Cells* **2022**, *11*, 525. [CrossRef] [PubMed]
5. Fan, X.G.; Yu, X.; Zhao, H.D.; Liu, B.D.; Cao, J.K.; Jiang, W.B. Improving fresh apricot (*Prunus armeniaca* L.) quality and antioxidant capacity by storage at near freezing temperature. *Sci. Hortic.* **2018**, *231*, 1–10. [CrossRef]
6. Chen, Y.; Hung, Y.C.; Chen, M.; Lin, H. Effects of acidic electrolyzed oxidizing water on retarding cell wall degradation and delaying softening of blueberries during postharvest storage. *LWT-Food Sci. Technol.* **2017**, *84*, 650–657. [CrossRef]
7. Cheng, D.W.; Ma, Q.Y.; Zhang, J.H.; Jiang, K.L.; Cai, S.J.; Wang, W.X.; Wang, J.; Sun, J.F. Cactus polysaccharides enhance preservative effects of ultrasound treatment on fresh-cut potatoes. *Ultrason. Sonochemistry* **2022**, *90*, 106205. [CrossRef] [PubMed]
8. Ranjith, F.H.; Adhikari, B.; Muhialdin, B.J.; Yusof, N.L.; Mohammed, N.K.; Ariffin, S.H.; Hussin, A.S.M. Peptide-based edible coatings to control postharvest fungal spoilage of mango (*Mangifera indica* L.) fruit. *Food Control* **2022**, *135*, 108789. [CrossRef]
9. Huang, J.; Wu, W.J.; Niu, B.; Fang, X.J.; Chen, H.J.; Wang, Y.H.; Gao, H.Y. Characterization of Zizania latifolia polysaccharide-corn starch composite films and their application in the postharvest preservation of strawberries. *LWT-Food Sci. Technol.* **2023**, *173*, 114332. [CrossRef]
10. Molassiotis, A.; Tanou, G.; Filippou, P.; Fotopoulos, V. Proteomics in the fruit treescience arena: New insights into fruit defense, development, and ripening. *Proteomics* **2013**, *13*, 1871–1884. [CrossRef]
11. Liu, R.L.; Wang, Y.Y.; Qin, G.Z.; Tian, S.P. iTRAQ-based quantitative proteomic analysis reveals the role of the tonoplast in fruit senescence. *J. Proteom.* **2016**, *146*, 80–89. [CrossRef] [PubMed]
12. Lin, M.; Fang, J.; Qi, X.; Li, Y.K.; Chen, J.Y.; Sun, L.M.; Zhong, Y.P. iTRAQ-based quantitative proteomic analysis reveals alterations in the metabolism of Actinidia arguta. *Sci. Rep.* **2017**, *7*, 5670. [CrossRef] [PubMed]
13. Xu, Y.; Huo, L.; Zhao, K.; Li, Y.; Zhao, X.; Wang, H.; Wang, W.; Shi, H. Salicylic acid delays pear fruit senescence by playing an antagonistic role toward ethylene, auxin, and glucose in regulating the expression of PpEIN3a. *Front. Plant Sci.* **2023**, *13*, 1096645. [CrossRef] [PubMed]
14. Pedreschi, R.; Hertog, M.; Robben, J.; Lilley, K.S.; Karp, N.A.; Baggerman, G.; Vanderleyden, J.; Nicolaï, B. Gel-based proteomics approach to the study of metabolic changes in pear tissue during storage. *J. Agric. Food Chem.* **2009**, *57*, 6997–7004. [CrossRef] [PubMed]
15. Yun, Z.; Jin, S.; Ding, Y.; Wang, Z.; Gao, H.; Pan, Z.; Xu, J.; Cheng, Y.; Deng, X. Comparative transcriptomics and proteomics analysis of citrus fruit, to improve understanding of the effect of low temperature on maintaining fruit quality during lengthy post-harvest storage. *J. Exp. Bot.* **2012**, *63*, 2873–2893. [CrossRef] [PubMed]
16. Ji, X.L.; Peng, Q.; Yuan, Y.P.; Shen, J.; Xie, X.Y.; Wang, M. Isolation, structures and bioactivities of the polysaccharides from jujube fruit (*Ziziphus jujuba* Mill.): A review. *Food Chem.* **2017**, *227*, 349–357. [CrossRef] [PubMed]
17. Yang, W.T.; Liu, Y.X.; Sang, Y.Y.; Ma, Y.Y.; Guo, M.R.; Bai, G.R.; Cheng, S.B.; Chen, G.G. Influences of ice-temperature storage on cell wall metabolism and reactive oxygen metabolism in Xinjiang (*Diaogan*) apricot. *Postharvest Biol. Technol.* **2021**, *180*, 111614. [CrossRef]
18. Win, N.M.; Yoo, J.; Kwon, S.; Watkins, C.B.; Kang, I. Characterization of fruit quality attributes and cell wall metabolism in 1-methylcyclopropene (1-MCP)-treated '*Summer King* 'and '*Green Ball*'apples during cold storage. *Front. Plant Sci.* **2019**, *10*, 1513. [CrossRef]
19. Hagerman, A.E.; Austin, P.J. Continuous spectrophotometric assay for plant pectin methyl esterase. *J. Agric. Food Chem.* **1986**, *34*, 440–444. [CrossRef]
20. Liu, B.D.; Zhao, H.D.; Fan, X.G.; Jiao, W.X.; Cao, J.K.; Jiang, W.B. Near freezing point temperature storage inhibits chilling injury and enhances the shelf-life quality of apricots following long-time cold storage. *J. Food Process. Preserv.* **2019**, *43*, e13958. [CrossRef]
21. Jia, C.; Lu, X.; Gao, J.; Wang, R.; Sun, Q.; Huan, J. TMT-labeled quantitative proteomic analysis to identify proteins associated with the stability of peanut milk. *J. Sci. Food Agric.* **2021**, *101*, 6424–6433. [CrossRef] [PubMed]
22. Ji, Y.; Hu, W.Z.; Liao, J.; Xiu, Z.L.; Jiang, A.L.; Guan, Y.G.; Yang, X.Z.; Feng, K. Ethanol vapor delays softening of postharvest blueberry by retarding cell wall degradation during cold storage and shelf life. *Postharvest Biol. Technol.* **2021**, *177*, 111538. [CrossRef]
23. Huang, P.; Ding, K.; Liu, C.Y.; Li, H.J.; Wang, C.L.; Lin, Y.Z.; Sameen, D.E.; Hossen, M.A.; Chen, M.R.; Yan, J.; et al. Konjac glucomannan/low-acyl gellant gum edible coating containing thymol microcapsule regulates cell wall polysaccharides disassembly and delays postharvest softening of blueberries. *Postharvest Biol. Technol.* **2023**, *204*, 112449. [CrossRef]

24. Wang, H.B.; Cheng, X.; Wu, C.; Fan, G.J.; Li, T.T.; Dong, C. Retardation of postharvest softening of blueberry fruit by methyl jasmone is correlated with altered cell wall modification and energy metabolism. *Sci. Hortic.* **2021**, *276*, 109752. [CrossRef]
25. Kosova, A.A.; Khodyreva, S.N.; Lavrik, O.I. Role of glyceraldehyde-3-phosphate dehydrogenase (GAPDH) in DNA repair. *Biochemistry* **2017**, *82*, 643–654. [CrossRef] [PubMed]
26. Zhang, L.H.; Wang, C.Z.; Jia, R.P.; Yang, N.X.; Jin, L.; Zhu, L.C.; Ma, B.Q.; Yao, Y.X.; Ma, F.W.; Li, M.J. Malate metabolism mediated by the cytoplasmic malate dehydrogenase gene Mdcy-MDH affects sucrose synthesis in apple fruit. *Hortic. Res.* **2022**, *9*, 194. [CrossRef]
27. Clara, I.M.; Jezrel, M.; Maarten, L.A.T.M.; Hertog, B.V.D.P.; Bart, M.N. Expression and protein levels of ethylene receptors, CTRs and EIN2 during tomato fruit ripening as affected by 1-MCP. *Postharvest Biol. Technol.* **2021**, *179*, 111573.
28. Reyes, J.S.; Fuentes-Lemus, E.; Figueroa, J.D.; Rojas, J.; Fierro, A.; Arenas, F.; Hägglund, P.M.; Davies, M.J.; López-Alarcón, C. Implications of differential peroxyl radical-induced inactivation of glucose 6-phosphate dehydrogenase and 6-phosphogluconate dehydrogenase for the pentose phosphate pathway. *Sci. Rep.* **2022**, *12*, 21191. [CrossRef]
29. Yuan, X.Z.; Wu, Z.M.; Li, H.; Wang, Y.D.; Liu, F.; Cai, H.; Newlove, A.A.; Wang, Y. Biochemical and proteomic analysis of 'Kyoho' grape (*Vitis labruscana*) berries during cold storage. *Postharvest Biol. Technol.* **2014**, *88*, 79–87. [CrossRef]
30. Giovannoni, J.J. Genetic regulation of fruit development and ripening. *Plant Cell* **2004**, *16*, 170–180. [CrossRef]
31. Li, T.T.; Yun, Z.; Zang, D.D.; Yang, C.W.; Zhu, H.; Jiang, Y.M.; Duan, X.W. Proteomic analysis of differentially expressed proteins involved in ethylene-induced chilling tolerance in harvested banana fruit. *Front. Plant Sci.* **2015**, *6*, 00845. [CrossRef]
32. Long, L.; Gao, W.; Xu, L.; He, X.; Yang, X.Y.; Zhang, X.L.; Zhu, L.F. GbMPK3, a mitogen-activated protein kinase from cotton, enhances drought and oxidative stress tolerance in tobacco. *Plant Cell* **2014**, *116*, 153–162. [CrossRef]
33. D'ambrosio, C.; Arena, S.; Rocco, M.; Verrillo, F.; Novi, G.; Viscosi, V.; Marra, M.; Scaloni, A. Proteomic analysis of apricot fruit during ripening. *J. Proteom.* **2013**, *78*, 39–57. [CrossRef]
34. Kortste, A.J.; Appeldoorn, N.J.G.; Ooetwijn, M.E.P.; Visser, R.G.F. Differences in regulation of carbohydrate metabolism during early fruit development between domesticated tomato and two wild relatives. *Planta* **2007**, *226*, 929–939. [CrossRef]
35. Ma, Y.Y.; Zhang, X.L.; Li, L.L.; Yang, W.T.; Zhang, W.D.; Cheng, S.B.; Guo, M.R.; Chen, G.G. Delaying fruit softening of 'France' prune (*Prunus domestica* L.) using near-freezing temperature storage. *LWT-Food Sci. Technol.* **2022**, *172*, 114165. [CrossRef]
36. Yu, F.; Shao, X.F.; Yu, L.; Xu, F.; Wang, H.F. Proteomic analysis of postharvest peach fruit subjected to chilling stress or non-chilling stress temperatures during storage. *Sci. Hortic.* **2015**, *197*, 72–89. [CrossRef]
37. Lorenzo, G.D.; D'Ovidio, R.; Cervone, F. The role of polygalacturonate inhibiting proteins (PGIPS) in defense against pathogenic fungi. *Annu. Rev. Phytopathol.* **2001**, *39*, 313–335. [CrossRef]
38. Kalunke, R.M.; Tundo, S.; Benedetti, M.; Cervone, F.; Lorenzo, G.D.; D'Ovidio, R. An update on polygalacturonate-inhibiting protein (PGIP), a leucine-rich repeat protein that protects crop plants against pathogens. *Front. Plant Sci.* **2015**, *6*, 146. [CrossRef]
39. Li, H.Y.; Smigocki, A.C. Wound induced Beta vulgaris polygalacturonate-inhibiting protein genes encode a longer leucine-rich repeat domain and inhibit fungal polygalacturonates. *Physio. Mol. Plant.* **2016**, *96*, 8–18. [CrossRef]
40. Li, R.J.; Yang, S.X.; Wang, D.; Liang, J.; Huang, T.Z.; Zhan, L.; Luo, A.W. Electron-beam irradiation delayed the postharvest senescence of kiwifruit during cold storage through regulating the reactive oxygen species metabolism. *Radiat. Phys. Chem.* **2021**, *189*, 109717. [CrossRef]
41. Soh, C.; Ali, Z.M.; Lazan, H. Characterization of an α-galactosidase with potential relevance to ripening related texture changes. *Phytochemistry* **2006**, *67*, 242–254. [CrossRef]

foods

MDPI

Article

Iron Competition as an Important Mechanism of Pulcherrimin-Producing *Metschnikowia* sp. Strains for Controlling Postharvest Fungal Decays on Citrus Fruit

Shupei Wang [1,2,3,†], Zhimei Tan [4,†], Chenshu Wang [4], Wenqing Liu [4], Fangxue Hang [4], Xuemei He [1,2], Dongqing Ye [1,2], Li Li [1,2] and Jian Sun [1,2,*]

1 Guangxi Academy of Agricultural Sciences, Nanning 530007, China; wshp2000@163.com (S.W.); xuemeihe1981@126.com (X.H.); yedongqing@gxaas.net (D.Y.); lili@gxaas.net (L.L.)
2 Guangxi Key Laboratory of Fruits and Vegetables Storage-Processing Technology, Nanning 530007, China
3 College of Environmental and Life Sciences, Nanning Normal University, Nanning 530001, China
4 College of Light Industry and Food Engineering, Guangxi University, Nanning 530004, China; tzhm0810@163.com (Z.T.); cswang0321@163.com (C.W.); wqliue@163.com (W.L.); hangfx@163.com (F.H.)
* Correspondence: jiansun@gxaas.net
† These authors contributed equally to this work.

Abstract: This study identified and tested fruit-isolated *Metschnikowia* yeasts against three major postharvest citrus pathogens, namely, *Penicillium digitatum*, *Penicillium italicum*, and *Geotrichum citri-aurantii*, and further evaluated the impact of $FeCl_3$ on the biocontrol efficiency of pulcherrimin-producing *M. pulcherrima* strains. Based on the characterization of the pigmented halo surrounding the colonies and the analysis of the D1/D2 domain of 26S rDNA, a total of 46 *Metschnikowia* sp. were screened and identified. All 46 *Metschnikowia* strains significantly inhibited the hyphal growth of *Penicillium digitatum*, *Penicillium italicum*, and *Geotrichum citri-aurantii*, and effectively controlled the development of green mold, blue mold and sour rot of citrus fruit. The introduction of exogenous $FeCl_3$ at certain concentrations did not significantly impact the pulcherriminic acid (PA) production of pigmented *M. pulcherrima* strains, but notably diminished the size of pigmented zones and the biocontrol efficacy against the three pathogens. Iron deficiency sensitivity experiments revealed that *P. digitatum* and *P. italicum* exhibited higher sensitivity compared to *G. citri-aurantii*, indicating that iron dependence varied among the three pathogens. These results suggested that *M. pulcherrima* strains, capable of producing high yields of PA, possessed great potential for use as biocontrol agents against postharvest citrus diseases. The biocontrol efficacy of these yeasts is mainly attributed to their ability to competitively deplete iron ions in a shared environment, with the magnitude of their pigmented halo directly correlating to their antagonistic capability. It is worth noting that the level of sensitivity of pathogens to iron deficiency might also affect the biocontrol effect of pulcherrimin-producing *M. pulcherrima*.

Keywords: *Metschnikowia pulcherrima*; pulcherrimin; iron competition; *P. digitatum*; *P. italicum*; *G. citri-aurantii*

1. Introduction

As one of the most commercially valuable fruits, citrus (*Citrus* spp.) is highly appreciated by consumers due to its unique flavor and rich nutrients [1]. Citrus fruits, however, are highly susceptible to spoilage fungi, including *Penicillium digitatum* (green mold), *Penicillium italicum* (blue mold), and *Geotrichum citri-aurantii* (sour rot) during postharvest, causing substantial economic losses [2]. With rising public concerns about the effects of synthetic chemical fungicides on food safety and environmental preservation, the pursuit of eco-friendly management techniques is intensifying. One of the most promising methods employs antagonistic yeasts as biological controls against postharvest decay in various fruits [3,4].

Metschnikowia pulcherrima is a ubiquitous yeast that exhibits strong ecological adaptability and is widely distributed across the globe. It is commonly found in flowers, nectars, and fresh and spoiled fruits [5]. This yeast typically produces pulcherrimin, a red pigment, which results in distinctive colony features [6]. *M. pulcherrima* possesses excellent antifungal properties. This was demonstrated by conducting competition assays. These involved using a sample of 40 yeasts on a diverse collection of 16 filamentous fungi, revealing *M. pulcherrima* to be the most potent antagonist [7]. Furthermore, when applied to freshly cut fruits, *M. pulcherrima* effectively inhibited the growth and proliferation of foodborne pathogens, including *Listeria monocytogenes* and *Salmonella enterica* [8,9]. In terms of controlling postharvest diseases, fruit-borne strains of *M. pulcherrima* exhibit strong antagonistic activity against a range of pathogens including *Penicillium* spp., *Botrytis cinerea*, *Alternaria alternata*, *Fusarium* spp., *Monilinia* spp., and *Pestalotiopsis vismiae*. Research has shown these strains to be highly effective in managing postharvest diseases of some fruits, such as apples, pears, sweet cherries, strawberries, lemons, grapes, loquats, and mangoes [10–16]. *M. pulcherrima* strains demonstrate the traits of an "ideal antagonist" because of their capacity to withstand harsh environmental circumstances, effectively outcompete for nutrients and space, and refrain from generating toxic substances [17,18]. These attributes, combined with its broad-spectrum antifungal qualities and biosafety, make *M. pulcherrima* a promising candidate for commercial application.

Iron (Fe) is an essential micronutrient for the survival and growth of almost all living organisms on earth. Competition for or depletion of iron ions are considered important mechanisms of action in the biocontrol of postharvest fruit diseases by pulcherrimin-producing *Metschnikowia* strains, including *M. pulcherrima*, *M.* andauensis, *M. fructicola*, *M. citriensis*, etc. [17,19]. These yeasts compete for iron ions with pathogens through the secretion of a siderophore—pulcherriminic acid (PA). This water-soluble PA is capable of spontaneously chelating four Fe^{3+} ions in a non-enzymatic manner, resulting in the formation of the red, insoluble pulcherrimin complex [20]. This process results in iron fixation in the coexistence environment, thereby depleting the availability of free iron and further inhibiting the growth of various bacteria and pathogenic fungi [6,17,21]. Therefore, pulcherrimin-producing *Metschnikowia* strains are regarded as highly promising for biocontrol agents [22].

Limited information suggests that *M. pulcherrima* has the potential to control green mold in lemons and mandarins [16,23]. However, there is little information available about the effectiveness of *M. pulcherrima* against citrus blue mold and sour rot. The aim of this research was to assess the biocontrol efficacy of fruit-sourced *Metschnikowia* yeasts against three postharvest citrus pathogens, and further to investigate the competitive ability of *M. pulcherrima* for iron acquisition against *P. digitatum*, *P. italicum*, and *G. citri-aurantii*. This study could provide new high-performance antagonistic yeast resources for the biocontrol of postharvest citrus diseases, as well as a certain theoretical basis for the investigation of iron competition mechanisms of *M. pulcherrima*.

2. Materials and Methods

2.1. Fruit Material

Commercially mature Olinda Valencia and Newhall navel oranges [*Citrus sinensis* (L.) Osbeck] were harvested from the commercial orchard located in Hepu and Fuchuan country, Guangxi, China, respectively. The selected fruits, which displayed consistent size, shape, and ripeness and showed no visible symptoms of disease or mechanical injury, were surface-sterilized via dipping in a 2% (*v*/*v*) sodium hypochlorite solution for 2 min. They were then thoroughly washed with tap water and left to air-dry at room temperature for the next experiment.

2.2. Pathogens Preparation

The strains of *P. digitatum*, *P. italicum* and *G. citri-aurantii* were identified, isolated from naturally decayed citrus fruits exhibiting disease-related symptoms and preserved in our laboratory. Before use, all isolates were cultivated on potato dextrose agar (PDA) at 25 °C for 5 days, and then the spores of the three pathogens were rinsed off with sterile water. The resulting cell numbers were calculated using a hemocytometer and adjusted to the desired concentrations.

2.3. Sampling and Yeast Isolation

The yeasts of the genus *Metschnikowia* were isolated from the different cultivars of table grapes collected in turn at 15 d intervals from agricultural wholesale markets in Xixiangtang (XX), Jiangnan (JN), Xingning (XN) and Wuming (WM) districts of Nanning, Guangxi, respectively, from April to October 2021. After labeling, the samples were stored in sterile self-sealing bags and transported back to the laboratory aseptically. Grapes were picked and rinsed four times with sterile water. Approximately 30–50 grapes from each sample were then randomly selected and incubated in a triangular flask containing 500 mL of sterile water at 28 °C for 3 h with shaking (180 rpm). The yeast suspension was then diluted 10-fold, and 50 μL of the dilution was plated on YPD + Fe plates (20 g L^{-1} peptone, 20 g L^{-1} dextrose, 10 g L^{-1} yeast extract, 20 g L^{-1} agar, 5 mg L^{-1} $FeCl_3$) containing 50 mg L^{-1} of kanamycin. All the media formulations were obtained from Solaibao Biological Technology Co., Ltd. (Beijing, China). After culturing at 28 °C for 48 h, the reddish colonies were picked based on pulcherrimin pigment production characteristics and investigated for cell morphology using an optical microscope. Afterwards, the single colonies were purified via re-streaking on YPD + Fe plates three times and the pure yeast cultures were subsequently stored in glycerol stocks at −80 °C.

2.4. Yeast Identification

The isolates were grown in YEPD medium (YPD without agar) at 28 °C (180 rpm) to reach the mid-log phase, after which the cells were collected for genomic DNA extraction using an Ezup column yeast genomic DNA purification kit according to the manufacturer's protocol (Sangon Biotech, Shanghai, China). The D1/D2 region of the 26S rDNA of the yeast isolates was amplified using the universal primers NL1 (5′-GCATATCAATAAG CGGAGGAAAAG-3′) and NL4 (5′-GGTCCGTGTTTCAAGACGG-3′), as previously described [24]. The PCR products were sequenced by Sanger sequencing (Sangon Biotech Co., Ltd., Shanghai, China), and the resulting sequences were blasted against the NCBI nucleotide database (http://blast.ncbi.nlm.nih.gov/, accessed on 11 December 2022) with BLASTn for species identification.

2.5. Antagonism Assays

2.5.1. In Vitro Biocontrol Assay on Agar Plates

The inhibitory effects of the *Metschnikowia* yeasts against *P. digitatum*, *P. italicum*, and *G. citri-aurantii* were tested on PDA plates, using the method of Wang et al. [25] with slight modifications. The yeast isolate cells (20 μL, 1×10^8 cells mL^{-1}) were streaked evenly along the central axis of a PDA plate, covering an area of 60 mm long × 3 mm wide. After incubation for 10 min, 5 μL conidial suspensions (1×10^6 spores mL^{-1}) of *P. digitatum*, *P. italicum* and *G. citri-aurantii* were inoculated at 30 mm on both sides of the coating position. The width of the inhibition zone (subtract the width of the yeast growth zone) was measured after 7 d of incubation at 25 °C. Three replicates of each treatment were performed, and each replicate contained six plates.

In addition, five strains of *M. pulcherrima* were used as controls: *M. pulcherrima* (CICC 33433, CICC 1467, CICC 32343, and CICC 33447) was purchased from the China Center of Industrial Culture Collection (CICC), and *M. pulcherrima* (CGMCC 2.3314) was obtained from the China General Microbiological Culture Collection Center (CGMCC). These five yeast strains were activated via inoculation in YPD medium for 48 h before use.

2.5.2. In Vivo Biocontrol Assay on Citrus Fruits

All yeast isolates were evaluated for biocontrol activity against the three major postharvest citrus diseases, following the method described by Liu et al. [26] with some modifications. The yeast strains were inoculated in 100 mL YEPD medium and incubated at 28 °C and 200 rpm for 16 h to reach the mid-log phase, and then centrifuged at 8000× *g* for 5 min to collect the cells. After being washed twice with sterile water, the concentration of yeast cells was counted and adjusted to 1 × 10^8 cells mL^{-1} for each treatment group. The prepared citrus fruits were artificially wounded (approximately 3 mm diameter and 3 mm deep) on opposite sides of their equators with a sterile 1000 µL pipette tip. Each wound was then inoculated with 20 µL of yeast cell suspension (1 × 10^8 cells mL^{-1}), and an equal amount of sterile water was used as a control. After 4 h, 10 µL spore suspensions of *P. digitatum* (1 × 10^4 spores mL^{-1}), *P. italicum* (1 × 10^4 spores mL^{-1}) and *G. citri-aurantii* (1 × 10^6 spores mL^{-1}) were added to each wound. After being air-dried, all treated fruits were individually wrapped in high-density polyethylene plastic bags and stored at 25 ± 2 °C with 85–90% relative humidity (RH). The disease incidence (DI) and lesion diameter (LD) were counted daily at 1 d intervals after the appearance of disease characteristics on the fruit. Each treatment was replicated three times, with each replicate consisting of 10 fruits.

2.6. *Test for Pulcherrimin Pigment Production of Yeast Isolates*

The pulcherrimin pigment produced by *Metschnikowia* yeasts was tested using PDA + Fe plates with a concentration of 10 mg L^{-1} $FeCl_3$. In total, 10 mL of molten medium was poured into Petri dishes and allowed to set. Next, a 20 µL yeast cells suspension (1 × 10^8 cells mL^{-1}) was evenly spread along the plate's central axis, covering an area of 60 mm in length and 30 mm in width. The width of the pigmented halos surrounding the yeast colonies was recorded following a 3-day incubation period at 28 °C. There were six replicate plates for each yeast isolate, and the experiment was reproduced twice.

2.7. *Iron Competition and Antagonism*

2.7.1. Iron Competition Assay on Agar Plates In Vitro

The effect of iron on the growth of pulcherrimin-producing *Metschnikowia* yeasts, inhibiting *P. digitatum*, *P. italicum*, and *G. citri-aurantii*, was tested using PDA plates with varying $FeCl_3$ concentrations (0, 5, 10, 15, 20, 25, 50 mg L^{-1}). This technique was referring to our previously published methods [27] with slight modifications. Overall, 10 mL PDA + Fe medium was poured into Petri plates and allowed to solidify; then, 20 µL of yeast cell suspension (1 × 10^8 cells mL^{-1}) was uniformly streaked along the plate's central axis, covering an area measuring 60 × 30 mm (length × width). After 10 min, the conidia suspensions (5 µL, 1 × 10^6 spores mL^{-1}) of *P. digitatum*, *P. italicum* and *G. citri-aurantii* were positioned at 30 mm on either side of the yeast strip location. The width of the inhibition zone was measured after incubation for 7 d at 25 °C. In addition, the PDA plates coated with yeast alone were placed in a 28 °C incubator for 3 days; then, the width of the pigmented circles was measured. Five plates were prepared for each treatment, and the experiments were performed twice.

2.7.2. Iron Competition on Citrus Fruits In Vivo

Citrus fruits with two artificial wounds were prepared as described in Section 2.5.2. The wounds were injected with 20 µL of (1) yeast cells suspension (1 × 10^8 cells mL^{-1}) and (2) sterile water as a control. After 4 h, 10 µL of spore suspensions of *P. digitatum* or *P. italicum* (1 × 10^4 spores mL^{-1}), and *G. citri-aurantii* (1 × 10^6 spores mL^{-1}) were applied to each wound. After drying, aliquots of 20 µL of $FeCl_3$ (0, 5, 15, 25 mg L^{-1}) were applied to each wound. After being bagged with the polyethylene bag, all fruits were stored at conditions of 25 °C and RH 85–90%. The DI and LD of each fruit was measured when the fruit showed symptoms of disease every 24 h. Each treatment was replicated 3 times with 10 fruits per repetition.

2.7.3. Test Iron on the Growth and PA Production of Selected Yeasts

The PDB (potato dextrose broth), supplemented with $FeCl_3$ at varying concentrations (0, 1.0, 5, 10, 15, 25, 50, 100 mg L^{-1}), was employed to test the effect of iron on the cell growth and intracellular PA production of *Metschnikowia* yeasts. The selected yeasts were incubated at 28 °C for 16 h with 180 rpm shaking and harvested at the mid-log phase. After washing and counting, the yeast cells were seeded into 50 mL of PDB + Fe medium at a final concentration of 1×10^6 cells mL^{-1}, and cultured for 30 h in a shaking incubator (180 rpm) at 28 °C. The cell growth and PA production of the yeasts was determined by reference to the previous study [22,28,29]. Briefly, 6 mL fermentation broth was taken and centrifuged at 8000× *g* for 10 min at 4 °C; then, the precipitated cells were washed twice with sterile water and resuspended in 0.1 mmol L^{-1} NaOH to dissolve the pulcherrimin. Following centrifugation at 10,000× *g* for 2 min, the pellet was resuspended in sterile water. Subsequently, the yeast growth was evaluated by measuring optical density at OD_{600}. Another 6 mL of yeast fermentation broth was harvested and centrifuged at 8000× *g* for 10 min. The resulting cell pellets were washed twice with deionized water and then resuspend in 2 mol L^{-1} NaOH to dissolve the PA. Subsequently, a centrifugation step (10,000× *g*, 3 min) was performed. The supernatant was then collected and the OD_{410} was measured using a UV-vis spectrophotometer (Tecan Infinite F-Plex plate reader) to determine the PA concentration. Each treatment was conducted in triplicate and each replicate was measured three times.

2.7.4. Deficiency Sensitivity Tests for *P. digitatum*, *P. italicum* and *G. citri-aurantii*

The test pathogens' sensitivity to iron deficiency was examined using tropolone (≥98%, Macklin, Shanghai, China), a chelating agent that exhibits a strong affinity for ferric ions [16]. An aliquot of 100 μL spore suspension of *P. digitatum*, *P. italicum*, and *G. citri-aurantii* at 1×10^6 spores·mL^{-1} was uniformly spread on PDA plates. The plates were left to stand for 30 min until the spore suspension was fully adsorbed by the culture medium. Subsequently, a 5 mm diameter well in the center of the plate was produced via a sterilized punch. Then, 50 μL tropolone solution (5 mg mL^{-1}) was added to the wells, and 50 μL sterile water was added as a control. After 3 days of incubation at 25 °C, the inhibition zone was measured. Six plates were prepared for each treatment, and the experiments were carried out twice.

2.8. Statistical Analysis

The statistical analysis was conducted utilizing SPSS 27.0 (SPSS Inc., Chicago, IL, USA). The data were presented as mean ± standard deviation ($n \geq 3$). One-way analysis of variance (ANOVA) was performed, followed by Duncan's multiple comparison tests, to ascertain any statistically significant distinctions ($p < 0.05$) among the various treatments.

3. Results

3.1. Identification of *Metschnikowia* Yeast Isolates

A total of 46 target yeast strains of the genus *Metschnikowia* were screened from 417 fresh table grape samples obtained from a wholesale market with different agricultural products in Nanning, Guangxi. All of the screened yeasts had a clear production of pulcherrimin-pigmented halo (Figure S1). Detailed molecular biological analysis revealed that all 46 yeast isolates were identified as *Metschnikowia* strains (Table S1). These were classified into three species: one isolate of *Metschnikowia* sp., one isolate of *M. shanxiensis*, and 44 isolates of *M. pulcherrima*. *M. pulcherrima* was the predominant species, comprising over 95% of the identified isolates.

3.2. Antagonistic Activity of Yeast Strains on Culture Media In Vitro

As shown in Table 1 and Figure S2, all the screened isolates of *Metschnikowia* significantly inhibited the growth of *P. digitatum*, *P. italicum* and *G. citri-aurantii* on PDA plates.

Table 1. In vitro mycelial growth inhibition of 51 yeast isolates of *Metschnikowia* against *P. digitatum*, *P. italicum* and *G. citri-aurantii* on PDA plates.

Isolate	Width of Inhibition Zone (mm)		
	P. digitatum	*P. italicum*	*G. citri-aurantii*
XX01	12.60 ± 055	11.00 ± 0.87	16.86 ± 0.69
XX02	12.50 ± 1.22	11.13 ± 0.64	16.75 ± 1.28
XX03	7.00 ± 0.89	5.71 ± 1.25	10.78 ± 0.83
XX04	11.60 ± 1.14	12.43 ± 1.27	16.17 ± 0.75
XX05	12.00 ± 1.00	11.33 ± 0.82	16.13 ± 0.64
XX06	5.43 ± 0.98	3.71 ± 1.38	9.57 ± 1.40
XX07	12.20 ± 1.30	11.29 ± 1.50	17.50 ± 0.55
XX08	10.17 ± 0.75	9.50 ± 1.87	15.57 ± 0.79
XX09	9.00 ± 0.58	8.17 ± 1.94	15.00 ± 0.82
XX10	9.83 ± 0.75	10.14 ± 1.77	15.75 ± 0.89
XX11	9.33 ± 0.52	9.57 ± 1.40	14.00 ± 1.10
XX12	12.40 ± 0.89	10.00 ± 1.53	15.33 ± 0.52
JN01	11.00 ± 0.58	11.17 ± 1.72	16.43 ± 0.53
JN02	10.29 ± 0.76	8.86 ± 1.21	15.14 ± 1.57
JN03	7.83 ± 0.98	8.00 ± 1.12	13.67 ± 1.03
JN04	9.67 ± 0.82	10.00 ± 0.89	15.44 ± 1.88
JN05	9.60 ± 0.55	9.14 ± 1.77	15.22 ± 1.48
JN06	9.20 ± 0.45	9.33 ± 1.37	14.00 ± 0.63
JN07	6.20 ± 0.45	7.88 ± 1.73	13.83 ± 1.47
JN08	10.00 ± 0.71	9.67 ± 0.82	13.67 ± 0.82
JN09	9.83 ± 0.75	9.71 ± 0.76	14.00 ± 1.26
JN10	10.17 ± 0.75	9.83 ± 0.75	15.50 ± 1.05
JN11	11.43 ± 0.79	11.14 ± 1.35	16.43 ± 1.13
JN12	11.14 ± 0.69	11.25 ± 1.49	16.00 ± 0.93
JN13	9.80 ± 1.10	10.38± 1.06	15.50 ± 1.05
JN14	7.60 ± 0.89	8.57 ± 1.40	15.00 ± 0.89
JN15	7.80 ± 1.64	8.71 ± 1.25	14.00 ± 0.63
XN01	8.33 ± 0.52	9.75 ± 1.16	15.14 ± 0.69
XN02	9.00 ± 0.71	10.13 ± 1.13	14.60 ± 0.55
XN03	9.86 ± 0.69	10.00 ± 1.07	15.33 ± 0.52
XN04	9.57 ± 0.53	10.29 ± 0.95	15.43 ± 1.51
XN05	11.13 ± 1.13	11.33 ± 0.82	16.00 ± 0.93
XN06	7.50 ± 1.41	10.14 ± 1.46	14.29 ± 0.76
XN07	7.71 ± 0.76	10.67 ± 0.82	14.67 ± 0.82
XN08	8.40 ± 0.55	10.83 ± 1.83	14.00 ±0.63
WM01	12.00 ± 0.71	9.33 ± 0.98	16.50 ± 0.84
WM02	11.86 ± 0.90	10.00 ± 1.26	17.00 ± 0.89
WM03	10.14 ± 1.21	9.50 ± 1.41	14.86 ± 0.38
WM04	11.00 ± 0.76	11.43 ± 1.51	16.29 ± 1.11
WM05	11.14 ±1.21	12.33 ± 0.52	16.00 ± 1.15
WM06	10.86 ± 1.07	10.00 ± 0.63	14.57 ± 1.27
WM07	10.43 ± 1.27	10.57 ± 0.98	15.43 ± 0.53
WM08	10.17 ± 0.98	10.50 ± 1.38	13.50 ± 1.07
WM09	12.33 ± 0.52	10.33 ± 1.03	14.89 ± 1.36
WM10	13.63 ± 1.19	11.57 ± 0.79	16.57 ± 0.98
WM11	12.43 ± 0.79	11.13 ± 1.25	16.71 ± 0.76
CICC33433	9.40 ± 1.34	9.57 ± 1.62	13.67 ± 1.21
CICC33447	0.86 ± 0.69	0.86 ± 0.69	1.14 ± 0.90
CICC32343	5.75 ± 0.89	6.17 ± 1.17	10.14 ± 1.35
CICC1467	6.83 ± 0.75	7.50 ± 1.22	10.43 ± 0.79
CGMCC2.3314	6.00 ± 1.14	8.33 ± 1.21	11.00 ± 1.20

Each value was presented as the mean ± standard deviation of three replicates.

To obtain yeasts that have more potent antagonistic effects against the three pathogens, further screening was performed using inhibitory zone widths of ≥11.00 mm for *Metschnikowia* sp. against *P. digitatum* and *P. italicum*, and of ≥16.00 mm for *Metschnikowia* sp. against *G.*

citri-aurantii, respectively. The resulting selected *Metschnikowia* yeast strains with excellent in vitro antimicrobial activity include: XX01, XX02, XX04, XX05, XX07, JN01, JN011, JN012, XN05, WM04, WM05, WM10 and WM11. All of these belong to the *Metschnikowia pulcherrima*.

3.3. Antagonistic Activity of Yeast Strains on Citrus Fruit In Vivo

A The DI and LD values (as shown in Figure 1 and Tables S2–S4) suggested that all 51 *Metschnikowia* yeasts tested had a significant inhibitory effect on the green and blue mold, and sour rot, which are caused by *P. digitatum*, *P. italicum*, and *G. citri-aurantii*, respectively, when compared to the control.

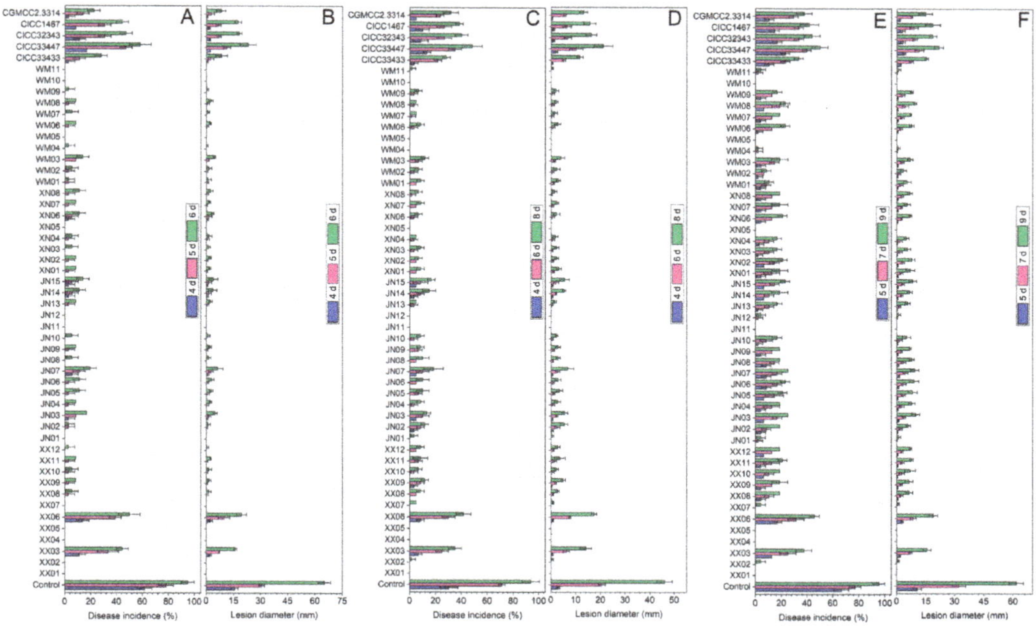

Figure 1. In vivo antagonism of 51 yeast isolates in the control of green mold (**A**,**B**), blue mold (**C**,**D**) and sour rot (**E**,**F**) caused by *P. digitatum*, *P. italicum* and *G. citri-aurantii* on citrus fruit. The columns represent the mean of three replicates and vertical bars show the standard error.

In general, among the 46 screened *Metschnikowia* sp. strains, except for the cases of XX03 and XX06, the ability to control the three major diseases in postharvest citrus fruits was superior to that of the five commercially purchased *M. pulcherrima* control strains. During the storage period, all 51 yeast strains, with the exception of XX03 and XX06, decreased the incidence of the three diseases by more than 70%. Among them, XX01, XX04, XX05, JN11, XN05, WM05, and WM10 completely controlled the occurrence of green mold, blue mold, and sour rot throughout the storage period, demonstrating outstanding biocontrol efficacy.

3.4. Pulcherrimin Pigment Production of Yeast Isolates

Except for CICC33447, the colony bands of 51 *Metschnikowia* yeasts exhibited red pigment production on PDA plates containing 10 mg L^{-1} $FeCl_3$, while CICC33447 appeared white (Figure 2). Additionally, except for XX03, XX06 and CICC33447, bands of pulcherrimin pigmentation of varying sizes formed around the 48 yeast colonies, with only small pigmented halos produced by CICC32343 and CICC1467.

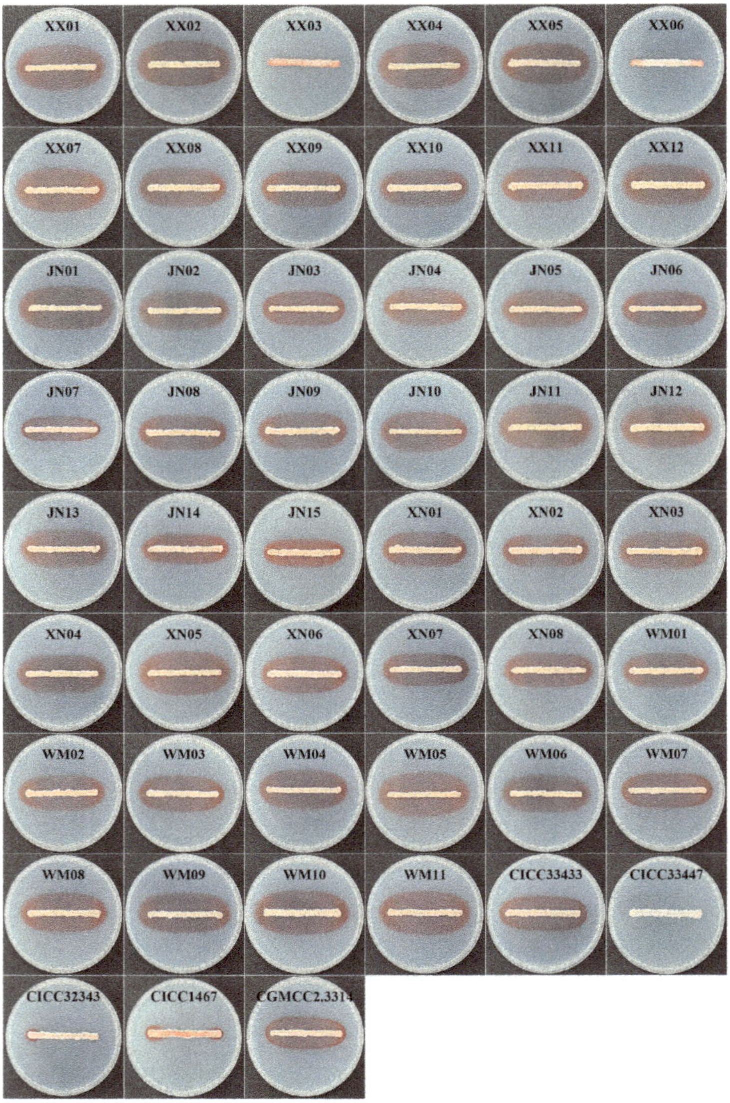

Figure 2. The pulcherrimin pigment produced by 51 *Metschnikowia* yeasts on PDA plate supplemented with 10 mg L^{-1} $FeCl_3$.

3.5. Iron Competition and Antagonism

For the subsequent experiments, we randomly selected three strains (XX01, XX05, WM05) from a pool of seven *M. pulcherrima* strains (XX01, XX04, XX05, JN11, XN05, WM05, and WM10). These strains have been proven to effectively manage the three major post-harvest citrus diseases during storage, as detailed in Section 3.3. Furthermore, the *M. pulcherrima* s strains XX06 and CICC33447 were also chosen as the control group in the experiment due to their limited and nearly negligible pigment production, respectively (Figure 2).

3.5.1. Subsubsection Effect of Iron on Pigment Production and Antagonism of Yeasts In Vitro

As presented in Figure 3, it could be observed that the increase in $FeCl_3$ concentration within the PDA medium resulted in a gradual reduction in the width of the pigment halos produced by the three pulcherrimin-producing *M. pulcherrima* strains. Additionally, the colony band transitioned from white to pink, and ultimately to dark red. At a $FeCl_3$ concentration of 50 mg L^{-1}, the pigmented halo widths of XX01, XX05, and WM05 coincided with their respective colony band widths. Furthermore, the colony bands exhibited a dark red color, and no pigment ring was observed surrounding the colony in any of the three strains. The *M. pulcherrima* XX06, characterized by its lower intensity of pigment production, produced a narrow-colored halo, and the production of additional pigment bands was halted when the exogenous $FeCl_3$ concentration was $\geq$5 mg L^{-1}. For the virtually unpigmented strain CICC33447, no pigment halos were formed under varying concentrations of $FeCl_3$.

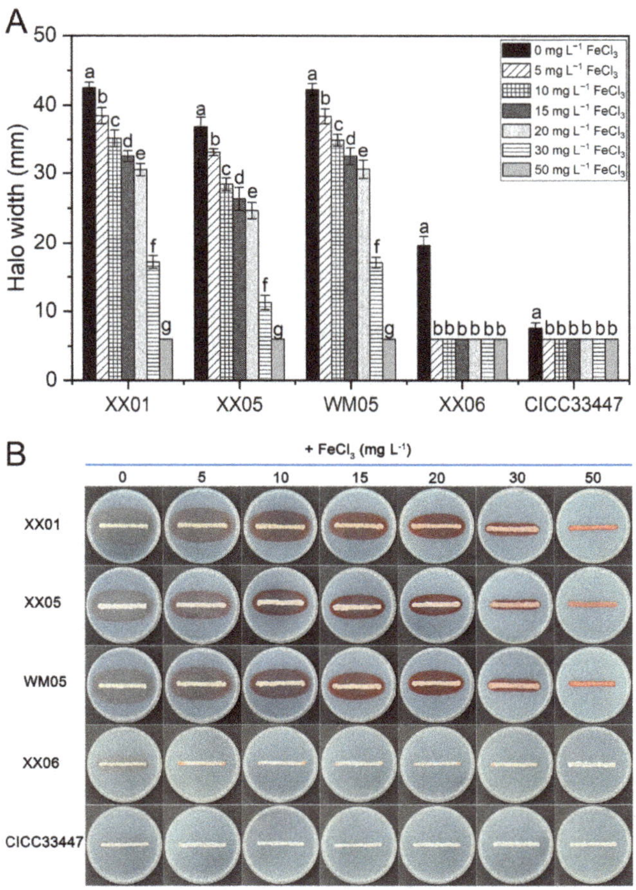

Figure 3. Effect of $FeCl_3$ concentration on the width of pulcherrimin-pigmented halos produced by *M. pulcherrima* strains on PDA plates. (**A**) Columns (treatments) represent the mean of three replicates and vertical bars show the standard error of the mean. Different letters indicate statistically significant differences based on Duncan's multiple range test ($p < 0.05$). (**B**) The photographs show the pigmentation halos of *M. pulcherrima* strains on the $FeCl_3$-supplemented PDA plates corresponding to the histogram.

Similarly, without the addition of exogenous $FeCl_3$, the pulcherrimin-producing *M. pulcherrima* yeasts XX01, XX05, WM05, and XX06 exhibited the widest inhibition zones against *P. digitatum*, *P. italicum*, and *G. citri-aurantii* (Figure 4). However, the size of these inhibition zones noticeably decreased as the concentration of $FeCl_3$ increased, and *M. pulcherrima* (XX01, XX05, and WM05) did not exhibit any inhibitory effect on the growth of these three pathogens when the $FeCl_3$ concentration was $\geq$30 mg L^{-1}. As for the strain XX06, characterized by its weaker pigment production capability, no inhibitory zones were observed when the $FeCl_3$ concentration $\geq$5 mg L^{-1}. CICC33447, which produces almost no pigment, exhibited no substantial inhibitory effects on all three citrus pathogenic fungus, whereas varied doses of $FeCl_3$ had no apparent impact on its biocontrol antagonistic efficacy.

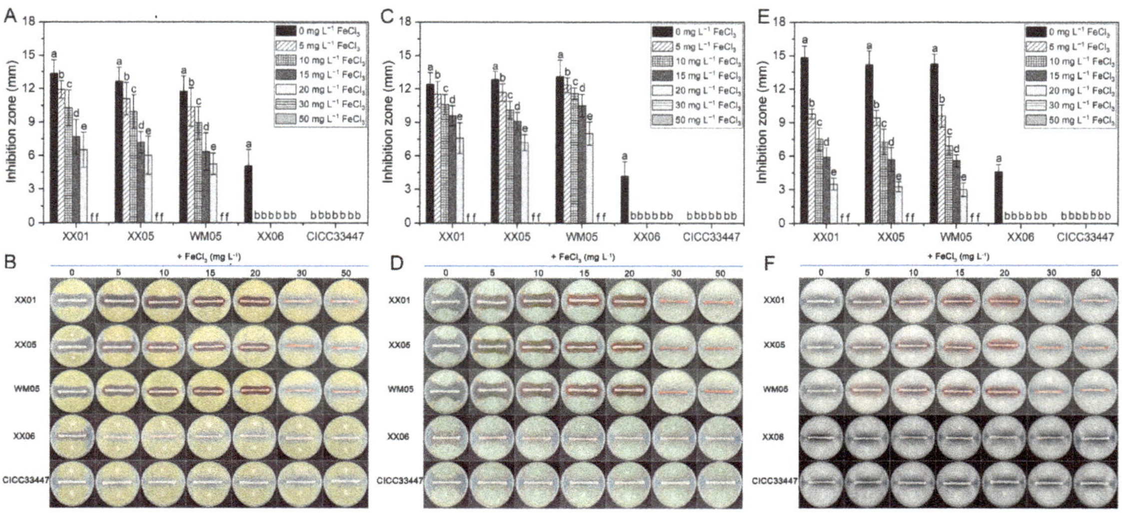

Figure 4. Effect of $FeCl_3$ concentration on the width of the inhibition zones by *M. pulcherrima* strains against *P. digitatum* (**A**,**B**), *P. italicum* (**C**,**D**) and *G. citri-aurantii* (**E**,**F**) on PDA plates. Columns (**A**,**C**,**E**) represent the mean of three replicates and vertical bars show the standard error of the mean. Different letters indicate statistically significant differences based on Duncan's multiple range test ($p < 0.05$). The photographs (**B**,**D**,**F**) show the inhibition areas of *M. pulcherrima* strains on the $FeCl_3$-supplemented PDA plates corresponding to the histogram.

3.5.2. Effect of Iron on Biocontrol Efficacy of Yeasts against Postharvest Citrus Pathogens In Vivo

Figure 5 shows the effect of iron on the antagonistic activity of *M. pulcherrima* against citrus green mold (A–C), blue mold (D–F), and sour rot (G–I) diseases. The in vivo experiments on citrus fruit showed similar outcomes to those obtained in in vitro studies (Figure 4). The addition of $FeCl_3$ to the wounds of fruit treated with three strains of high pigment-producing *M. pulcherrima* resulted in a significant increase in the disease incidence and lesion diameter of green mold, blue mold and sour rot. Within the tested range of 0 to 50 mg L^{-1}, the biocontrol activities of the three *M. pulcherrima* strains (XX01, XX05 and WM05) against both citrus green and blue mold significantly decreased as the concentration of $FeCl_3$ increased (Figure 5A–F). However, when the $FeCl_3$ concentration ranged from 5.0 to 50 mg L^{-1}, there was no significant impact on the development of citrus sour rot (Figure 5G–I). This implied that *P. digitatum* and *P. italicum* require a greater iron consumption compared to *G. citri-aurantii* during their growth and infection processes. Different concentrations of $FeCl_3$ did not significantly affect the occurrence and development of XX06 and CICC33447 for controlling of citrus green and blue mold, as well as sour rot decay.

However, for XX06, the addition of certain concentration of $FeCl_3$ ($\geq$5 mg L^{-1}) resulted in higher disease incidence and lesion diameter statistics than the control (0 mg L^{-1} of $FeCl_3$).

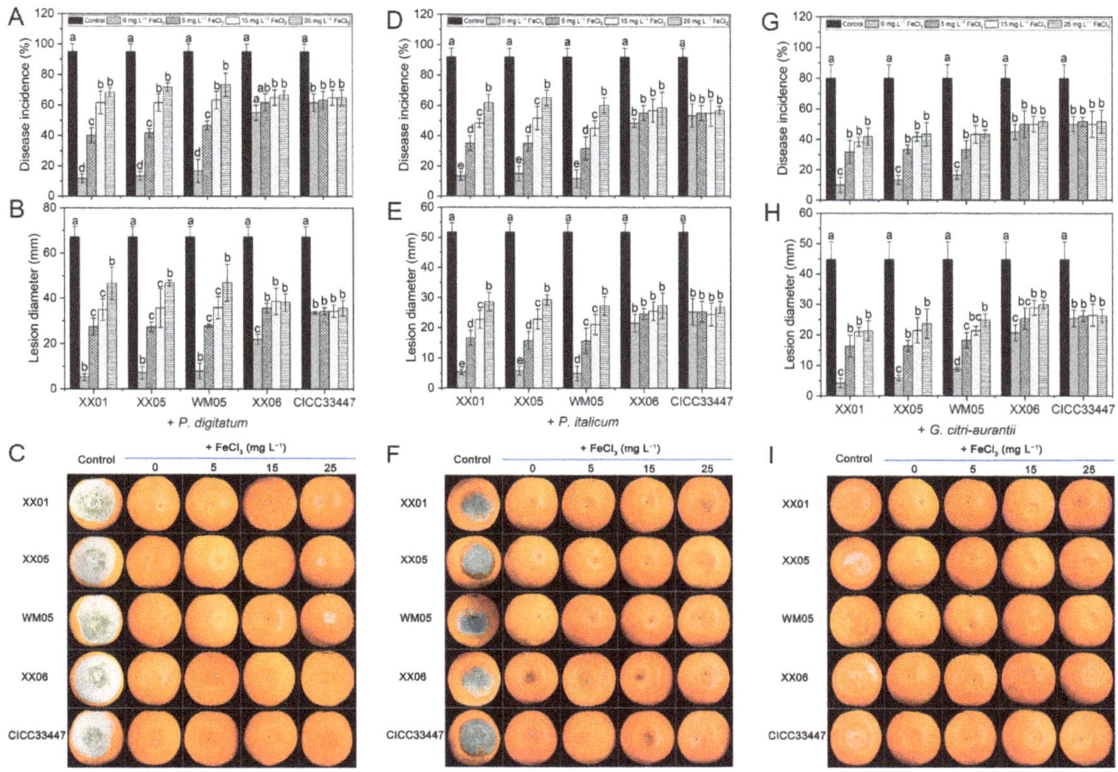

Figure 5. Effect of iron on efficacy of *M. pulcherrima* strains in controlling green mold (**A–C**), blue mold (**D–F**) and sour rot (**G–I**) on citrus fruit. Statistical analysis of disease incidence and lesion diameters are illustrated following the 6 d (infection with *P. digitatum*, **A**,**B**), 8 d (infection with *P. italicum*, **D**,**E**) and 9 d (infection with *G. citri-aurantii*, **G**,**H**) storage, respectively. The pictures showed the biocontrol performance of different treatment groups on the 6th (**C**), 8th (**F**), and 9th (**I**) day after inoculation. The columns represent the mean of three replicates and vertical bars show the standard error. Different letters indicate statistically significant differences based on Duncan's multiple range test ($p < 0.05$).

3.5.3. Effect of Iron on the Growth and the PA Production of Selected Yeasts

As illustrated in Figure 6A, supplementing the medium with $FeCl_3$ enhanced the cell growth of the five *M. pulcherrima* strains. Notably, the addition of 5–50 mg L^{-1} $FeCl_3$ markedly stimulated yeast proliferation. PA is a secondary metabolite. $FeCl_3$ at the tested concentrations (0–100 mg L^{-1}) had no significant effect on PA production by the five yeast strains (Figure 6B).

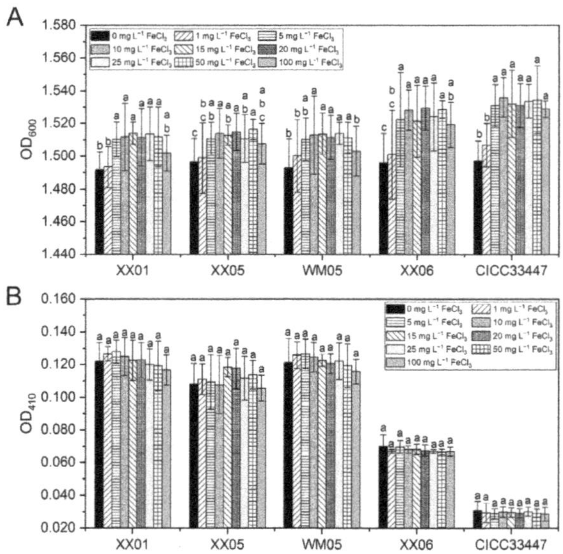

Figure 6. Effect of $FeCl_3$ on the cell densities (**A**) and the pulcherriminic acid yield (**B**) of selected *M. pulcherrima* strains. The OD_{600} and OD_{410} of each strain was measured after culture for 30 h. The columns represent the mean of three replicates and vertical bars show the standard error. Different letters indicate statistically significant differences based on Duncan's multiple range test ($p < 0.05$).

3.5.4. Sensitivity of *P. digitatum*, *P. italicum* and *G. citri-aurantii* to Iron Starvation

We selected tropolone, known for its potent affinity for iron ions [11,13,16], for evaluating the susceptibility of *P. digitatum*, *P. italicum*, and *G. citri-aurantii* to iron deprivation. As illustrated in Figure 7, all three pathogens exhibited sensitivity to iron-deficient conditions. Among them, *P. digitatum* exhibited the highest sensitivity, followed by *P. italicum* and *G. citri-aurantii*. These findings aligned closely with the results from the fruit experiment.

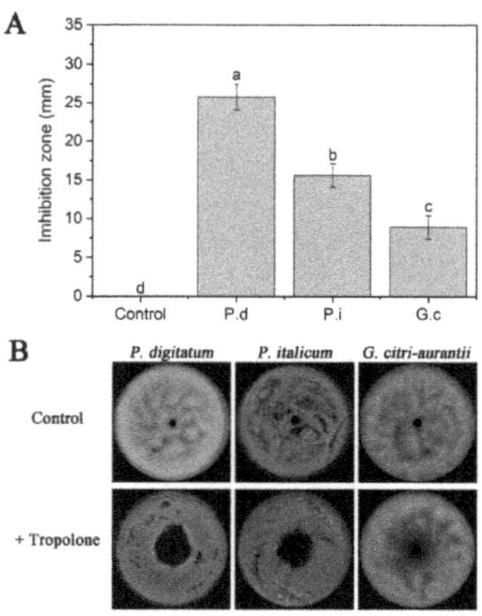

Figure 7. Inhibition of *P. digitatum*, *P. italicum* and *G. citri-aurantii* by tropolone. Columns (**A**) represent the mean of three replicates and vertical bars show the standard error of the mean. Different letters indicate statistically significant differences based on Duncan's multiple range test ($p < 0.05$). The photograph (**B**) shows the inhibition zones of tropolone activity against *P. digitatum*, *P. italicum* and *G. citri-aurantii*. P.d, *P. digitatum*; P.i, *P. italicum*; G.c, *G. citri-aurantii*.

4. Discussion

In general, PA or pulcherrimin-producing strains can generally serve as ideal candidates for the development of antibacterial and anti-plant-pathogenic fungal agents [20,22]. When tested for pigment production, some *Metschnikowia* species are found to produce characteristic reddish halos known as pulcherrimin and their antagonistic capacity appears well correlated with the size of the red-pigmented zones [17]. In this study, *Metschnikowia* yeasts, originating from the surface of table grapes, were screened in an iron-containing medium based on the characteristics of their colonies that produced the red pulcherrimin pigment. A total of 46 *Metschnikowia* strains were screened and identified. Notably, all 46 strains exhibited distinct pulcherrimin-pigmented halos surrounding their colonies. Among them, *M. pulcherrima* had the largest number of strains, with 44 strains. This was consistent with the previous finding that *M. pulcherrima* strains were frequently found in grapes [5]. On PDA plates containing 10 $mg \cdot L^{-1}$ $FeCl_3$, 51 *Metschnikowia* strains were observed to be able to produce pigment circles of different sizes (Figure 2), suggesting that yeasts belonging to the genus *Metschnikowia* display strain-specific differences in pulcherrimin production [18]. All 51 *Metschnikowia* yeasts were highly effective against *P. digitatum*, *P. italicum* and *G. citri-aurantii* in vitro and in vivo on citrus fruits compared with the control (Tables 1 and S2–S4, Figures 1 and S2), implying that *Metschnikowia* yeasts with pigment-producing capabilities could potentially be applied in the prevention and control of postharvest diseases of citrus.

A comparison of yeast strains revealed that those with a larger pigment-producing ability (e.g., XX01, XX02, XX04, XX05, XX07, JN01, JN11, JN12, XN05, WM01, WM02, WM04, WM05, WM10, and WM11) showed greater effectiveness in controlling *P. digitatum*, *P. italicum*, and *G. citri-aurantii*. In contrast, yeast strains with less pigment production (e.g., XX03, XX06, and CICC33447) were significantly less antagonistic to these three pathogens (Figures 1 and S2, Tables S2–S4). Closer inspection revealed that there seemed to be a consistent correlation between the boundaries of pigmentation zones and the edges of inhibitory areas. Similar results were also found by Horváth et al. [19].

Iron is a vital element necessary for the growth, proliferation, and pathogenicity of pathogens. It is required as an essential cofactor for numerous enzymes and proteins implicated in pivotal biological processes, such as cellular metabolism and energy generation, etc. [30,31]. Competition for iron has long been recognized as an important biocontrol mechanism employed by antagonists against post-harvest pathogens [32]. Pulcherrimin, an insoluble specific siderophore formed by PA bound Fe^{3+} [33], cannot be recognized and therefore utilized by *P. digitatum*, *P. italicum*, and *G. citri-aurantii*. This leads indirectly to an inhibitory effect on the three pathogens due to a competitive advantage in Fe^{3+} depletion. Upon the introduction of exogenous $FeCl_3$ into the PDA culture medium, the pigmentation halos around the yeast colony notably diminished and eventually disappeared with increasing $FeCl_3$ concentration. Simultaneously, the inhibitory zone generated against the pathogens also significantly reduced and gradually vanished (Figures 3 and 4). This is due to the fact that as the concentration of Fe^{3+} increases, although the amount of PA produced by the *M. pulcherrima* remains constant (Figure 6), the PA secreted into the extracellular space however cannot diffuse further before it binds with the abundant iron to form stable pulcherrimin near the yeast colonies, leading to progressively narrower bands of pigmentation. Simultaneously, the increasing amount of free Fe^{3+} in the medium was sufficient to meet requirements for the spore germination and mycelial growth of *P. digitatum*, *P. italicum*, and *G. citri-aurantii*. Thus, the inhibitory effect on the pathogens began to decrease gradually until disappearing. For yeast strains that produce weaker pigments (XX06 and CICC33447), lower levels of PA production mean that a lower concentration of iron is sufficient to completely deprive the yeast of its ability to produce the pulcherrimin-pigmented halo and inhibit pathogens (Figures 4 and 5). It is not difficult to perceive that a positive correlation appears to exist between the antagonistic capacity of *M. pulcherrima* and the size of the red pigmented halo. Furthermore, it was noted that even when high concentrations of iron were added to the fruit wounds, the *M. pulcherrima* did not completely lose its effectiveness in controlling green mold, blue mold, and sour rot (Figure 5). This suggested that the biocontrol of citrus diseases by the pulcherrimin-producing *Metschnikowia* yeast involves various mechanisms, in which the competition for iron may play a pivotal role.

In addition, the addition of exogenous Fe^{3+} significantly reduced the effectiveness of *M. pulcherrima* in controlling three postharvest citrus diseases. This is evident from the significant increases in the DI and LD of green mold, blue mold and sour rot with the increase in iron concentration (Figure 5). Moreover, the magnitude of the increase in DI of green and blue mold was higher than that of sour rot, indicating that there are greater requirements for iron in the growth and infection processes of *P. digitatum* and *P. italicum* compared to *G. citri-aurantii*. Oztekin and Karbancioglu-Guler [16] also observed the higher dependency of spore germination and mycelial growth of *P. digitatum* on iron ions compared to *P. expansum*. Different pathogens exhibit varying sensitivities to iron-deficient environments, and our experiments confirmed that *P. digitatum* and *P. italicum* were more sensitive to iron deficiency compared to *G. citri-aurantii* (Figure 7). However, the molecular mechanisms underlying the differences in competition for iron ions between the three pathogens and *M. pulcherrima* need to be further explored.

5. Conclusions

In this study, 46 strains of *Metschnikowia* yeasts from the surface of table grapes were screened and identified. The results of antagonism assays showed that all the 46 *Metschnikowia* yeast strains could effectively control the citrus green mold, blue mold and sour rot caused by *P. digitatum*, *P. italicum* and *G. citri-aurantii*, respectively. The pulcherrimin-producing strains of *M. pulcherrima* showed stronger antagonistic properties against the three postharvest citrus pathogens, suggesting that this species has the potential to be utilized as a biocontrol agent. The production of pulcherrimin and the biocontrol efficacy of *M. pulcherrima* strains were directly influenced by the iron concentration, and there appeared to be a positive correlation between antagonistic capacity and the size of the red pigmented halo. By competing for iron ions, pulcherrimin-producing *M. pulcherrima* strains inhibited the growth of *P. digitatum*, *P. italicum* and *G. citri-aurantii* in vitro and in vivo. However, the inhibitory effect may vary depending on the sensitivity of the three pathogens to iron starvation. Additionally, the mechanism of how iron starvation, triggered by iron competition, inhibits the growth of the three pathogens needs to be further investigated.

Supplementary Materials: The following supporting information can be downloaded at: https://www.mdpi.com/article/10.3390/foods12234249/s1, Table S1. Identification of the yeast species isolated from the surface of table grape from fruit market in Nanning. Table S2. In vivo antagonism of 51 yeast isolates against green mold caused by *P. digitatum* on citrus fruit. Table S3. In vivo antagonism of 51 yeast isolates against blue mold caused by *P. italicum* on citrus fruit. Table S4. In vivo antagonism of 51 yeast isolates in the control of sour rot caused by *G. citri-aurantii* on citrus fruit. Figure S1: Morphological characteristics of screened *Metschnikowia* yeasts on YPD + Fe medium; Figure S2: Effects of the *Metschnikowia* yeast isolates on the mycelial growth of *P. digitatum*, *P. italicum* and *G. citri-aurantii*.

Author Contributions: Conceptualization, S.W. and J.S.; methodology, D.Y., L.L. and X.H.; software, C.W. and W.L.; data curation, F.H., D.Y., L.L. and X.H.; writing—original draft preparation, S.W. and Z.T.; writing—review and editing, S.W., Z.T., W.L. and J.S.; visualization, W.L., C.W. and Z.T.; supervision, J.S. and F.H.; funding acquisition, J.S. and S.W. All authors have read and agreed to the published version of the manuscript.

Funding: This work was financially supported by China Postdoctoral Science Foundation (Grant No. 2022MD713737), Guangxi Natural Science Foundation (Grant No. 2022GXNSFBA035561) and the National Natural Science Foundation of China (Grant No. 32302164).

Data Availability Statement: Data is contained within the article or Supplementary Material.

Conflicts of Interest: The authors declare no conflict of interest.

References

1. Chhikara, N.; Kour, R.; Jaglan, S.; Gupta, P.; Gat, Y.; Panghal, A. Citrus medica: Nutritional, phytochemical composition and health benefits—A review. *Food Funct.* **2018**, *9*, 1978–1992. [CrossRef] [PubMed]
2. Xu, L.; Feng, L.; Sun, J.; Mao, L.; Li, X.; Jiang, Y.; Duan, X.; Li, T. Antifungal activities of a natural trisaccharide ester against sour rot in mandarin fruit. *Postharvest Biol. Technol.* **2022**, *191*, 111981. [CrossRef]
3. Zhang, X.; Li, B.; Zhang, Z.; Chen, Y.; Tian, S. Antagonistic yeasts: A promising alternative to chemical fungicides for controlling postharvest decay of Fruit. *J. Fungi* **2020**, *6*, 158. [CrossRef]
4. Wang, Z.; Sui, Y.; Li, J.; Tian, X.; Wang, Q. Biological control of postharvest fungal decays in citrus: A review. *Crit. Rev. Food Sci. Nutr.* **2020**, *62*, 861–870. [CrossRef] [PubMed]
5. Morata, A.; Loira, I.; Escott, C.; del Fresno, J.M.; Bañuelos, M.A.; Suárez-Lepe, J.A. Applications of *Metschnikowia pulcherrima* in wine biotechnology. *Fermentation* **2019**, *5*, 63. [CrossRef]
6. Kregiel, D.; Nowacka, M.; Rygala, A.; Vadkertiová, R. Biological activity of pulcherrimin from the *Meschnikowia pulcherrima* Clade. *Molecules* **2022**, *27*, 1855. [CrossRef]
7. Hilber-Bodmer, M.; Schmid, M.; Ahrens, C.H.; Freimoser, F.M. Competition assays and physiological experiments of soil and phyllosphere yeasts identify *Candida subhashii* as a novel antagonist of filamentous fungi. *BMC Microbiol.* **2017**, *17*, 4. [CrossRef]
8. Leverentz, B.; Conway, W.S.; Janisiewicz, W.; Abadias, M.; Kurtzman, C.P.; Camp, M.J. Biocontrol of the food-borne pathogens *Listeria monocytogenes* and *Salmonella enterica* serovar Poona on fresh-cut apples with naturally occurring bacterial and yeast antagonists. *Appl. Environ. Microbiol.* **2006**, *72*, 1135–1140. [CrossRef]
9. Fadahunsi, I.F.; Olubodun, S. Antagonistic pattern of yeast species against some selected food-borne pathogens. *Bull. Natl. Res. Cent.* **2021**, *45*, 34. [CrossRef]
10. Tian, Y.Q.; Li, W.; Jiang, Z.T.; Jing, M.M.; Shao, Y.Z. The preservation effect of *Metschnikowia pulcherrima* yeast on anthracnose of postharvest mango fruits and the possible mechanism. *Food Sci. Biotechnol.* **2018**, *27*, 95–105. [CrossRef]
11. Saravanakumar, D.; Ciavorella, A.; Spadaro, D.; Garibaldi, A.; Gullino, M.L. *Metschnikowia pulcherrima* strain MACH1 outcompetes *Botrytis cinerea, Alternaria alternata* and *Penicillium expansum* in apples through iron depletion. *Postharvest Biol. Technol.* **2008**, *49*, 121–128. [CrossRef]
12. Yang, H.; Wang, L.; Li, S.; Gao, X.; Wu, N.; Zhao, Y.; Sun, W. Control of postharvest grey spot rot of loquat fruit with *Metschnikowia pulcherrima* E1 and potential mechanisms of action. *Biol. Control* **2021**, *152*, 104406. [CrossRef]
13. Sipiczki, M. *Metschnikowia* strains isolated from botrytized grapes antagonize fungal and bacterial growth by iron depletion. *Appl. Environ. Microbiol.* **2006**, *72*, 6716–6724. [CrossRef] [PubMed]
14. Pawlikowska, E.; James, S.A.; Breierova, E.; Antolak, H.; Kregiel, D. Biocontrol capability of local *Metschnikowia* sp. isolates. *Antonie Van. Leeuwenhoek* **2019**, *112*, 1425–1445. [CrossRef] [PubMed]
15. Parafati, L.; Vitale, A.; Restuccia, C.; Cirvilleri, G. Biocontrol ability and action mechanism of food-isolated yeast strains against *Botrytis cinerea* causing post-harvest bunch rot of table grape. *Food Microbiol.* **2015**, *47*, 85–92. [CrossRef] [PubMed]
16. Oztekin, S.; Karbancioglu-Guler, F. Bioprospection of *Metschnikowia* sp. isolates as biocontrol agents against postharvest fungal decays on lemons with their potential modes of action. *Postharvest Biol. Technol.* **2021**, *181*, 111634. [CrossRef]
17. Sipiczki, M. *Metschnikowia pulcherrima* and related pulcherrimin-producing yeasts: Fuzzy species boundaries and complex antimicrobial antagonism. *Microorganisms* **2020**, *8*, 29. [CrossRef] [PubMed]
18. Pawlikowska, E.; Kolesińska, B.; Nowacka, M.; Kregiel, D. A new approach to producing high yields of pulcherrimin from *Metschnikowia* yeasts. *Fermentation* **2020**, *6*, 114. [CrossRef]
19. Horváth, E.; Dályai, L.; Szabó, E.; Barna, T.; Kalmár, L.; Posta, J.; Sipiczki, M.; Csoma, H.; Miklós, I. The antagonistic *Metschnikowia andauensis* produces extracellular enzymes and pulcherrimin, whose production can be promoted by the culture factors. *Sci. Rep.* **2021**, *11*, 10593. [CrossRef]
20. Yuan, S.; Yong, X.; Zhao, T.; Li, Y.; Liu, J. Research progress of the biosynthesis of natural bio-antibacterial agent pulcherriminic acid in *Bacillus*. *Molecules* **2020**, *25*, 5611. [CrossRef]
21. Kregiel, D.; Czarnecka-Chrebelska, K.H.; Schusterová, H.; Vadkertiová, R.; Nowak, A. The *Metschnikowia pulcherrima* clade as a model for assessing inhibition of *Candida* spp. and thetoxicity of its metabolite, pulcherrimin. *Molecules* **2023**, *28*, 5064. [CrossRef] [PubMed]
22. Wang, S.; Wang, H.; Zhang, D.; Li, X.; Zhu, J.; Zhan, Y.; Cai, D.; Wang, Q.; Ma, X.; Wang, D.; et al. Multistep metabolic engineering of *Bacillus licheniformis* to improve pulcherriminic acid production. *Appl. Environ. Microbiol.* **2020**, *86*, e03041-19. [CrossRef] [PubMed]
23. Öztekin, S.; Karbancioglu-Guler, F. Biological control of green mould on mandarin fruit through the combined use of antagonistic yeasts. *Biol. Control* **2023**, *180*, 105186. [CrossRef]
24. Rodriguez Assaf, L.A.; Pedrozo, L.P.; Nally, M.C.; Pesce, V.M.; Toro, M.E.; Castellanos de Figueroa, L.I.; Vazquez, F. Use of yeasts from different environments for the control of *Penicillium expansum* on table grapes at storage temperature. *Int. J. Food Microbiol.* **2020**, *320*, 108520. [CrossRef] [PubMed]
25. Wang, S.; Zhang, H.; Ruan, C.; Yi, L.; Deng, L.; Zeng, K. *Metschnikowia citriensis* FL01 antagonize *Geotrichum citri-aurantii* in citrus fruit through key action of iron depletion. *Int. J. Food Microbiol.* **2021**, *357*, 109384. [CrossRef] [PubMed]
26. Liu, Y.; Yao, S.; Deng, L.; Ming, J.; Zeng, K. Different mechanisms of action of isolated epiphytic yeasts against *Penicillium digitatum* and *Penicillium italicum* on citrus fruit. *Postharvest Biol. Technol.* **2019**, *152*, 100–110. [CrossRef]

27. Wang, S.; Ruan, C.; Yi, L.; Deng, L.; Yao, S.; Zeng, K. Biocontrol ability and action mechanism of *Metschnikowia citriensis* against *Geotrichum citri-aurantii* causing sour rot of postharvest citrus fruit. *Food Microbiol.* **2020**, *87*, 103375. [CrossRef]
28. Wang, D.; Zhan, Y.; Cai, D.; Li, X.; Wang, Q.; Chen, S. Regulation of the synthesis and secretion of the iron chelator cyclodipeptide pulcherriminicacid in *Bacillus licheniformis*. *Appl. Environ. Microbiol.* **2018**, *84*, e00262-18. [CrossRef]
29. Wang, S.; Zhang, H.; Qi, T.; Deng, L.; Yi, L.; Zeng, K. Influence of arginine on the biocontrol efficiency of *Metschnikowia citriensis* against *Geotrichum citri-aurantii* causing sour rot of postharvest citrus fruit. *Food Microbiol.* **2022**, *101*, 103888. [CrossRef]
30. Zhang, J.; Liu, B.; Li, M.; Feng, D.; Jin, H.; Wang, P.; Liu, J.; Xiong, F.; Wang, J.; Wang, H.-B. The bHLH transcription factor bHLH104 Interacts with IAA-LEUCINE RESISTANT3 and modulates iron homeostasis in Arabidopsis. *Plant Cell* **2015**, *27*, 787–805. [CrossRef]
31. Qin, Y.; He, Y.; She, Q.; Larese-Casanova, P.; Li, P.; Chai, Y. Heterogeneity in respiratory electron transfer and adaptive iron utilization in a bacterial biofilm. *Nat. Commun.* **2019**, *10*, 3702. [CrossRef] [PubMed]
32. Dukare, A.S.; Paul, S.; Nambi, V.E.; Gupta, R.K.; Singh, R.; Sharma, K.; Vishwakarma, R.K. Exploitation of microbial antagonists for the control of postharvest diseases of fruits: A review. *Crit. Rev. Food Sci. Nutr.* **2019**, *59*, 1498–1513. [CrossRef] [PubMed]
33. Krause, D.J.; Kominek, J.; Opulente, D.A.; Shen, X.X.; Zhou, X.; Langdon, Q.K.; DeVirgilio, J.; Hulfachor, A.B.; Kurtzman, C.P.; Rokas, A.; et al. Functional and evolutionary characterization of a secondary metabolite gene cluster in budding yeasts. *Proc. Natl. Acad. Sci. USA* **2018**, *115*, 11030–11035. [CrossRef] [PubMed]

foods

MDPI

Article

Volatile Organic Compounds of *Scheffersomyces spartinae* W9 Have Antifungal Effect against *Botrytis cinerea* on Strawberry Fruit

Xiurong Zou [1,2], Yingying Wei [2], Jianhua Zhu [1], Jincai Sun [3,*] and Xingfeng Shao [2,*]

[1] College of Food Science and Technology, Guangdong Provincial Key Laboratory of Utilization and Conservation of Food and Medicinal Resources in Northern Region, Shaoguan University, Shaoguan 512005, China
[2] State Key Laboratory for Managing Biotic and Chemical Threats to the Quality and Safety of Agro-Products, Zhejiang-Malaysia Joint Research Laboratory for Agricultural Product Processing and Nutrition, College of Food and Pharmaceutical Sciences, Ningbo University, Ningbo 315800, China
[3] Faculty of Food Science, Zhejiang Pharmaceutical University, Ningbo 315500, China
* Correspondence: sunjingcai66@126.com (J.S.); shaoxingfeng@nbu.edu.cn (X.S.)

Abstract: This study aims to evaluate the antifungal effects of volatile organic compounds (VOCs) produced by a marine biocontrol yeast, *Scheffersomyces spartinae* W9. The results showed that the VOCs from the yeast inhibited the growth of *Botrytis cinerea* mycelium and spore germination by 77.8% and 58.3%, respectively. Additionally, it reduced the disease incidence and lesion diameter of gray mold on the strawberry fruit surface by 20.7% and 67.4%, respectively. Electronic micrographs showed that VOCs caused damage to the morphology and ultrastructure of the hyphae. Based on headspace solid-phase microextraction gas chromatography–mass spectrometry (HS-SPME/GC-MS), *S. spartinae* W9 emitted 18 main VOCs, and the pure substance of VOCs, such as 3-methyl-1-butanol, 2-methyl-1-butanol, 2-phenylethanol, and isoamyl acetate, showed antifungal effects against *B. cinerea* mycelium growth. Among them, 2-phenylethanol exhibited the strongest antifungal activity. It has been concluded that VOCs are the key antifungal mechanism of *S. spartinae* W9, and a promising strategy for controlling gray mold on strawberry fruit.

Keywords: antagonistic yeast; biocontrol; gray mold; volatile organic compounds (VOCs); 2-phenylethanol

1. Introduction

So far, a lot of antagonistic yeasts have been discovered to suppress fruit and vegetable diseases caused by phytopathogenic fungi, such as gray mold, green mold, brown rot, and root rot [1]. The biocontrol mechanisms exhibited by antagonistic yeasts are diverse and depend on species to species. They primarily include competition for nutrition and space, parasitism, biofilm formation, production of diffusible or volatile antimicrobial substances, and induction of plant host resistance [1,2].

During recent years, volatile organic compounds (VOCs) have been proven to be one of the most crucial biocontrol mechanisms of antagonistic yeasts [2]. VOCs can not only inhibit the mycelial growth and spore germination of pathogenic fungi, but also limit the production of mycotoxins. For instance, Farbo et al. [3] found that VOCs produced by yeast remarkably suppressed the growth, sporulation, and ochratoxin A production in *Aspergillus* spp. Furthermore, yeast VOCs have demonstrated efficacy in controlling gray mold in strawberry fruit [4], green mold and blue mold in mandarin fruit [5], green mold in lemons [6], and brown rot in cherry, peach, and apricot fruit [7]. Notably, yeast VOCs have also been found to enhance the disease resistance of plants. Di Francesco et al. [7] observed an increase in the concentration of membrane proteins, as well as the biosynthesis of cuticle and wax, in stone fruits after treatment with VOCs derived from *Aureobasidium pullulans*. They speculated that VOCs enhanced the disease resistance by changing the fruit's structural composition.

VOCs refer to a category of organic compounds with a molecular weight of less than 300 Da, high vapor pressure, low boiling point, and low water solubility [8]. They mainly include alcohols, ketones, aldehydes, phenols, acids, benzene rings, heterocycles, mercaptans, and thioesters [9]. The volatility of VOCs allows them to diffuse in soil particles and the air, enabling them to function without direct physical contact with pathogenic microorganisms. The compound of VOCs varies with the yeast species, the presence of other microorganisms, culture time, nutritional conditions, ambient temperature, etc. [10]. Nasanit et al. [11] found that the emitted VOCs were distinct among five different species of antagonistic yeast, and even among different strains of the same species, the produced VOCs varied. Due to their individual and complicated content, it is vital to explore the composition, functions, and mechanisms of biocontrol yeast VOCs for their further practical application.

In a previous study, a strain of antagonistic yeast, *Scheffersomyces spartinae* W9, was isolated from a tropical marine sea. It has exhibited excellent biocontrol effects against *Botrytis cinerea*, a common fungal pathogen of stored fruit and vegetables [12], both in vitro and in strawberry fruit, and the VOCs produced by *S. spartinae* W9 have been found to inhibit the growth of *B. cinerea* mycelium [13]. However, there have been limited reports on the biocontrol abilities of *S. spartinae* W9 and its VOCs. Therefore, this study aims to further explore the antifungal effects of VOCs secreted by *S. spartinae* W9 on stored strawberry fruit. Specifically, this study includes four parts: (i) evaluating the biocontrol effects of VOCs on the mycelium growth and spore germination of *B. cinerea*, as well as gray mold in strawberry fruit; (ii) analyzing the effects of VOCs on the morphology and ultrastructure of *B. cinerea* mycelium; (iii) determining the composition of VOCs; and (iv) investigating the antifungal effects of the main substances present in the VOCs.

2. Materials and Methods

2.1. Fruit

Strawberry (*Fragaria ananassa* Duch. cv. Hongyan) fruit was picked early in the morning from a local orchard in Ningbo, China. All fruits were of similar maturity (with uniform color and size). The fruits were free from any mechanical, pathological, or pest damage.

2.2. Pathogen and the Yeast

B. cinerea (strain No. ACCC 36028) was purchased from the Agricultural Culture Collection of China in Beijing and stored in glycerol at −80 °C. To rejuvenate *B. cinerea*, it was inoculated on potato dextrose agar medium (PDA, 20% fresh potato, 2% glucose, 2% agar). Then, the plates were incubated at 25 °C for 5 d, and this process was repeated twice before use.

To prepare the spore suspension of *B. cinerea*, a 10-day-old *B. cinerea* culture was rinsed with sterile distilled water. The resulting liquid was filtered through two layers of sterile 300-mesh silk cloth to remove mycelia. The concentration of *B. cinerea* spore suspension was then adjusted at 1×10^5 spores mL^{-1} using a hemocytometer.

S. spartinae W9 was isolated from marine sediment and identified through molecular biology analysis in our previous study [13]. To prepare the yeast suspension, a ring of yeast colony from a glycerol tube stored at −80 °C was selected. The colony was then inoculated into nutrient yeast dextrose broth (NYDB, 0.8% beef paste, 0.5% yeast extract, 1% glucose). The culture was incubated on a shaker at 28 °C and a shaking speed of 180 rpm for 48 h. After one subculture, the culture medium was centrifuged to obtain yeast pellets. The pellets were rinsed twice with sterile distilled water. Finally, the yeast pellets were suspended again in sterile distilled water, with the concentration adjusted to 1×10^8 cells mL^{-1}.

2.3. Effects of VOCs Produced by S. spartinae W9 on Mycelium Growth and Spore Germination of B. cinerea

In a dual culture assay, the effects of VOCs produced by *S. spartinae* W9 on pathogen mycelium growth were assessed, following the method described by Zou et al. [14]. Briefly, a 100 μL yeast suspension of *S. spartinae* W9 (1×10^8 cells mL^{-1}) was inoculated and spread evenly on plates of nutrient yeast dextrose agar (NYDA: 0.8% beef paste, 0.5% yeast extract, 1% glucose, 2% agar). An equal volume of sterile distilled water was used as a control. The yeast plates were then incubated at 28 °C for 48 h.

Mycelium growth evaluation: a 9 mm mycelium block, obtained from a 5-day-old culture of *B. cinerea*, was laid on a PDA plate. The NYDA plate previously inoculated with *S. spartinae* W9 was placed on the top of the PDA plate. The two based plates were sealed with parafilm and incubated at 25 °C for 5 d. At the end of the incubation period, the diameter of *B. cinerea* colony was measured. Each treatment had three biological replicates, and the experiment was repeated twice.

Spore germination assessment: a spore suspension (1×10^5 spores mL^{-1}, 1 mL) was pipetted into a 3.5 cm sterile petri dish. The dish's cover was removed, and the dish was placed in the center of an NYDA plate previously incubated with *S. spartinae* W9 for 48 h. The dish was then sealed and incubated at 25 °C for 14~19 h. An NYDA plate inoculated with sterile distilled water served as a control. The spore germination per 100 spores was observed with a 40× optical microscope. Conidia germination was recorded when the length of the germ tube reached or exceeded half the diameter of the conidia. The spore germination rate (%) = number of germinated conidia/total number of conidia × 100. Each treatment was replicated three times, and the experiment was conducted two times.

2.4. Effect of VOCs on Strawberry Gray Mold

A 3 × 3 mm wound was pricked at the equator of strawberry fruits using a sterile iron nail. After the wound was naturally dried, 10 μL of a 1×10^5 spores mL^{-1} spore suspension of *B. cinerea* was inoculated onto the wound and allowed to dry naturally. Ten strawberries with artificial inoculation were placed onto a 15 cm glass plate. An NYDA plate coated with 500 μL of yeast cell suspension (1×10^8 cells mL^{-1}) was laid upside down on the glass plate. Another NYDA plate spread with 500 μL of sterile distilled water was used as a control. The plates were sealed with two layers of parafilm to create a closed environment and stored at 90% relative humidity and 20 °C for 4 d. After four days of storage, the sealing film was torn off, leaving a small gap for ventilation. The strawberries were further stored for two more days. On the 6th day of storage, the disease incidence rate (percentage of infected strawberries) and lesion diameter were measured. The experiment was repeated three times.

2.5. Microscopic Observation of Hyphae Morphology and Ultrastructure

B. cinerea hyphae were exposed to VOCs of *S. spartinae* W9, as described in 2.3. After incubation for five days, the mycelium block of the *B. cinerea* culture was removed and fixed in 2.5% glutaraldehyde in 0.1 M PBS (pH 7.2) overnight at 4 °C. For the hyphae morphology observation, further experimental procedures were conducted following the method described in Xu et al. [15]. The fixed samples were first rinsed with 0.1 M PBS (pH 7.2) three times, each for 15 min. Then, they were dehydrated using ethanol gradients (30%, 50%, 70%, 80%, 90%) for 15 min at each concentration. Absolute ethanol was used for dehydration twice, also for 15 min each time. The samples were further dehydrated with anhydrous ethanol: tert-butanol mixture (in ratio of 3:1, 1:1, 3:1) for 15 min each. Pure tert-butanol was then replaced for 20 min. After centrifugation, the supernatant was removed, and an appropriate amount of tert-butanol was added. The sample was then placed in a freeze-dryer for further drying. Finally, the samples were observed using scanning electron microscopy (SEM) (Hitachi S-3400N, Tokyo, Japan) after coating with a thin layer of gold.

For ultrastructure observing, the samples were prepared and observed following the method described by Shao et al. [16]. Briefly, the fixed samples were washed three times in PBS (pH 7.2, 0.1 M) to remove any residual fixative. The washed samples were further post-fixed with 1% osmium tetroxide for 2 h, then rinsed three times with PBS, and dehydrated using a series of dehydration solution gradients (30%, 50%, 70%, 90%), each time for 15 min. The samples were then dehydrated with 90% acetone once, followed by four dehydrations with anhydrous acetone, each time for 15 min. After embedding, polymerization, sectioning, and double staining with uranyl acetate and lead citrate, the sections were finally observed using transmission electron microscopy (TEM) (Hitachi H-7650, Tokyo, Japan).

2.6. Identification of VOC Components

The preparation method of yeast VOC samples was slightly modified according to Yalage Don et al. [17]. A 20 mL glass extraction bottle was sterilized by autoclaving, and the bottle cap was sterilized using ultraviolet light for 2 h. Five mL of sterile NYDA medium was pipetted into the sterilized glass extraction bottle to create a ramp. A cell suspension of *S. spartinae* W9 (1×10^8 cells mL^{-1}, 30 μL) was transferred onto the ramp. The suspension was spread evenly on the ramp surface. The glass extraction bottle was sealed and then incubated at 28 °C for 48 h. The control group consisted of a glass extraction bottle inoculated with an equal volume of sterile distilled water instead of the yeast suspension. After the cultivation, the VOCs produced by *S. spartinae* W9 were detected using headspace solid-phase microextraction gas chromatography–mass spectrometry (HS-SPME/GC-MS), according to a previous method [14].

The extraction and determination process of volatile components was as follows: the sample was first shaken and equilibrated at 50 °C for 30 min, and then, VOCs were extracted by the SPME fiber with 50/30 μm, DVB/CAR/PDMS (Supelco Co., Bellefonte, PA, USA;) coating for 30 min at 50 °C. Once adsorption was completed, the SPME fiber was inserted into the GC (7890B–7000C, Agilent Technologies Inc., Santa Clara, CA, USA) inlet and desorbed at 250 °C for 7 min. A DB-5 chromatographic column (30 m × 0.25 mm × 0.25 μm) was utilized to separate the VOC samples, and the separation program was set as follows: 35 °C for 2 min; the temperature was then increased to 100 °C at a rate of 5 °C min^{-1} and held for 1 min. Next, the temperature was further increased to 230 °C at a rate of 7 °C min^{-1} and kept for 12 min. The mass spectrometer utilized a 70 eV EI ion source, with a filament emission current of 200 μA and an ion source temperature of 230 °C. The scanning mass range was set as 29–500 m/z. The n-alkanes (C_7 to C_{40}, o2si, Charleston, SC, USA) were run using the same conditions to calculate [18]. The compositions of VOCs were identified by comparing the calculated LRI with the literature LRI and the mass spectrum in the NIST spectral library. Components that were only produced or increased in the treatment group were considered VOCs produced by *S. spartinae* W9. Each treatment consisted of five biological replicates.

2.7. Effect of Selected Individual VOCs on Mycelial Growth of *B. cinerea*

The antifungal activity of four main single components of yeast VOCs was evaluated following the method described by Lyu et al. [19]. 3-methyl-1-butanol (AR, ≥98.5%), 2-methyl-1-butanol (AR, ≥98%), 2-phenylethanol (AR, ≥99%), isoamyl acetate (AR, ≥99%) were purchased from Shanghai Macklin Biochemical Technology Co., Ltd. (Shanghai, China). A mycelium block with a diameter of 6 mm was taken from the edge of a 5-day-old *B. cinerea* culture and positioned at the center of a PDA plate. A sterile filter paper with a 3 cm diameter was placed on the bottom of the petri dish, and a specific volume of 3-methyl-1-butanol, 2-methyl-1-butanol, 2-phenylethanol, isoamyl acetate was dropped onto the filter paper, resulting in fumigation concentrations of 50, 100, and 150 μL L^{-1}, respectively. As a control, 100 μL of sterile water was added to the filter paper. The plate with *B. cinerea* block was placed on top of the cover plate with filter paper, and the petri dish was sealed with two layers of parafilm. The sealed plate was then incubated at 25 °C for 5 d.

The diameter of *B. cinerea* colony was measured daily, and the inhibition rate was calculated. Each treatment was replicated three times, and the experiment was repeated twice.

2.8. Statistical Analysis

Data analysis was performed using SPSS (version 22.0, SPSS Inc., Chicago, IL, USA). The data are presented as mean ± standard deviation (SD). The differences between the data were assessed using Duncan's multiple range tests of one-way analysis of variance (ANOVA), or an independent samples *t*-test. A significance level of $p < 0.05$ was used to indicate the data were significantly different.

3. Results

3.1. VOCs Produced by S. spartinae W9 Inhibited Mycelium Growth and Spore Germination of B. cinerea

After five days of cultivation, the colony diameter of the *S. spartinae* W9 VOCs treatment group was 16.0 mm, which was 77.8% smaller than that of the control group, demonstrating a significant difference ($p < 0.05$) (Figure 1a). As shown in Figure 1b, the germination rate of the VOCs-treated group was significantly lower than that of the control group after 14 h and 19 h of culture ($p < 0.05$). When the incubation time reached 19 h, the spore germination rates of the control group and the VOCs treatment group were 78.0% and 32.5%, respectively, indicating a 58.3% reduction compared to the control treatment. These indicate that the VOCs produced by *S. spartinae* W9 inhibit the mycelial growth and spore germination of *B. cinerea*.

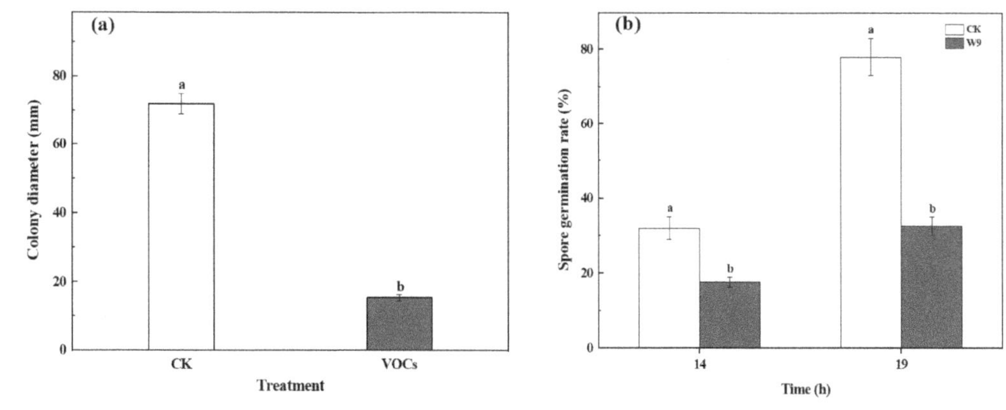

Figure 1. Antifungal effects of VOCs emitted by *S. spartinae* W9 on *B. cinerea*. (**a**) Mycelium growth of *B. cinerea*. The hyphae were cultured in PDA at 25 °C for 5 d under control and yeast VOCs treatment conditions. (**b**) Spore germination rate of *B. cinerea*. The spore suspensions were treated under control and yeast VOCs treatment conditions, and incubated at 25 °C for 14~19 h, respectively. Columns with different letters are significantly different according to an independent samples t-test at $p < 0.05$.

3.2. VOCs Produced by S. spartinae W9 Suppressed Strawberry Gray Mold

As shown in Figure 2, after 6 d of storage, the yeast VOCs-treated group exhibited a significantly lower incidence rate and lesion diameter compared to the control group ($p < 0.05$). The control group had a 100% incidence rate and a lesion diameter of 26.1 mm, whereas the VOCs treatment group showed an incidence rate of 79.3% and a lesion diameter of 8.5 mm, making a reduction of 20.7% and 67.4%, respectively, compared to the control group. These data indicate that the VOCs produced by *S. spartinae* W9 effectively inhibit the occurrence and development of strawberry gray mold, and VOCs serve as a vital biocontrol mechanism of *S. spartinae* W9.

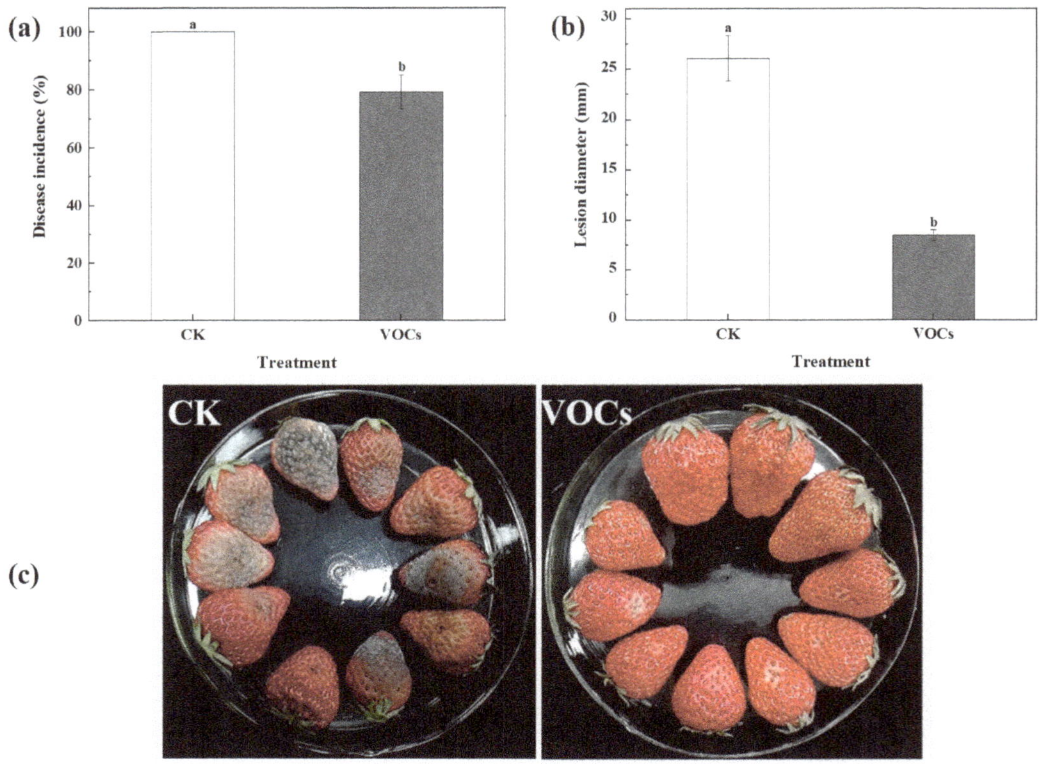

Figure 2. Biocontrol effects of gray mold by VOCs produced from *S. spartinae* W9 on artificially infected strawberry fruits after 6 d of storage. Disease incidence (**a**), lesion diameter (**b**), and pictures of representative strawberry samples recorded on the sixth day of storage (**c**). Columns with different letters are significantly different according to an independent samples t-test at $p < 0.05$.

3.3. VOCs produced by S. spartinae W9 Destroyed the Mycelium Surface Morphology of B. cinerea

After being cultured at 25 °C for 5 d, the hyphae of the control group appeared rough, with smooth and plump surface morphology, while some hyphae exhibited slight wrinkles (Figure 3a). In contrast, the mycelia of *B. cinerea* treated with VOCs from *S. spartinae* W9 were smaller and displayed more severe surface shriveling than the control (Figure 3b). These observations show that the VOCs produced by *S. spartinae* W9 cause damage [17] to the mycelial morphology of *B. cinerea*.

3.4. VOCs Produced by S. spartinae W9 Damaged the Cell Ultrastructure of B. cinerea

TEM observations revealed differences in the hyphae cell walls between the control group and the *S. spartinae* W9 VOCs-treated group. In the control group, the hyphae exhibited clear boundaries, and organelles such as liposomes, mitochondria, nuclei, and nucleoli were clearly visible. Additionally, the presence of few small vacuoles was observed (Figure 4a–c). Conversely, in the *S. spartinae* W9 VOCs-treated group, the hyphal cells formed larger vacuoles, some cells showed slight plasmolysis (Figure 4d), mitochondria appeared swelled (Figure 4e,f), and the matrix electron density decreased (Figure 4f). These observations indicate that the VOCs produced by *S. spartinae* W9 induce damage in the internal structure of *B. cinerea* hyphae.

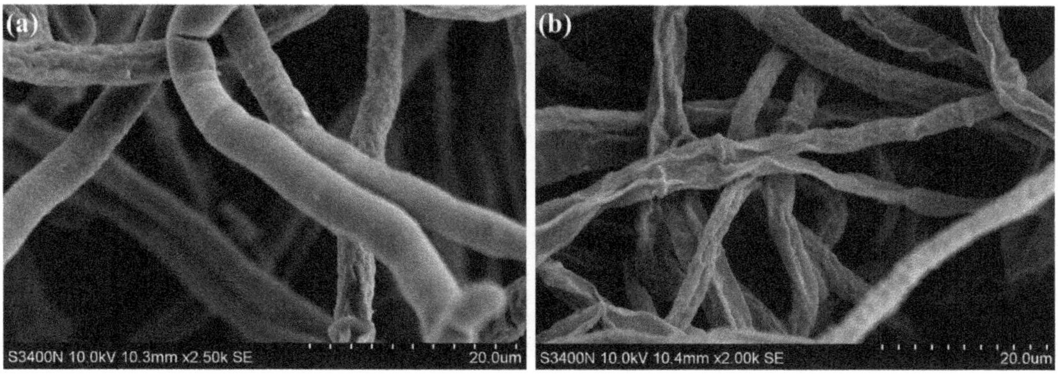

Figure 3. Scanning electron microscope observation of the hyphae morphology of *B. cinerea* treated with VOCs produced from *S. spartinae* W9. (**a**) Control group; (**b**) VOCs treatment group. The hyphae were cultured in PDA at 25 °C for 5 d under control and yeast VOCs treatment, respectively. Magnification: ×2500 (**a**); ×2000 (**b**).

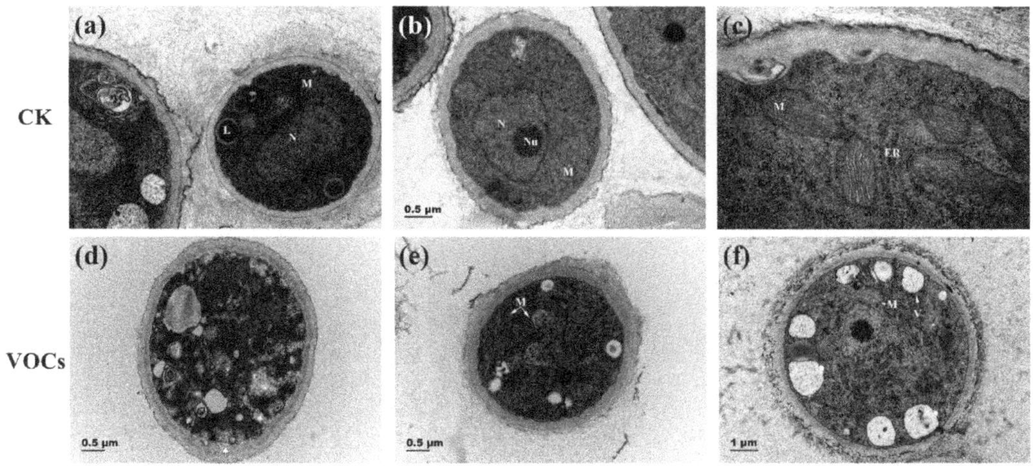

Figure 4. Transmission electron microscope observation of *B. cinerea* hyphae ultrastructure treated with VOCs produced by *S. spartinae* W9. Images (**a**–**c**) represent the control group, while images (**d**–**f**) represent the yeast VOCs treatment group. The arrow in (**d**) indicates slight plasmolysis. L: lipid body; M: mitochondria; N: nuclei; Nu: nucleus; ER: endoplasmic reticulum; V: vacuole. Magnification: ×30,000 (**a**,**b**); ×80,000 (**c**); ×25,000 (**d**,**e**); ×15,000 (**f**).

3.5. Composition of VOCs Produced by S. spartinae W9

As shown in Table 1, *S. spartinae* W9 produced 18 main VOCs, including 4 alcohols, 4 esters, 3 heterocycles, 2 alkanes, 1 acid, 1 amide, 1 amine, 1 benzene, and 1 ether. Among them, 3-methyl-1-butanol exhibited the highest abundance, with a relative peak area of 51.29%. This was followed by 2-methyl-1-butanol (11.74%), 2-phenylethanol (9.02%), formamide (8.40%), dimethyl ether (5.26%), isoamyl acetate (4.32%), dimethylamine (2.90%), formic acid (1.44%), and ethyl hexanoate (1.07%). The remaining VOCs had a relative peak area of less than 1.0%.

Table 1. List of major VOCs produced by *S. spartinae* W9 after 48 h of incubation on NYDA medium at 28 °C.

Volatile Compound	CAS	Molecular Formula	RT [a] (min)	RA [b] (%)	$LRI_{(calculated)}$ [c]	$LRI_{(lit)}$ [d]
Alcohols						
3-Methyl-1-butanol	123-51-3	$C_5H_{12}O$	4.223	51.29	726	727
2-Methyl-1-butanol	137-32-6	$C_5H_{12}O$	4.611	11.74	742	743
2-phenylethanol	60-12-8	$C_8H_{10}O$	17.016	9.02	1110	1110
3-Cyclohexen-1-ol	822-66-2	$C_6H_{10}O$	19.557	0.44	1179	1182
Esters						
Isoamyl acetate	123-92-2	$C_7H_{14}O_2$	8.321	4.32	877	878
Ethyl hexanoate	123-66-0	$C_8H_{16}O_2$	12.796	1.07	999	999
Octanoic acid, ethyl ester	106-32-1	$C_{10}H_{20}O_2$	20.185	0.50	1196	1193
Decanoic acid, ethyl ester	110-38-3	$C_{12}H_{24}O_2$	26.941	0.88	1394	1391
Heterocycles						
Furan, 2-pentyl-	3777-69-3	$C_9H_{14}O$	12.393	0.41	988	989
Pyrazine, 3-ethyl-2,5-dimethyl-	13360-65-1	$C_8H_{12}N_2$	15.629	0.63	1074	1077
Pyrazine, 2, 5-dimethyl-3-(3-methylbutyl)-	18433-98-2	$C_{11}H_{18}N_2$	24.177	0.29	1310	1310
Alkanes						
Pentane, 3-ethyl-	617-78-7	C_7H_{16}	3.598	0.55	696	687
Dodecane	112-40-3	$C_{12}H_{26}$	20.334	0.37	1200	1200
Acids						
Formic acid	64-18-6	CH_2O_2	2.315	1.44	- [e]	-
Amides						
Formamide	75-12-7	CH_3NO	2.079	8.40	-	-
Amines						
Dimethylamine	124-40-3	C_2H_7N	2.139	2.90	-	426
Benzenes						
o-Cymene	527-84-4	$C_{10}H_{14}$	13.722	0.49	1023	1027
Ethers						
Dimethyl ether	115-10-6	C_2H_6O	1.755	5.26	-	-

[a] RT: Retention time. [b] RA: Relative peak area. [c] $LRI_{(calculated)}$: Linear retention index, calculated via analysis of n-alkanes. [d] $LRI_{(lit)}$: Linear retention index from NIST Standard Reference Database. [e] "-": the LRI of the compound was either not calculated or found in the NIST database.

3.6. Antifungal Effect of Main VOCs Produced by S. spartinae W9

3-methyl-1-butanol, 2-methyl-1-butanol, 2-phenylethanol, and isoamyl acetate from the main VOCs of *S. spartinae* W9 were investigated for their antifungal effects on *B. cinerea* mycelium growth. As shown in Table 2, all these pure substances exhibited varying inhibitory effects on the growth of *B. cinerea* mycelia. The inhibition rate increased with the increase in the concentration of the substances and decreased with the increase in the incubation time. On the second day, the inhibitory rates of the pure compounds at a concentration of 150 $\mu L\ L^{-1}$ on mycelial growth ranged from 49% to 100%. On the fifth day, the inhibitory rates of the compound at the same concentration ranged from 21% to 92%. Notably, 2-phenylethanol showed the most potent antifungal effect, with an inhibition rate of 100% on the second day and 92.9% on the fifth day of incubation. These were significantly higher than the inhibition rates observed in other treatment groups ($p < 0.05$). These findings indicate that the main VOCs produced by *S. spartinae* W9 possess antifungal activity against the mycelium growth of *B. cinerea*, with 2-phenylethanol exhibiting the strongest inhibition effect.

Table 2. Effects of selected individual of VOCs produced by *S. spartinae* W9 on mycelial growth of *B. cinerea*.

Volatile Compound	Inhibition Rate (%)					
	Day 2			Day 5		
	50 μL/L	100 μL/L	150 μL/L	50 μL/L	100 μL/L	150 μL/L
3-Methyl-1-butanol	25.6 ± 12.3 [c]	43.2 ± 3.7 [c]	49.1 ± 9.4 [cd]	4.3 ± 4.5 [b]	14.4 ± 5.2 [b]	21.4 ± 8.2 [c]
2-Methyl-1-butanol	29.2 ± 6.1 [bc]	49.9 ± 3.2 [bc]	56.7 ± 3.3 [c]	5.7 ± 6.4 [b]	19.7 ± 8.4 [b]	27.4 ± 6.4 [bc]
2-Phenylethanol	76.6 ± 13.4 [a]	87.4 ± 13.4 [a]	100 ± 0.0 [a]	22.5 ± 11.6 [a]	83.4 ± 2.7 [a]	92.9 ± 7.9 [a]
Isoamyl acetate	40.3 ± 2.8 [b]	53.9 ± 5.4 [b]	90.7 ± 7.2 [b]	17.2 ± 1.3 [a]	22.1 ± 6.4 [b]	25.5 ± 4.9 [c]

Note: Different alphabetic superscripts in the same column indicate significant difference ($p < 0.05$) based on Independent Samples *t*-test or Duncan's Multiple Range test.

4. Discussion

The VOCs with antimicrobial characteristics have been documented in bacteria [19–21], yeasts [22–24], and filamentous fungi [25,26]. The release of VOCs has emerged as a critical mode of action of biocontrol microorganisms. With the trait of being volatile and rarely leaving residue, applying antagonistic microorganism VOC products in a closed non-contact environment is a new field of biological control applications [9]. In addition to exerting direct antimicrobial activity, microbial VOCs may play a role in complex intra- and inter-species interactions, such as intercellular communication, cell differentiation, morphogenesis, plant growth promotion, and induction of plant resistance [9,10,27].

It is indeed true that not all VOCs produced by yeasts have antifungal effects. While the literature has reported that certain VOCs from yeasts exhibit antimicrobial activity, the antifungal effects can vary among different species and strains. For instance, a study conducted in Thailand isolated 366 yeast strains from field crops, of which only 49 strains exhibited antifungal ability against the mycelium growth of *Aspergillus* spp., and the degree of fungal growth inhibition also varied among different species [28]. Some specific examples of yeasts and their VOCs with antifungal effects include: *A. pullulans* [7,17], *Hanseniaspora uvarum* [23,29], *Hanseniaspora opuntiae* [30], *Candida sake* [22], *Yarrowia lipolytica, Kluyveromyces marxianus* and *Pichia kudriavzevii* [31], and *Wickerhamomyces anomalus, Metschnikowia pulcherrima* and *Saccharomyces cerevisiae* [4]. These yeast VOCs have been reported to exhibit antifungal effects against various fungal species, including *Pestalotiopsis vismiae, B. cinerea, Aspergillus* spp., *Penicillium spp., Monilinia spp.*, and *Alternaria alternata.*

In the present study, the strong antifungal activity of VOCs produced by *S. spartinae* W9 on the mycelium growth and spore germination of *B. cinerea* in vitro (Figure 1) was observed and demonstrated, as well as a decrease in disease incidence and the lesion diameter of gray mold in strawberry fruit (Figure 2). This may be the first report describing the fumigation of *S. spartinae* VOCs to suppress the growth of gray mold in strawberry fruit. Similar results have been reported previously; for instance, VOCs produced by *H. uvarum* [23,29], *W. anomalus, M. pulcherrima,* and *S. cerevisiae* [4] reduced *B. cinerea* growth in vitro and strawberry gray mold incidence, and even prolonged the storage time and shelf life of the fruit [23]. The present study confirmed that *S. spartinae* W9 VOCs damaged the morphology and internal structure of *B. cinerea* hyphae (Figures 3 and 4). These results are consistent with those of Farbo et al. [3], who found that yeast VOCs emitted by *Lachancea thermotolerans* caused hyphal swollen and lysis in *Aspergillus* spp. The volatiles emitted by *S. cerevisiae* resulted in hyphal deformities, damage, and reduced size in *Aspergillus flavus* [32]. Xing et al. [20] also discovered that VOCs of *Streptomyces fimicarius* caused morphological disruption in hyphae, sporangiophores, and sporangia of *Peronophythora litchi*, and even led to thickened cell walls and cell membranes, as well as damage in vacuoles and mitochondria.

Using the HS-SPME/GC-MS technology, we identified 18 main VOCs produced by *S. spartinae* W9. The top three VOCs in terms of relative content were 3-methyl-1-butanol, 2-methyl-1-butanol, and 2-phenylethanol, accounting for more than 72% of the total VOCs

(Table 1). These VOCs, along with isoamyl acetate, are commonly found in yeast [33–35]. Di Francesco et al. [33] studied the effects of the main volatile substances produced by *A. pullulans* and found that 2-phenylethanol, 3-methyl-1-butanol, and 2-methyl-1-butanol inhibited the spore germination of *B. cinerea*. Among them, 2-phenylethanol exhibited the strongest inhibitory effect, with an EC_{50} of 0.57 $\mu L\ mL^{-1}$. Ando et al. [36] demonstrated that isoamyl acetate and 3-methyl-1-butanol inhibited the germination of spores from 12 and 15 different filamentous fungi, respectively. Similar results were also observed in the present study regarding the inhibitory effects of four VOCs produced by *S. spartinae* W9, namely 3-methyl-1-butanol, 2-methyl-1-butanol, 2-phenylethanol, and isoamyl acetate, on the hyphal growth of *B. cinerea*. Among them, 2-phenylethanol showed the strongest antifungal effect (Table 2).

In recent years, there has been research interest in 2-phenylethanol, a common VOC secreted by many yeasts. Hua et al. [37] reported that 2-phenylethanol is the main volatile substance emitted by *Pichia anomala*. They found it inhibited mycelium growth, conidia germination, toxin production, and aflatoxin gene expression in *A. flauvs*. Yalage Don et al. [17] discovered that ethanol and 2-phenylethanol are key substances of VOCs emitted by *A. pullulans*. These substances inhibited the fungal growth of *B. cinerea* and *A. alternata*. Lu et al. [38] revealed that 2-phenylethanol is the third-most abundant component in the VOCs of *Candida quercitrusa*. They also found that it blocked the oxidative phosphorylation pathway of *Phytophthora infestans*. Zou et al. [14] also identified 2-phenylethanol as the key compound in the VOCs of *Candida pseudolambica* W16, and observed that 2-phenylethanol inhibited *B. cinerea* growth by causing cell membrane damage and inducing reactive oxygen species (ROS) stress [39].

Based on the findings of this paper, we speculate the production of several antifungal volatile substances may serve as a crucial biocontrol mechanism of *S. spartinae* W9. The antifungal performance of the yeast VOCs is probably due to an additive and synergistic action of several compounds, not only the above four compounds, but also other volatile components with lower concentrations. The mixtures of VOCs from *S. spartinae* W9 possibly elicit multicomponent modes of action, such as cell membrane permeability and disruption, organelle damage, electrolyte loss, and oxidative stress [20,34]. The profound cytological impact implies that VOCs from *S. spartinae* W9 cause significant changes in gene expression in *B. cinerea*, ultimately inhibiting its growth. The detailed antifungal mechanisms of *S. spartinae* W9 VOCs need to be further explored.

5. Conclusions

In conclusion, *S. spartinae* W9 produced 18 main VOCs, some of which exhibited antifungal activities, resulting in microscopic damage to *B. cinerea* and displaying biocontrol ability in vitro and in strawberry fruits during storage. The VOCs produced by *S. spartinae* W9 offer a promising biocontrol strategy against gray mold in strawberries. For example, fumigation with yeast VOCs or specific antifungal volatile compounds. However, further investigations are required to explore the synergistic effects of these VOCs and their broad-spectrum antimicrobial effects on other microorganisms. Additionally, practical applications of these VOCs need to be explored.

Author Contributions: Conceptualization, X.Z. and X.S.; methodology, X.Z., Y.W. and J.S.; software, J.Z.; validation, X.Z. and X.S.; writing—original draft, X.Z.; writing—review & editing, Y.W., J.Z., J.S. and X.S. All authors have read and agreed to the published version of the manuscript.

Funding: This work was funded by the National Natural Science Foundation of China (grant no. 32202122), the Ningbo Agricultural Research Project of China (grant no. 2022S154), the Shaoguan Science and Technology Plan Project of China (grant no. 210717174530674), and the Natural Science Foundation of Shaoguan University (grant no. SZ2022JK03).

Data Availability Statement: The data presented in this study are included in the article. Further inquiries can be directed to the corresponding author.

Conflicts of Interest: The authors declare no conflict of interest.

References

1. Dukare, A.S.; Paul, S.; Nambi, V.E.; Gupta, R.K.; Singh, R.; Sharma, K.; Vishwakarma, R.K. Exploitation of microbial antagonists for the control of postharvest diseases of fruits: A review. *Crit. Rev. Food Sci. Nutr.* **2019**, *59*, 1498–1513. [CrossRef] [PubMed]
2. Freimoser, F.M.; Rueda-Mejia, M.P.; Tilocca, B.; Migheli, Q. Biocontrol yeasts: Mechanisms and applications. *World J. Microbiol. Biotechnol.* **2019**, *35*, 154. [CrossRef] [PubMed]
3. Farbo, M.G.; Urgeghe, P.P.; Fiori, S.; Marcello, A.; Oggiano, S.; Balmas, V.; Hassan, Z.U.; Jaoua, S.; Migheli, Q. Effect of yeast volatile organic compounds on ochratoxin A-producing *Aspergillus carbonarius* and *A. ochraceus*. *Int. J. Food Microbiol.* **2018**, *284*, 1–10. [CrossRef] [PubMed]
4. Oro, L.; Feliziani, E.; Ciani, M.; Romanazzi, G.; Comitini, F. Volatile organic compounds from *Wickerhamomyces anomalus*, *Metschnikowia pulcherrima* and *Saccharomyces cerevisiae* inhibit growth of decay causing fungi and control postharvest diseases of strawberries. *Int. J. Food Microbiol.* **2018**, *265*, 18–22. [CrossRef] [PubMed]
5. Parafati, L.; Vitale, A.; Restuccia, C.; Cirvilleri, G. Performance evaluation of volatile organic compounds by antagonistic yeasts immobilized on hydrogel spheres against gray, green and blue postharvest decays. *Food Microbiol.* **2017**, *63*, 191–198. [CrossRef] [PubMed]
6. Pereyra, M.M.; Garmendia, G.; Rossini, C.; Meinhardt, F.; Vero, S.; Dib, J.R. Volatile organic compounds of *Clavispora lusitaniae* AgL21 restrain citrus postharvest pathogens. *Biol. Control* **2022**, *174*, 105025. [CrossRef]
7. Di Francesco, A.; Di Foggia, M.; Baraldi, E. *Aureobasidium pullulans* volatile organic compounds as alternative postharvest method to control brown rot of stone fruits. *Food Microbiol.* **2020**, *87*, 103395. [CrossRef]
8. Rios-Navarro, A.; Gonzalez, M.; Carazzone, C.; Celis Ramirez, A.M. Learning about microbial language: Possible interactions mediated by microbial volatile organic compounds (VOCs) and relevance to understanding *Malassezia* spp. metabolism. *Metabolomics* **2021**, *17*, 39. [CrossRef]
9. Oufensou, S.; Ul Hassan, Z.; Balmas, V.; Jaoua, S.; Migheli, Q. Perfume guns: Potential of yeast volatile organic compounds in the biological control of mycotoxin-producing fungi. *Toxins* **2023**, *15*, 45. [CrossRef]
10. Tilocca, B.; Cao, A.C.; Migheli, Q. Scent of a killer: Microbial volatilome and its role in the biological control of plant pathogens. *Front. Microbiol.* **2020**, *11*, 41. [CrossRef]
11. Nasanit, R.; Jaibangyang, S.; Onwibunsiri, T.; Khunnamwong, P. Screening of volatile organic compound-producing yeasts and yeast-like fungi against aflatoxigenic *Aspergillus flavus*. *Microbiol. Biotechnol. Lett.* **2022**, *50*, 202–210. [CrossRef]
12. Fan, L.; Wei, Y.; Chen, Y.; Jiang, S.; Xu, F.; Zhang, C.; Wang, H.; Shao, X. Epinecidin-1, a marine antifungal peptide, inhibits *Botrytis cinerea* and delays gray mold in postharvest peaches. *Food Chem.* **2023**, *403*, 134419. [CrossRef] [PubMed]
13. Zou, X.; Wei, Y.; Dai, K.; Xu, F.; Wang, H.; Shao, X. Yeasts from intertidal zone marine sediment demonstrate antagonistic activities against *Botrytis cinerea* in vitro and in strawberry fruit. *Biol. Control* **2021**, *158*, 104612. [CrossRef]
14. Zou, X.; Wei, Y.; Jiang, S.; Cao, Z.; Xu, F.; Wang, H.; Zhan, P.; Shao, X. Volatile organic compounds and rapid proliferation of *Candida pseudolambica* W16 are modes of action against gray mold in peach fruit. *Postharvest Biol. Technol.* **2022**, *183*, 111751. [CrossRef]
15. Xu, Y.; Wei, J.; Wei, Y.; Han, P.; Dai, K.; Zou, X.; Jiang, S.; Xu, F.; Wang, H.; Sun, J.; et al. Tea tree oil controls brown rot in peaches by damaging the cell membrane of *Monilinia fructicola*. *Postharvest Biol. Technol.* **2021**, *175*, 111474. [CrossRef]
16. Shao, X.; Cao, B.; Xu, F.; Xie, S.; Yu, D.; Wang, H. Effect of postharvest application of chitosan combined with clove oil against citrus green mold. *Postharvest Biol. Technol.* **2015**, *99*, 37–43. [CrossRef]
17. Yalage Don, S.M.; Schmidtke, L.M.; Gambetta, J.M.; Steel, C.C. *Aureobasidium pullulans* volatilome identified by a novel, quantitative approach employing SPME-GC-MS, suppressed *Botrytis cinerea* and *Alternaria alternata* in vitro. *Sci. Rep.* **2020**, *10*, 4498. [CrossRef] [PubMed]
18. Yan, J.; Liu, X.-B.; Zhu, W.-W.; Zhong, X.; Sun, Q.; Liang, Y.-Z. Retention Indices for Identification of Aroma Compounds by GC: Development and Application of a Retention Index Database. *Chromatographia* **2014**, *78*, 89–108. [CrossRef]
19. Lyu, A.; Yang, L.; Wu, M.; Zhang, J.; Li, G. High efficacy of the volatile organic compounds of *Streptomyces yanglinensis* 3–10 in suppression of *Aspergillus contamination* on peanut kernels. *Front. Microbiol.* **2020**, *11*, 142. [CrossRef]
20. Xing, M.; Zheng, L.; Deng, Y.; Xu, D.; Xi, P.; Li, M.; Kong, G.; Jiang, Z. Antifungal activity of natural volatile organic compounds against litchi downy blight pathogen *Peronophythora litchii*. *Molecules* **2018**, *23*, 358. [CrossRef]
21. Wang, Z.; Zhong, T.; Chen, K.; Du, M.; Chen, G.; Chen, X.; Wang, K.; Zalán, Z.; Takács, K.; Kan, J. Antifungal activity of volatile organic compounds produced by *Pseudomonas fluorescens* ZX and potential biocontrol of blue mold decay on postharvest citrus. *Food Control* **2021**, *120*, 107499. [CrossRef]
22. Arrarte, E.; Garmendia, G.; Rossini, C.; Wisniewski, M.; Vero, S. Volatile organic compounds produced by Antarctic strains of *Candida sake* play a role in the control of postharvest pathogens of apples. *Biol. Control* **2017**, *109*, 14–20. [CrossRef]
23. Qin, X.; Xiao, H.; Cheng, X.; Zhou, H.; Si, L. *Hanseniaspora uvarum* prolongs shelf life of strawberry via volatile production. *Food Microbiol.* **2017**, *63*, 205–212. [CrossRef] [PubMed]
24. Lemos, W.J.F.; Binati, R.L.; Felis, G.E.; Slaghenaufi, D.; Ugliano, M.; Torriani, S. Volatile organic compounds from *Starmerella bacillaris* to control gray mold on apples and modulate cider aroma profile. *Food Microbiol.* **2020**, *89*, 103446.
25. Elsherbiny, E.A.; Amin, B.H.; Aleem, B.; Kingsley, K.L.; Bennett, J.W. *Trichoderma* volatile organic compounds as a biofumigation tool against late blight pathogen *Phytophthora infestans* in postharvest potato tubers. *J. Agric. Food Chem.* **2020**, *68*, 8163–8171. [CrossRef] [PubMed]

26. Rajani, P.; Rajasekaran, C.; Vasanthakumari, M.M.; Olsson, S.B.; Ravikanth, G.; Uma Shaanker, R. Inhibition of plant pathogenic fungi by endophytic *Trichoderma* spp. through mycoparasitism and volatile organic compounds. *Microbiol. Res.* **2021**, *242*, 126595. [CrossRef] [PubMed]
27. Schmidt, R.; Cordovez, V.; de Boer, W.; Raaijmakers, J.; Garbeva, P. Volatile affairs in microbial interactions. *ISME J.* **2015**, *9*, 2329–2335. [CrossRef] [PubMed]
28. Jaibangyang, S.; Nasanit, R.; Limtong, S. Biological control of aflatoxin-producing *Aspergillus flavus* by volatile organic compound-producing antagonistic yeasts. *BioControl* **2020**, *65*, 377–386. [CrossRef]
29. Ruiz-Moyano, S.; Hernández, A.; Galvan, A.I.; Córdoba, M.G.; Casquete, R.; Serradilla, M.J.; Martín, A. Selection and application of antifungal VOCs-producing yeasts as biocontrol agents of grey mould in fruits. *Food Microbiol.* **2020**, *92*, 103556. [CrossRef]
30. Galvan, A.I.; Hernandez, A.; Cordoba, M.G.; Martin, A.; Serradilla, M.J.; Lopez-Corrales, M.; Rodriguez, A. Control of toxigenic *Aspergillus* spp. in dried figs by volatile organic compounds (VOCs) from antagonistic yeasts. *Int. J. Food Microbiol.* **2022**, *376*, 109772. [CrossRef]
31. Delali, K.I.; Chen, O.; Wang, W.; Yi, L.; Deng, L.; Zeng, K. Evaluation of yeast isolates from kimchi with antagonistic activity against green mold in citrus and elucidating the action mechanisms of three yeast: *P. kudriavzevii*, *K. marxianus*, and *Y. lipolytica*. *Postharvest Biol. Technol.* **2021**, *176*, 111495. [CrossRef]
32. Natarajan, S.; Balachandar, D.; Senthil, N.; Velazhahan, R.; Paranidharan, V. Volatiles of antagonistic soil yeasts inhibit growth and aflatoxin production of *Aspergillus flavus*. *Microbiol. Res.* **2022**, *263*, 127150. [CrossRef] [PubMed]
33. Di Francesco, A.; Ugolini, L.; Lazzeri, L.; Mari, M. Production of volatile organic compounds by *Aureobasidium pullulans* as a potential mechanism of action against postharvest fruit pathogens. *Biol. Control* **2015**, *81*, 8–14. [CrossRef]
34. Yalage Don, S.M.; Schmidtke, L.M.; Gambetta, J.M.; Steel, C.C. Volatile organic compounds produced by *Aureobasidium pullulans* induce electrolyte loss and oxidative stress in *Botrytis cinerea* and *Alternaria alternata*. *Res. Microbiol.* **2020**, *172*, 103788. [CrossRef] [PubMed]
35. Contarino, R.; Brighina, S.; Fallico, B.; Cirvilleri, G.; Parafati, L.; Restuccia, C. Volatile organic compounds (VOCs) produced by biocontrol yeasts. *Food Microbiol.* **2019**, *82*, 70–74. [CrossRef] [PubMed]
36. Ando, H.; Hatanaka, K.; Ohata, I.; Yamashita-Kitaguchi, Y.; Kurata, A.; Kishimoto, N. Antifungal activities of volatile substances generated by yeast isolated from Iranian commercial cheese. *Food Control* **2012**, *26*, 472–478. [CrossRef]
37. Hua, S.S.; Beck, J.J.; Sarreal, S.B.; Gee, W. The major volatile compound 2-phenylethanol from the biocontrol yeast, *Pichia anomala*, inhibits growth and expression of aflatoxin biosynthetic genes of *Aspergillus flavus*. *Mycotoxin Res.* **2014**, *30*, 71–78. [CrossRef]
38. Lu, J.; Li, J.; Li, L.; Qi, L.; Wang, Y.; Yang, S.; Xu, G.; Dou, D.; Wang, X.; Liu, J. Natural product 2-Phenylethanol inhibits ATP synthesis of *P. infestans* by blocking the oxidative phosphorylation pathway to prevent potato late blight. *Postharvest Biol. Technol.* **2023**, *199*, 112310. [CrossRef]
39. Zou, X.; Wei, Y.; Jiang, S.; Xu, F.; Wang, H.; Zhan, P.; Shao, X. ROS stress and cell membrane disruption are the main antifungal mechanisms of 2-phenylethanol against *Botrytis cinerea*. *J. Agric. Food Chem.* **2022**, *70*, 14468–14479. [CrossRef]

foods MDPI

Article

Pullulan-Based Active Coating Incorporating Potassium Metabisulfite Maintains Postharvest Quality and Induces Disease Resistance to Soft Rot in Kiwifruit

Yiming Tian [1], Lamei Li [1], Rui Wang [1,*], Ning Ji [1], Chao Ma [1], Jiqing Lei [1], Wenqiang Guan [2] and Xu Zhang [1]

[1] College of Food and Pharmaceutical Engineering, Guiyang University, Guiyang 550005, China; tym764656@163.com (Y.T.); llmei2020@126.com (L.L.); jining552100@163.com (N.J.); gyuchaoma@163.com (C.M.); jiqinglei@gmail.com (J.L.); z810156401@163.com (X.Z.)

[2] Tianjin Key Laboratory of Food Biotechnology, College of Biotechnology and Food Science, Tianjin University of Commerce, Tianjin 300134, China

* Correspondence: wangrui__gyu@163.com

Abstract: Soft rot is a severe postharvest disease of kiwifruit that causes enormous economic losses annually. In this study, we aimed to explore an effective pullulan-based active coating, incorporating food additives to reduce soft rot and extend the shelf life of cold-stored kiwifruit. The results showed that 1 g/L potassium metabisulfite could completely inhibit the mycelial growth of *Diaporthe* sp., *Botryosphaeria dothidea*, *Phomopsis* sp. and *Alternaria* sp., which were the primary pathogens of kiwifruit soft rot. Furthermore, the pullulan coating, combined with a 10 g/L potassium metabisulfite group, had a decay rate 46% lower than the control (CK) group and maintained fruit quality at the end of shelf life. The retention of physicochemical properties such as soluble solid content (SSC), firmness, weight loss and respiration rate also confirmed the efficacy of the treatment. In addition, at the end of shelf life, pullulan coating, combined with potassium metabisulfite, increased the accumulation of total phenolic content (37.59%) and flavonoid content (9.28%), maintained a high energy charge (51.36%), and enhanced superoxide dismutase (SOD) (6.27%), peroxidase (POD) (62.50%), catalase (CAT) (84.62%) and phenylalanine ammonia lyase (PAL) (24.61%) enzyme activities as well as initiating the upregulation of their gene expression levels. As a result, the disease resistance of fruit was improved, and the occurrence of soft rot was delayed. Overall, this study demonstrated that using the pullulan-based active coating incorporating potassium metabisulfite treatment effectively controlled soft rot and retarded the senescence of postharvest kiwifruit.

Keywords: active coating; food additives; soft rot; disease resistance

1. Introduction

'Hong Yang' Kiwifruit (*Actinidia chinensis* cv. Hongyang) is one of the most commercially planted varieties in China and is well known for its nutritious properties and unique flavor [1]. However, during the planting and harvesting period, it can be easily infected by multiple diseases such as black spot, gray mold, soft rot, blue mold and anthracnose [1,2]. Kiwifruit is most susceptible to soft rot, which is primarily caused by *Diaporthe* sp., *Alternaria alternata*, *Botryosphaeria dothidea*, *Phomopsis* sp., *Botrytis cinerea* and *Cryptosporiopsis actinidiae* [3,4]. This manifests primarily during the process of transportation, storage and marketing. When this occurs, it causes huge economic losses annually. Currently, the use of chemical fungicides remains the primary measure to control diseases of kiwifruit and other fruits [5]. However, public concern about the detrimental effects of pesticides on human health and the environment, along with the induction of disease resistance in target pathogenic fungi, has raised questions about the application of fungicides for decay control after harvest. Consequently, it is necessary to develop an effective strategy for the inhibition of the infection and growth of pathogens in fresh foods and fruits.

Food additives such as natamycin, sodium dehydroacetate, sodium silicate and potassium sorbate have gained much attention in the postharvest field due to their broad-spectrum antimicrobial activity and food safety [6]. In a previous study, sodium dehydroacetate treatment was found to inhibit the growth of *Geotrichum citri-aurantii* and reduce sour rot in citrus fruits [7]. The same result was also observed in citrus fruits with the use of potassium metabisulfite, which could restrain diseases caused by *Penicillium digitatum*, *Penicillium italicum* and *Geotrichum citri-aurantii* [5]. In another study, the combination of potassium sorbate with heat treatment reduced the incidence of soft rot by 13% at 4 days by inhibiting the mycelial growth of *Alternaria alternata* compared to the control group [8].

Edible coatings have been used for fruit and vegetable preservation due to their advantages of easy use, good selective permeability to gas, being food-safe and effective prevention. Besides controlling fruit disease, the practice of coating can also improve fruit safety and extend fruit shelf life and so has been recognized as a safe, feasible and environmentally friendly method to control fruit yield losses [9]. As a bio-polysaccharide produced by *Aureobasidium pullulans*, pullulan can be used to make an edible coating with several advantages, such as being colorless, tasteless and biodegradable, owing to its possession of a good oxygen-barrier performance and it exhibiting no reaction with food ingredients [10]. For example, Pobiega et al. [11] found that pullulan coating with an ethanol extract of propolis reduced the numbers of *Penicillium chrysogenum*, *Fusarium solani* and *Botrytis cinerea* by about 1.9, 2.2 and 1.0 log CFU g^{-1}. This accounted for only 48%, 59% and 81% of the totals in their respective CK groups after 7 days in artificially inoculated cherry tomatoes.

Therefore, in the present study, an edible pullulan coating combined with food additives was considered a potentially useful strategy with which to control soft rot for kiwifruit and extend shelf life. The objectives of the study were as follows: (a) to identify the dominant fungal pathogens of 'Hong Yang' kiwifruit via high-throughput sequencing, as well as the optimal isolation and identification techniques to use; (b) to screen effective food additives with antifungal activity (in vitro); and (c) to study the effects of a pullulan coating combined with food additives on soft rot, defense-related enzymes and their gene expression, total phenolic and flavonoid content, energy metabolism and the physicochemical properties of 'Hong Yang' kiwifruit. Finally, we proposed and verified an environmentally friendly strategy for fruit preservation.

2. Materials and Methods

2.1. Fruit Materials

A series of experiments were conducted in 2020 and 2021. Before sampling, 25 fruit trees that had been followed for 7 years were selected and marked in a specific orchard. 'Hong Yang' kiwifruit was harvested from a local orchard in the town of Miluo (105.00° E, 26.37° W, Shuicheng County, Liupanshui City, China).

2.2. Experiment Design

In 2020, kiwifruits were harvested on 10 August, delivered to the laboratory within 4 h, precooled at 2 ± 0.5 °C for 24 h, packed into a polyethylene bag (60 × 40 cm, 20 μm thick) and stored at a low temperature of 2 ± 0.5 °C for 60 days. Then, the experiment was performed in three stages: (1) The decayed kiwifruit was picked for the isolation and identification of pathogens. (2) The isolated pathogen was identified using morphological and molecular techniques. (3) The optimal food additive and fungicidal concentration for the inhibition of the soft rot of kiwifruit (in vitro and vivo) was selected.

In 2021, the experiment was carried out in two stages: (1) In order to detect the fungi community composition of kiwifruit, 30 samples were harvested on 1 August 2021 and stored at 25 ± 1 °C (85 ± 5% RH) for 144 h. Then, the fruits with soft rot symptoms were washed with sterile water and peeled to obtain fruit pulp for use in high-throughput sequencing analysis (Shanghai OE Biotech). (2) Based on the identification and results of the in vitro experiment in 2020 and the high-throughput sequencing results from 2021,

1200 kiwifruits were harvested on 25 August 2021, coated by repeating a dip-coating process, and then consistently dried in a draught cupboard for 24 h. Groups of 30 fruits were each packed randomly into a polyethylene bag (dimensions 60 × 40 cm, 20 μm thick). Subsequently, kiwifruits were stored at 2 ± 0.5 °C for 60 days. Subsequently, both the treated and control groups were removed from refrigerated conditions and analyzed after 0, 2, 4, 6 and 8 days of storage at 25 ± 1 °C (70 ± 5% RH).

2.3. Isolation, Purification and Identification of Pathogens

The separation of plant lesion tissue was used to isolate and purify the pathogens. Prior to inoculation, fruits were disinfected with 2% (*v*/*v*) NaClO for 3 min, rinsed with sterile water and air-dried for 30 min. Then, the flesh tissue was cut using sterile scissors, inoculated on a PDA medium and placed in a biochemical incubator at 28 °C. After 2 days, the mycelium at the edge of the colony was picked with an inoculation needle and inoculated on a new PDA medium at 28 °C for 5 to 7 d, purified 3-4 times and used. Fungi were identified according to the Koch postulates. Then, rDNA-ITS sequences were used to identify fungal isolates combined with their morphological characteristics and similarities. As a result, *Diaporthe* sp., *Botryosphaeria dothidea*, *Phomopsis* sp. and *Alternaria* sp. were isolated from kiwifruits.

2.4. High-Throughput DNA Sequencing and Analysis

The high-throughput experiments were performed according to the method described in the report of Zhang et al. [12]; the fragments were amplified with the universal fungal primers ITS1F (CTTGGTCATTTAGAGGAAGTAA) and ITS2 (GCTGCGTTCTTCATCGATGCITS). The PCR program for 18S rDNA amplicon had parameters of 5 min, 94 °C and 26 cycles (30 s, 94 °C; 30 s at 56 °C; 30 s 72 °C), with a 10 min final extension at 72 °C using a PCR Thermocycler 580BR10905 (Bio-Rad).

2.5. In Vitro Antifungal Activity

In the primary screening, the effects of 18 food additives, namely nisin, ε-polylysine, ethyl p-hydroxy benzoate, natamycin, sodium diacetate, sodium propionate, potassium sorbate, stabilized chlorine dioxide, 2,4-dichlorophenoxy acetic acid, sodium benzoate, citric acid, sodium sulfite, ethoxy quin, ascorbic acid, sodium dehydroacetate, potassium metabisulfite, sodium hyposulfite and lysozyme, on the mycelial growth of pathogens (*Diaporthe* sp., *Botryosphaeria dothidea*, *Phomopsis* sp. and *Alternaria* sp. (Figure S2)) were tested.

The method of assessing growth rate was used to determine the effects of 18 food additives on the growth of pathogenic fungi at a concentration of 1 g/L. A medium containing a pathogenic fungus with a diameter of 4 mm was inserted into a PDA medium with food additives using a sterile hole punch, while the same medium without food additives was used for the control. Then, the cultures proceeded to grow in an incubator maintained at a constant temperature of 28 °C. When the hyphae of the control group almost covered the entire PDA medium after inoculation for 2~8 days, the colony diameters were recorded. Mycelial growth rates were assessed according to the recommendations of Duan et al. [13]. When mycelial growth was 100% inhibited in the food-additive-treated group, the treatment was considered an effective antimicrobial additive and used in the subsequent screened concentration experiments.

After the primary screening, sterile solutions at concentrations of 0, 50, 250, 500, 750 and 1000 mg/L of sodium dehydroacetate (SD), potassium metabisulfite (PM), sodium hyposulfite (SM) and lysozyme were prepared by diluting each with sterile water. Referring to the method of Duan et al. [13], the minimum inhibitory concentration (MIC) and minimal fungicidal concentration (MFC) values of four pathogens were determined.

2.6. Coating Preparation and Application

Pullulan coating solution was prepared with the following chemical composition (g/L): 2% pullulan (Shanghai Yuanye Bio-Technology Co., Ltd., Shanghai, China) (*v*/*v*), 1%

glycerol (v/v) and distilled water. The components were mixed at 45 °C with a magnetic stirrer (600 rpm) for 20 min. The pullulan + potassium metabisulfite (PM) coating was prepared by mixing (different concentrations of PM, 1% glycerol (v/v), 1% tween 80 (v/v)) in a magnetic stirrer (600 rpm) for an hour at 45 °C. Each coating dissolved completely and had a uniform texture.

2.7. In Vivo Antifungal Activity

Kiwifruit was inoculated with *Diaporthe* sp., *Botryosphaeria dothidea*, *Phomopsis* sp. or *Alternaria* sp. It was then soaked for 60 s in the corresponding solution: (1) sterile water, CK; (2) pullulan, PU; (3) PU + 1 g/L PM, T1; (4) PU + 5 g/L PM, T2; (5) PU + 10 g/L PM, T3. The fruits were treated, dried and incubated at 28 °C with 95% relative humidity (RH) for 8 days. Subsequently, a vernier caliper was used to measure soft rot diameters. Each fruit measurement and treatment was repeated three times.

2.8. Kiwifruit Treatment for Postharvest Study

During the coating experiment, kiwifruits with no visual defects were randomly divided separated into five groups, with 300 kiwifruits each. The five groups were, respectively, treated as follows: (1) sterile water, CK; (2) pullulan, PU; (3) PU + 1 g/L PM, T1; (4) PU + 5 g/L PM, T2; (5) PU + 10 g/L PM, T3. Firstly, the kiwifruits were coated in a coating solution (CK, PU, T1, T2, T3) for 2 min to ensure uniform coverage and then consistently dried in a draught cupboard for 24 h. Groups of 30 fruits were each randomly packed into a polyethylene bag (dimensions 60 × 40 cm, 20 μm thick). Subsequently, four groups of kiwifruits were stored at 2 ± 0.5 °C for 60 days. Then, both the treated and control groups were removed from refrigerated conditions and analyzed after 0, 2, 4, 6 and 8 days of storage at 25 ± 1 °C (70 ± 5% RH). All the experiments were carried out in triplicates, and the average data were used for analysis.

2.9. Index Determination

2.9.1. Quality and Physiological Parameter Indexes

The decay incidence was measured using the method described by Shah et al. [6] Fruits showing symptoms of mycelial development and severe softening were considered as decayed and were expressed as a percentage of the total fruit number. SSC was measured using a refractometer (PAL-1, ATAGO, Tokyo, Japan). The respiration rate of the kiwifruit was determined using the method of Kou et al. [14] with slight modifications. Before measurement, three replicates of six 'Hong Yang' kiwifruits were randomly chosen and placed in 1 L sealed plastic jars at 25 °C for 2 h. Then, 2 mL headspace gas was taken from each jar and analyzed with infrared O_2 and CO_2 analyzers (Checkpoint II Portable residual oxygen meter, Dansensor, Rinsted, Denmark). In this study, the respiratory rate was expressed as mg/kg/h CO_2.

The firmness was determined as described by Burdon et al. [15] with slight modifications. A penetrometer (TA. XT. Plus, SMS, London, UK) equipped with a puncture probe (2mm diameter) was used, with an experiment speed of 5 mm/s, a penetration depth of 5 mm and a trigger force of 5 g.

The weight loss ratio was calculated using the following formula:

$$\text{Weight loss ratio } (\%) = (W_0 - W_n)/(W_n) \times 100, \quad (1)$$

where W_0 is the initial (0 day) sample weight and W_n is the sample weight on day n. In total, 30 kiwifruits were used for each treatment.

The ethanol concentration of fruit was estimated using headspace gas chromatography according to the method described in the report of Maratab Ali et al. [16] with slight modifications: 2.0 g frozen sample was mixed with 10 mL deionized water and centrifuged at 10,000× *g* for 25 min at 4 °C. Ultrasound treatment was performed in an ice water bath for 30 min, and the volume was fixed to 10 mL. Then, 1.5 mL was centrifuged at

12,000 r/min for 2 min and stored at −20 °C until analysis. The ethanol contents were detected by GC (Agilent 6890 Gas Chromatograph).

According to Liu et al. [17], an LF-NMR measurement was performed. The transverse relaxation measurements were performed on an NMI20-015V-I NMR analyzer with a resonance frequency of 21.0 MHz at 32 °C. The Carr–Purcell–Meiboom–Gill pulse sequence was used to measure the sample's transverse relaxation time.

2.9.2. Total Phenolic and Flavonoid Contents

The total phenolic content and flavonoid content were determined based on a previous study [18] with slight modifications: a 5 g kiwifruit sample was homogenized in 20 mL of a 1% HCl-methanol solution and centrifuged at 14,000× *g* for 20 min at 4 °C. The total phenolic content of the sample was determined by measuring the absorbance of the supernatant at 280 nm. The flavonoid contents were determined by measuring absorbance at 325 nm. The total phenolic content and flavonoid content were expressed in mg/(100 g) and mg/mL, respectively.

2.9.3. The SOD, POD, CAT and PAL Enzyme Activity and Gene Expression

The activity of the SOD enzyme in kiwifruit was measured using an SOD assay kit (WST-1 method) (Nanjing Jian Cheng Bioengineering Institute, Nanjing, China NJBI). A sample of 0.4 g, accurate to 0.02 g, was weighed and added to 3.6 mL of 100 mm phosphate buffer (pH 7~7.4), which was mixed well. Then, the material was centrifuged at 3500 rpm (2380× *g*) for 10 min, and the supernatant was collected. The enzyme activity was recorded according to the manufacturer's instructions. In the reaction system, when the inhibition rate of SOD reached 50%, a unit of SOD activity was determined (U/g FW).

The activity of POD was measured using the methods described by Wang et al. [19]. The POD activity was calculated as U/(g·min) fresh weight.

The activity of CAT was measured using methods described by Lin et al. [20]. CAT activity was expressed in units of U/(g·min·FW).

The activity of the PAL enzyme in kiwifruit was measured using an assay kit (spectrophotometric assays) (Nanjing Jian Cheng Bioengineering Institute, Nanjing, China NJBI). We conducted the assay according to the manufacturer's instructions. Units of PAL activity were expressed as U/g.

RNA was extracted via a Prime Script™ RT reagent Kit with a gDNA Eraser (Perfect Real Time) (Ta Ka Ra, #RR047A) to synthesize cDNA through reverse transcription. Operation procedures were performed in accordance with the kit's instructions. Real-time quantitative PCR (RT-PCR) TaqMan assays were used with the BIO-RAD CFX Connect™ Real-time PCR Detection System. Actin was used as an internal reference gene for the quantitative PCR detection of kiwifruit resistance genes. The primer sequences used were listed in Table 1. Subsequently, cDNA was used as a template for RT-qPCR analysis. The PCR reaction conditions were set as follows: denaturation at 95 °C for 3 min, followed by 40 cycles of 95 °C for 10 s, 55 °C for 20 s, 72 °C for 20 s and 75 °C for 5 s.

Table 1. Sequences for primers used in real-time fluorescent quantitative PCR.

Gene	Gene Bank Accession Number	Forward Primer (5′-3′)	Reference
Actin	FG520231	F: GCAGGAATCCATGAGACTACC R: GTCTGCGATACCAGGGAACAT	[21]
SOD	FG471220	F: CACAAGAAGCACCACCAGAC R: TCTGCAATTTGACGACGGTG	[21]
POD	FJ422811	F: TCTGTCGTCTTCTGTTTGTATGG R: CTCCTCCTTTGAGAGGGTTATTG	[22]
CAT	FG470670	F: GCTTGGACCCAACTATCTGC R: TTGACCTCCTCATCCCTGTG	[21]

Table 1. *Cont.*

Gene	Gene Bank Accession Number	Forward Primer (5′-3′)	Reference
PAL	AAC18870	F: AAACGACAACCCCTTGATTG R: ACAAGCTCCGAAATTTGTGC	[23]

2.9.4. Determination of ATP, ADP and AMP Content and Energy Charge

The content of ATP, ADP and AMP was measured using the methods described by Liu et al. [24] with slight modifications: 4 g kiwifruit samples were weighed, and 3 mL of 0.6 mol/L perchloric acid solution was added. The samples were extracted for 1 min in an ice bath and centrifuged at 3000 r/min for 15 min at 4 °C, and 3 mL of the supernatant was neutralized with 1 mol/L potassium hydroxide solution to pH 6.5~6.8. Potassium perchlorate was precipitated for 30 min under ice-bath conditions and then centrifuged at 4 °C and 3000 r/min for 10 min. The supernatant was filtered through a 0.45 μm filter membrane and stored at 4 °C for testing. The supernatant obtained was analyzed with an HPLC system (UltiMate 3000, Thermo Fisher Scientific, Shanghai, China) coupled with an Agilent ZORBAX SB-C18 column (250 mm × 4.6 mm, 5 μm; Agilent Technologies Inc., CA, USA) via binary linear gradient elution. Results were expressed as mg/kg fresh weight. The energy charge (EC) was calculated using the following formula:

$$EC = ([ATP] + 1/2[ADP])/([ATP] + [ADP] + [AMP]), \quad (2)$$

2.10. Data Analysis

Data analyses were performed using SPSS. 26. 0 Software, and statistical differences between experimental groups were calculated using the Duncan test ($p < 0.05$) and ANOVA.

3. Results

3.1. Identification and Microbial Community Diversity Analysis

In the present study, we isolated four pathogens from the obtained kiwifruits in August 2020. These were *Diaporthe* sp., *Botryosphaeria dothidea*, *Phomopsis* sp. and *Alternaria* sp. (Figure S1). During the following year, high-throughput sequencing was used to evaluate the diversity of fungal communities, and the four pathogens mentioned above were still detected, in proportions of *Alternaria* sp. (22.11%), *Diaporthales* sp. (5.22%), *Phomopsis* sp. (2.17%) and *Botryosphaeria* sp. (0.60%), respectively. Therefore, four pathogens were used in further experiments.

3.2. In Vitro Antifungal Activity

Through preliminary screening, we found that SD, PM, SM and lysozyme showed strong antifungal activity against the tested pathogens among 18 food additives at the concentration of 1 g/L (Figure S2). Furthermore, these four food additives inhibited the mycelial growth of four pathogens in a concentration-dependent manner (Table 2). Compared with SD and lysozyme, both PM and SM displayed higher inhibitory activities. The MIC values of both PM and SM against *Diaporthe* sp., *Botryosphaeria dothidea*, *Phomopsis* sp. and *Alternaria* sp. were determined to be 0.25, 0.75, 0.5 and 0.5 g/L, and the MFC values were 0.5, 0.75, 1 and 1 g/L, respectively. Increased dietary sodium intake and decreased dietary potassium intake were associated with higher blood pressure. To decrease the risk of hypertension, heart disease and stroke, the 2010 Dietary Guidelines recommend reducing sodium intake and choosing foods that contain potassium. In addition, PM was reported to possess better antifungal activity compared with SM in citrus [5]. Therefore, PM was chosen for subsequent use in vivo antifungal experiments.

Table 2. Effect of different food additive concentrations on the mycelial growth of *Diaporthe* sp., *Botryosphaeria dothidea*, *Phomopsis* sp. and *Alternaria* sp. incubated at 25 ± 2 °C for 2~8 d.

Food Additive	Concentration (g/L)	Inhibition of *Diaporthe* sp. (%)		Inhibition of *Botryosphaeria dothidea* (%)		Inhibition of *Phomopsis* sp. (%)		Inhibition of *Alternaria* sp. (%)	
		Day2	Day8	Day2	Day8	Day2	Day8	Day2	Day8
SD	0.05	73.71 ± 0.43^{d}	20.09 ± 0.1^{e}	94.73 ± 0.06^{b}	61.40 ± 0.08^{d}	88.36 ± 0.37^{c}	16.54 ± 0.52^{e}	100 ± 0.00^{a}	11.71 ± 0.43^{b}
	0.25	93.27 ± 0.28^{c}	60.52 ± 0.33^{d}	100 ± 0.00^{a}	89.83 ± 0.05^{c}	94.30 ± 0.21^{b}	52.82 ± 0.22^{d}	100 ± 0.00^{a}	100 ± 0.00^{a}
	0.5	96.09 ± 0.18^{b}	71.38 ± 0.14^{c}	100 ± 0.00^{a}	97.25 ± 0.52^{b}	100 ± 0.00^{a}	84.63 ± 0.36^{c}	100 ± 0.00^{a}	100 ± 0.00^{a}
	0.75	100 ± 0.00^{a}	92.28 ± 0.15^{b}	100 ± 0.00^{a}	100 ± 0.00^{a}	100 ± 0.00^{a}	94.21 ± 0.41^{b}	100 ± 0.00^{a}	100 ± 0.00^{a}
	1	100 ± 0.00^{a}	100 ± 0.00^{a}	100 ± 0.00^{a}	100 ± 0.00^{a}	100 ± 0.00^{a}	100 ± 0.00^{a}	100 ± 0.00^{a}	100 ± 0.00^{a}
PM	0.05	63.83 ± 0.63^{b}	10.32 ± 0.21^{c}	44.46 ± 0.33^{d}	11.81 ± 0.18^{d}	34.28 ± 0.29^{c}	0.00 ± 0.00^{d}	88.05 ± 0.19^{c}	12.63 ± 0.22^{e}
	0.25	100 ± 0.00^{a}	18.59 ± 0.37^{b}	74.33 ± 0.26^{c}	20.76 ± 0.24^{c}	72.41 ± 0.44^{b}	0.00 ± 0.00^{d}	98.47 ± 0.11^{b}	38.38 ± 0.28^{d}
	0.5	100 ± 0.00^{a}	100 ± 0.00^{a}	80.03 ± 0.20^{b}	79.33 ± 0.65^{b}	100 ± 0.00^{a}	63.62 ± 0.34^{c}	100 ± 0.00^{a}	56.69 ± 0.28^{c}
	0.75	100 ± 0.00^{a}	100 ± 0.00^{a}	100 ± 0.00^{a}	100 ± 0.00^{a}	100 ± 0.00^{a}	97.18 ± 0.25^{b}	100 ± 0.00^{a}	63.57 ± 1.79^{b}
	1	100 ± 0.00^{a}	100 ± 0.00^{a}	100 ± 0.00^{a}	100 ± 0.00^{a}	100 ± 0.00^{a}	100 ± 0.00^{a}	100 ± 0.00^{a}	100 ± 0.00^{a}
SM	0.05	52.38 ± 0.32^{b}	11.33 ± 0.32^{c}	33.19 ± 0.33^{d}	8.93 ± 0.32^{d}	30.38 ± 0.33^{c}	9.83 ± 0.19^{e}	88.49 ± 0.16^{c}	10.21 ± 0.44^{e}
	0.25	100 ± 0.00^{a}	33.43 ± 0.33^{b}	46.74 ± 0.46^{c}	15.07 ± 0.24^{c}	70.69 ± 0.39^{b}	15.63 ± 0.39^{d}	99.31 ± 0.09^{b}	34.92 ± 0.29^{d}
	0.5	100 ± 0.00^{a}	100 ± 0.00^{a}	73.41 ± 0.00^{b}	69.12 ± 0.00^{b}	100 ± 0.00^{a}	36.56 ± 0.00^{c}	100 ± 0.00^{a}	74.57 ± 0.47^{c}
	0.75	100 ± 0.00^{a}	100 ± 0.00^{a}	100 ± 0.00^{a}	100 ± 0.00^{a}	100 ± 0.00^{a}	79.85 ± 0.00^{b}	100 ± 0.00^{a}	84.42 ± 0.44^{b}
	1	100 ± 0.00^{a}	100 ± 0.00^{a}	100 ± 0.00^{a}	100 ± 0.00^{a}	100 ± 0.00^{a}	100 ± 0.00^{a}	100 ± 0.00^{a}	100 ± 0.00^{a}
Lysozyme	0.05	73.89 ± 0.22^{d}	20.62 ± 0.26^{e}	96.68 ± 0.34^{b}	50.41 ± 0.43^{c}	68.51 ± 0.20^{d}	16.82 ± 0.26^{e}	100 ± 0.00^{a}	100 ± 0.00^{a}
	0.25	93.52 ± 0.42^{c}	52.66 ± 0.32^{d}	100 ± 0.00^{a}	94.41 ± 0.36^{b}	93.93 ± 0.24^{c}	53.70 ± 0.27^{d}	100 ± 0.00^{a}	100 ± 0.00^{a}
	0.5	97.52 ± 0.32^{b}	69.44 ± 0.32^{c}	100 ± 0.00^{a}	100 ± 0.00^{a}	96.54 ± 0.30^{b}	75.88 ± 0.40^{c}	100 ± 0.00^{a}	100 ± 0.00^{a}
	0.75	100 ± 0.00^{a}	85.55 ± 0.37^{b}	100 ± 0.00^{a}	100 ± 0.00^{a}	100 ± 0.00^{a}	95.60 ± 0.22^{b}	100 ± 0.00^{a}	100 ± 0.00^{a}
	1	100 ± 0.00^{a}	100 ± 0.00^{a}	100 ± 0.00^{a}	100 ± 0.00^{a}	100 ± 0.00^{a}	100 ± 0.00^{a}	100 ± 0.00^{a}	100 ± 0.00^{a}

Note: Data are the means of three replicates ± standard deviation. In columns, different superscript letters (a–e) at the same shelf time indicate a significant difference according to Duncan's test ($p < 0.05$).

3.3. *In Vivo Antifungal Activity*

Different PM concentrations (1 × MFC, 5 × MFC, 10 × MFC) of pullulan were selected for vivo experiments based on vitro experiments. Figure 1 shows the effects of coating on the lesion diameter (A) and visual appearances (B) of kiwifruits inoculated with four pathogens when stored at 25 °C for 8 days. On the 8th day, PM had a significant effect on the control of lesion extension compared to CK and PU treatments. The higher the concentration was, the more obvious the inhibitory effect became. It was noteworthy that pullulan treatment alone promoted the growth of kiwifruit lesions. This may be because pullulan as a biopolysaccharide provided carbon for the growth of pathogenic fungi. As shown in Figure 1, pullulan + 10 g/L PM (T3) coating proved to be the most effective option for significantly inhibiting the growth of the four tested pathogens with respect to the other treatments.

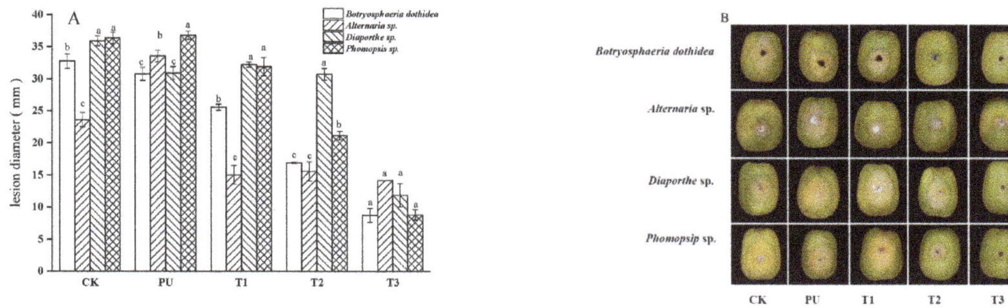

Figure 1. Effect of coating on the lesion diameter (**A**) and visual appearance (**B**) of kiwifruits inoculated with four pathogens and stored at 25 °C for 8 days (CK: sterile water, PU: pullulan, T1: PU + 1 g/L PM, T2: PU + 5 g/L PM, T3: PU + 10 g/L PM). Differences are indicated by different letters according to Duncan's multiple range test ($p < 0.05$).

3.4. *Kiwifruit Quality*

3.4.1. Incidence of Decay and Visual Appearance

The appearance of kiwifruit samples was evaluated in order to observe the rot symptoms of the fruit more clearly. For this purpose, the fruit was photographed on the 8th day after peeling (Figure 2). After 60 days of storage at a low temperature, groups treated with pullulan coating showed bright appearances, with no significant black spots on the surface, compared with the CK and PU groups (60 days). In the CK group, visible decay appeared after four days of shelf life, whereas black spots appeared in the pullulan + PM coating treatment group after only eight days of shelf life. On the 8th day, the CK and pullulan treatment groups showed obvious disease symptoms. As shown in Figure 3, all treatments appeared to decay after 60 days of storage at a low temperature and further showed an increasing trend throughout the period of shelf-life storage. An increase in the decay rate of kiwifruit was significantly inhibited by coating treatment. On the contrary, the PU treatment alone increased the decay rate in kiwifruit. This may be because pullulan, as a biological polysaccharide, provided carbon for the growth of pathogenic bacteria and increased the incidence of soft rot. In 8 days of shelf-life, CK decay incidence increased from 38.27% to 67.79%, and PU decay incidence increased from 47.95% to 77.89%, whereas pullulan + 10g/L PM (T3) coating treatment decay incidence only increased from 16.13% to 21.12%. As a result, pullulan + PM coating treatment was effective at controlling soft rot of kiwifruits, while PU treatment had no practical significance for the storage of kiwifruit. Therefore, the physical and chemical indexes of the PU group were not determined in subsequent experiments.

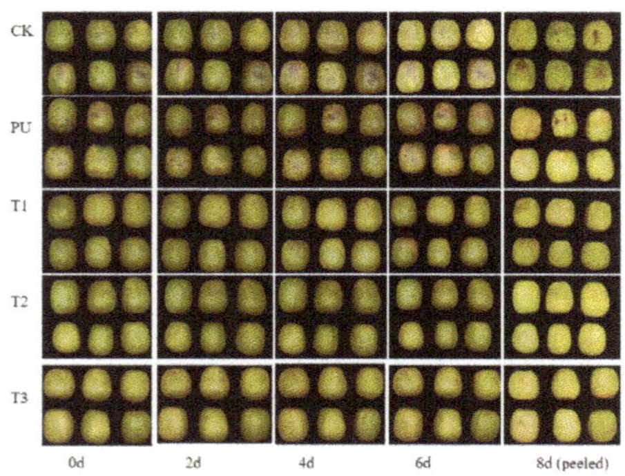

Figure 2. Visual appearance of kiwifruits without and with pullulan coating during storage at 2 ± 0.5 °C for 60 days and 8 days at 25 ± 1 °C (CK: sterile water, PU: pullulan, T1: PU + 1 g/L PM, T2: PU + 5 g/L PM, T3: PU + 10 g/L PM).

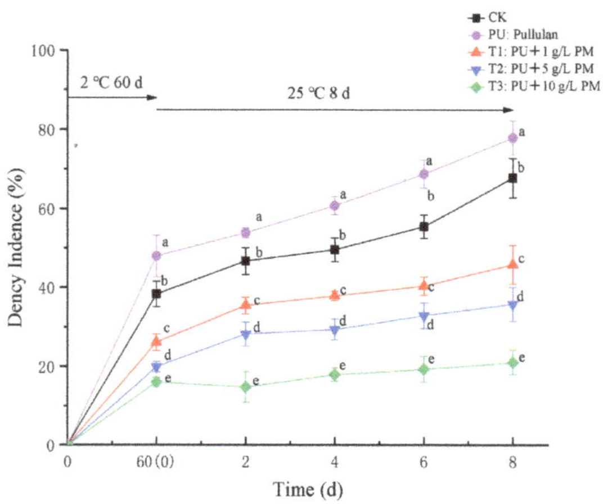

Figure 3. Decay incidence in uncoated and coated kiwifruits during storage at 2 ± 0.5 °C for 60 days and 8 days at 25 ± 1 °C. Differences are indicated by different letters according to Duncan's multiple range test ($p < 0.05$).

3.4.2. Effect of Pullulan Coating Combined with PM on Firmness, Respiration Rate, SSC, Weight Loss, Ethanol Content and Water Mobility in Harvested Kiwifruit

Firmness is one of the important indicators in evaluating the quality of kiwifruit. As shown in Figure 4A, after 60 days of cold storage, the firmness was reduced from an initial 39.6 N to 0.49, 0.71, 0.83 and 0.99 N in CK, T1, T2 and T3, respectively. When the fruit was transferred to room temperature storage, the fruit firmness still decreased with the extension of shelf life. Firmness was retained significantly better by pullulan with different concentrations of PM coating, achieving 8 days of shelf-life storage compared to CK ($p < 0.05$). However, there were no significant differences among the concentrations of PM.

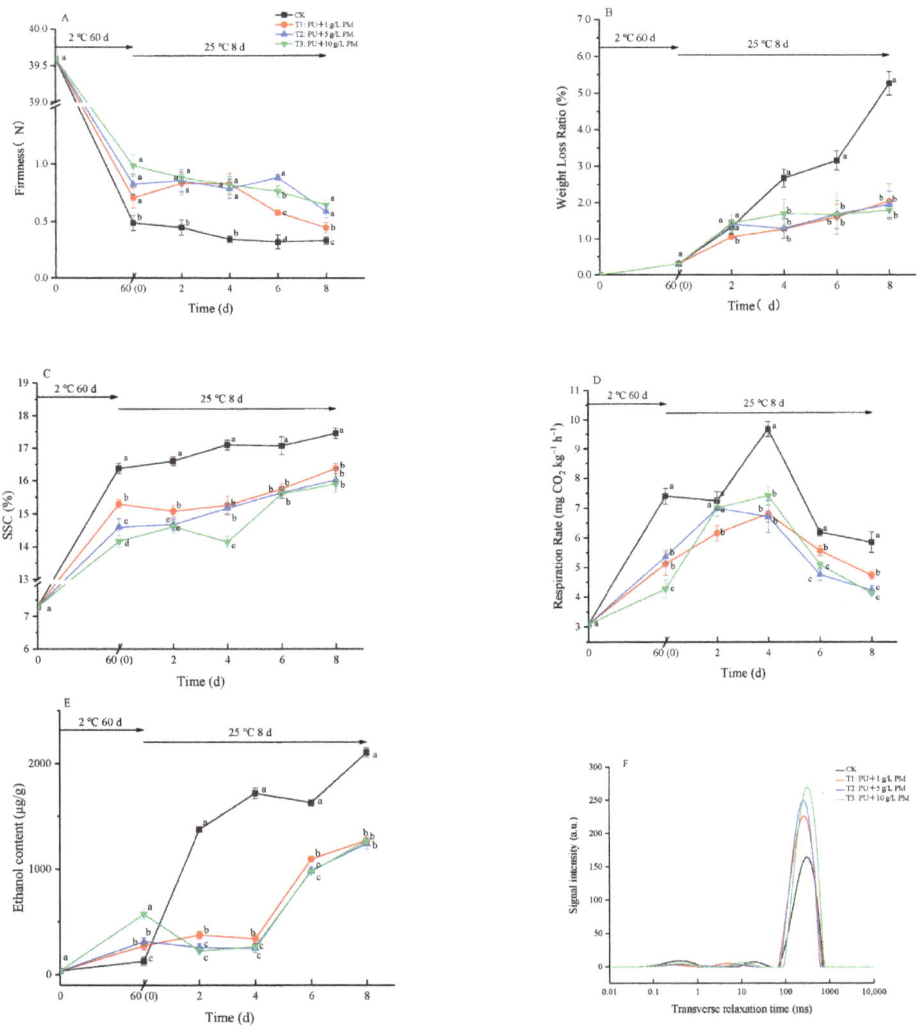

Figure 4. Effect of PU + 1 g/L PM, PU + 5 g/L PM, PU + 10 g/L PM and CK on firmness (**A**), weight loss (**B**), SSC (**C**), respiration rate (**D**) and ethanol content (**E**) during storage at 2 ± 0.5 °C for 60 days and 8 days at 25 ± 1 °C, and on water mobility (**F**) on day 8 of shelf life at 25 °C after being stored up to 60 days at 2 ± 0.5 °C. Differences are indicated by different letters according to Duncan's multiple range test ($p < 0.05$).

Weight loss is considered as being among the important reasons for the decline in fruit appearance and quality, issues that are primarily caused by water loss and nutrition consumption during transpiration and respiration. As shown in Figure 4B, weight loss showed an upward trend with the extension of storage time, which was consistent with the decay incidence. The weight loss of fruit began to increase rapidly on the 4th day. At the same time, the fruit surface of the CK group showed a shrinkage phenomenon (Figure 2). From the 4th to 8th day of shelf life, the weight loss of the CK group was significantly higher than that of the PM coating treatment group. However, there were no significant differences among the various concentration groups.

In Figure 4C, SSC is shown to continuously increase in all groups as storage time progresses. For the CK group, SSC increased rapidly, and it reached its highest level (17.46%) on the 8th day. This was significantly higher than that of the coating treatments ($p < 0.05$), indicating that pullulan + PM coating treatment delays fruit ripening. Similar to the results of firmness and weight loss, there were no significant differences between various concentration groups.

It was reported that fruits with lower respiration rates had higher storage lives than those with higher respiration rates [25]. After 60 days of refrigeration, the respiration rate of the CK group was much higher than that of the coating treatment group. During the shelf-life storage, the respiration rate of kiwifruit first increased and then decreased. All groups reached a peak on the 4th day, and the peaks were 9.68 (CK), 6.82 (pullulan + 1 g/L PM), 7.13 (pullulan + 5 g/L PM) and 7.42 (pullulan + 10 g/L PM) mg kg^{-1} h^{-1}. The coating treatment significantly reduced the peak of respiration in kiwifruit (Figure 4D).

The excessive accumulation of ethanol results in off-flavor formation, which reduces consumers' acceptance [26]. According to Figure 4E, the ethanol content in the CK group increased rapidly after the products were removed from refrigerated conditions, and the content reached a maximum value of 2101.86 μg/g on the 8th day. As for the coating treatment group, the ethanol content remained relatively stable for the first 4 days and began to rise on the 6th day. After 8 days of shelf life, the ethanol contents of the coating treatment groups were 1240.52~1271.55 μg/g and were significantly lower than the level of the CK group (2101.86 μg/g). These results indicated that coating treatment could significantly reduce the ethanol content compared with the CK group.

For kiwifruit, the moisture content and existing state are important indexes of quality. The distribution curve of transverse relaxation time on the 8th day of kiwifruit growth was determined and is shown in Figure 4F. The CK group had the lowest value of maximum amplitude, which was 164.55MS, i.e., significantly lower than that of pullulan coating combined with PM treatment. As PM concentrations increased, the value of maximum amplitude continued to change, with values of 219.41 (T1), 237.94 (T2) and 269.67 (T3) ms for samples coated with pullulan + PM. Therefore, pullulan + PM coating can delay the decline in free water and preserve the quality of kiwifruit. The higher the concentration is, the more obvious the effect will be.

3.4.3. Effect of Pullulan Coating Combined with PM Treatment on Total Phenolic Content and Flavonoid Content in Harvested Kiwifruit

As shown in Figure 5A,B, after 60 days of refrigeration, the content of total phenolics and flavonoids in kiwifruit increased significantly. On the 0th day of shelf life, the total phenolic contents of the pullulan + 5 g/L PM and pullulan + 10 g/L PM coating were significantly higher than those of the CK and pullulan + 1 g/L PM coating group. On the 2nd day, the total phenols of all treatments sharply decreased and maintained a downward trend throughout the shelf life. The coating treatment significantly inhibited this downward trend. On the 8th day, the total phenolic content of the pullulan + 10 g/L PM coating group was 37.59% higher than that of the CK group. Similarly, a significant decrease in the content of flavonoids was also observed in all treatments during shelf life. On the 8th day, the flavonoid content of the pullulan + 10 g/L PM coating group was 9.28% higher than that of the CK group.

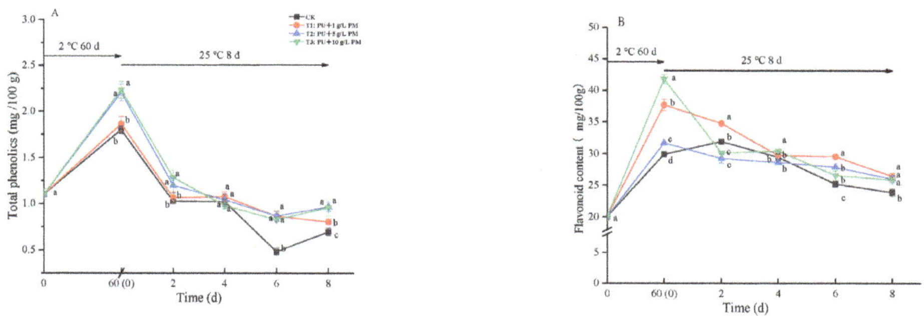

Figure 5. Effect of PU + 1 g/L PM, PU + 5 g/L PM, PU + 10 g/L PM and CK on total phenolic content (**A**) and flavonoid content (**B**) during storage at 2 ± 0.5 °C for 60 days and 8 days at 25 ± 1 °C. Differences are indicated by different letters according to Duncan's multiple range test ($p < 0.05$).

3.4.4. Effect of Pullulan Coating Combined with PM Treatment on Defense-Related Enzymes in Kiwifruit

Various biotic and abiotic stresses can cause plants to produce defense-related enzymes that increase their resistance to pathogens [27]. The activities of SOD, POD, CAT and PAL during kiwifruit sample storage at 2 ± 0.5 °C for 60 days and 25 ± 1 °C for 8 days are shown in Figure 6. After 60 days of cold storage, SOD, POD, CAT and PAL activity significantly increased ($p < 0.05$).

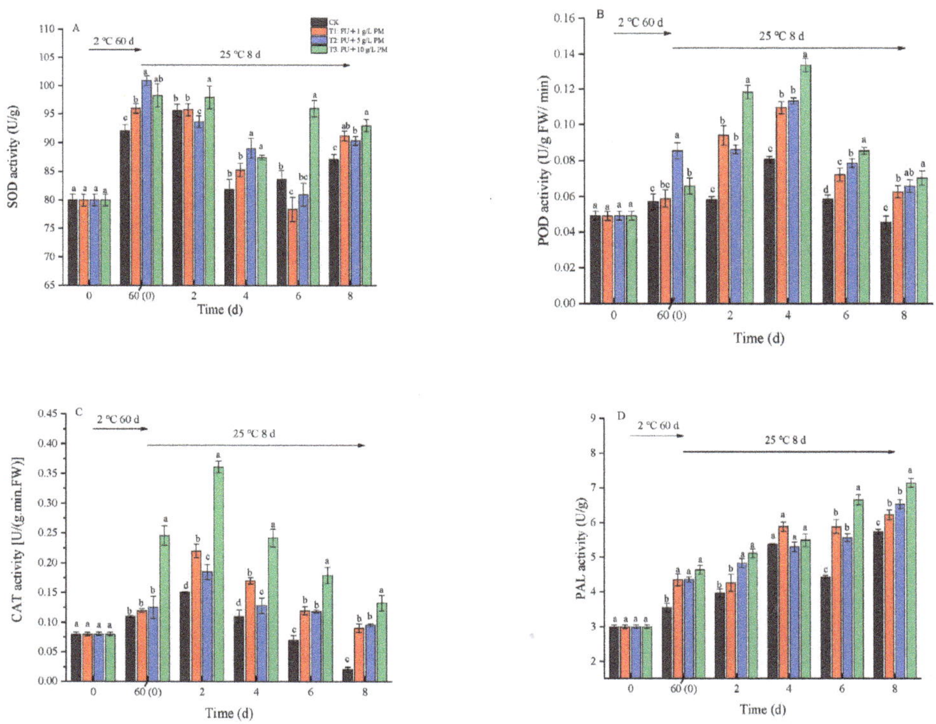

Figure 6. Effect of activities of SOD (**A**), POD (**B**), CAT (**C**) and PAL (**D**) during storage at 2 ± 0.5 °C for 60 days and 8 days at 25 ± 1 °C. Differences are indicated by different letters according to Duncan's multiple range test ($p < 0.05$).

During the shelf life, the SOD activity fluctuated over time (Figure 6A). However, pullulan + PM coating treatment remained at a higher level compared to the CK group during the whole experiment period in a concentration-dependent manner, except for on the 4th day. As shown in Figure 6B, the POD activity exhibited an increasing and then decreasing trend during shelf life. However, the pullulan + PM coating treatment significantly increased POD activity compared to the CK group. The POD activity of kiwifruit reached its peak on the 4th day, and the POD activity of the pullulan + 10 g/L PM coating group (0.13 U/g FW/min) was 62.50% higher than that of the CK group (0.08 U/g FW/min).

The CAT activity followed a similar trend to POD (Figure 6C). The initial CAT activity of the fruit was 0.08 U/g FW/min. Thereafter, the CK group attained peak values (0.15 U/g FW/min) on day 2 of shelf life, whereas pullulan + 10 g/L PM-coated fruit expressed the highest CAT activities of 0.36 U/g FW/min on the 2nd day. At the end of product shelf life, the CAT activity declined to 0.02, 0.09, 0.10 and 0.13 U/g FW/min for CK, pullulan + 1 g/L PM, pullulan + 5 g/L PM and pullulan + 10 g/L PM, respectively. In addition, the coating treatment also significantly increased the PAL activity of kiwifruit during storage compared to the control group (Figure 6D). The activity of the pullulan + 10 g/L PM coating treatment group continued to increase throughout the whole storage period. On the 8th day, the PAL activity of the pullulan + 10g/L PM coating group (7.14 U/g) was 24.61% higher than that of the CK group (5.73 U/g).

3.4.5. Effect of Pullulan Coating Combined with PM Treatment on Gene Expression of SOD, POD, CAT and PAL in Harvested Kiwifruit

In order to better understand the disease resistance of kiwifruit treated with a pullulan + PM coating in response to pathogens, we investigated the expression level of defense-related enzyme metabolism genes using RT-qPCR. The results showed that coating treatment had a positive effect on the gene expression of these defense-related enzymes. As shown in Figure 7, kiwifruits treated with pullulan + PM exhibited higher levels of transcript accumulation of SOD, POD, CAT and PAL compared to the CK group.

During the shelf life, the expression levels of SOD and POD fluctuated as time went by. The highest transcript level of SOD was recorded on the 2nd day of shelf life in the fruits coated with pullulan + 10g/L PM, increasing to 2.13 times that in the CK group (Figure 7A). The gene expression level of POD was primarily affected by PM concentration. In other words, a higher PM concentration resulted in a greater expression level (Figure 7B). The gene expression level of CAT in all groups showed a trend of gradual increase with the advance of the storage period. At the end of storage, the level of SOD transcripts increased by 1.93-fold (CK), 2.07-fold (pullulan + 1 g/L PM), 2.03-fold (pullulan + 5 g/L PM) and 2.13-fold (pullulan + 10 g/L PM) (Figure 7C). The expression level of the PAL gene showed a trend of increasing at first and then decreasing throughout the storage process. The coating treatment groups maintained high levels on day 8, which were 2.13 (pullulan + 1 g/L PM), 3.52 (pullulan + 5 g/L PM) and 3.91 (pullulan + 10 g/L PM) times that of the CK group (Figure 7D). These results were generally consistent with enzyme activity (Figure 6).

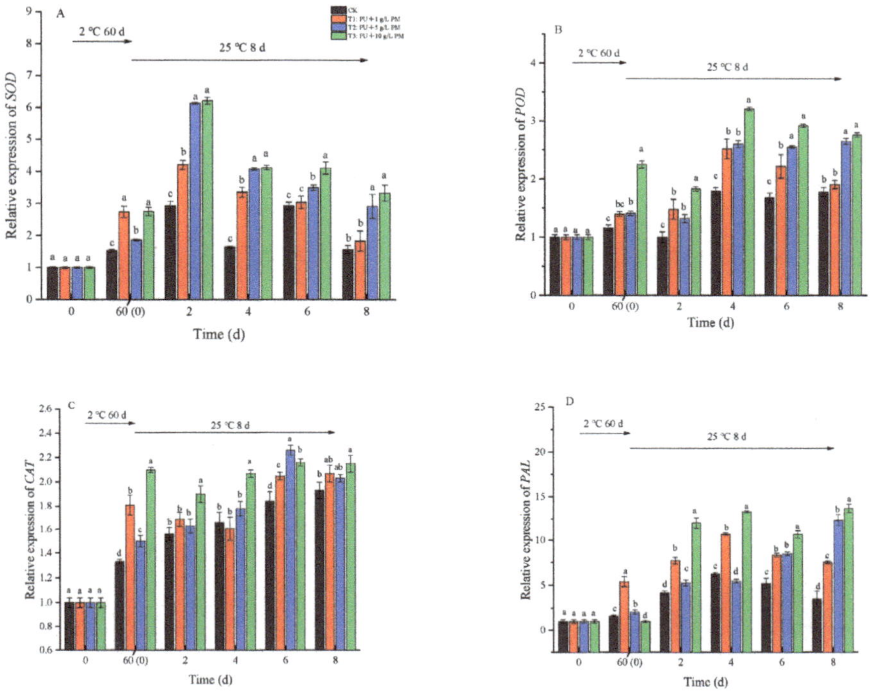

Figure 7. Effect of gene expression of *SOD* (**A**), *POD* (**B**), *CAT* (**C**) and *PAL* (**D**) during storage at 2 ± 0.5 °C for 60 days and 8 days at 25 ± 1 °C. Differences are indicated by different letters according to Duncan's multiple range test ($p < 0.05$).

3.5. ATP, ADP and AMP Contents and Energy Charge (EC)

A high level of energy status helps to maintain the integrity of the cell structure, thereby enhancing fruits' resistance to diseases. As shown in Figure 8A, with the prolongation of storage time, the ATP content in kiwifruit increased gradually, while it decreased rapidly on the 4th day of shelf life. The pullulan + PM coating treatment could significantly promote the accumulation of ATP in the early stage of storage and delayed the decline in ATP content in the later stage. In addition, the contents of ADP and AMP fluctuated over time (Figure 8B,C). The pullulan + PM coating treatment inhibited the increase in ADP and AMP in the early stage of storage and delayed the decline in ADP content in the later stage. After 60 days of cold storage, the energy charge of the CK group decreased slightly, while that of the pullulan + PM coating treatment increased. During the shelf life, the EC of all groups remained essentially stable and decreased rapidly on the 8th day. The pullulan + PM coating treatment significantly delayed this downward trend, with values that were 13.57% (pullulan + 1 g/L PM), 48.36% (pullulan + 5 g/L PM), and 51.36% (pullulan + 10 g/L PM) higher than that of CK group (Figure 8D).

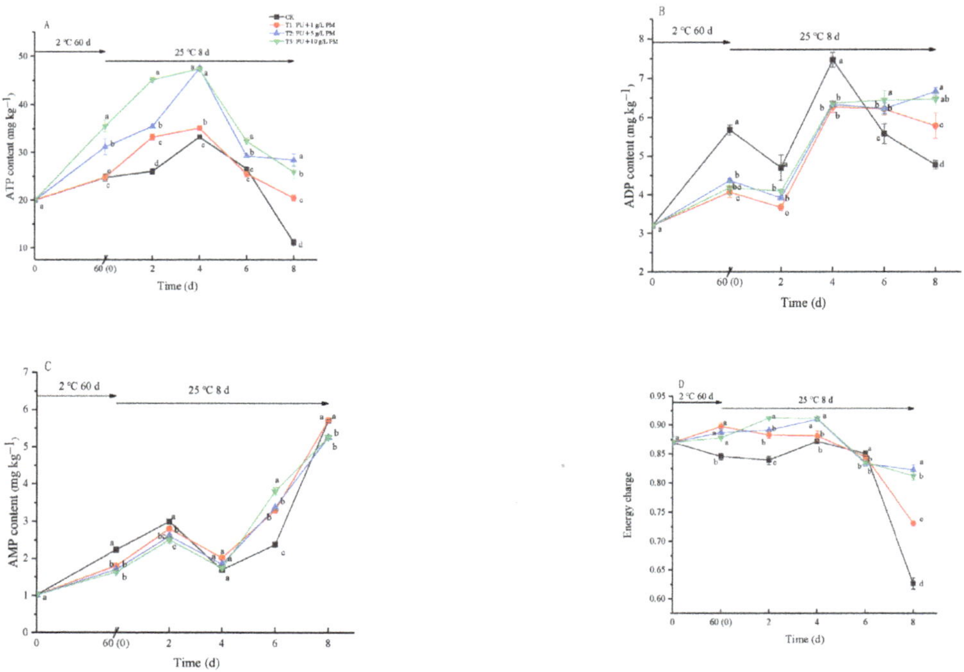

Figure 8. Effect of ATP (**A**), ADP (**B**), AMP (**C**) and energy charge (**D**) during storage at 2 ± 0.5 °C for 60 days and 8 days at 25 ± 1 °C. Differences are indicated by different letters according to Duncan's multiple range test ($p < 0.05$).

4. Discussion

Soft rot is a major concern for kiwifruit producers because it significantly reduces the quality of fruit, affecting consumers' acceptance [4]. In this study, *Diaporthe* sp., *Botryosphaeria dothidea*, *Phomopsis* sp. and *Alternaria* sp. were isolated and identified from decayed kiwifruit. They had previously been proven to be the primary pathogens causing kiwifruit soft rot [3,4].

In in vitro experiments with 18 food additives, PM proved to be the most effective fungicide for soft rot on kiwifruit. The MIC values of PM against *Diaporthe* sp., *Botryosphaeria dothidea*, *Phomopsis* sp. and *Alternaria* sp. were determined to be 0.25, 0.75, 0.5 and 0.5 g/L, and the MFC values were 0.5, 0.75, 1 and 1 g/L, respectively. It was also reported that PM could inhibit pathogenic fungi on other fruits, such as *Penicillium digitatum*, *Penicillium italicum* and *Geotrichum* citri-aurantii on citrus [5] or *Botryodiplodia theobromae*, *Colletotrichum gloeosporioides* and *Gliocephalotrichum microchlamydosporum* on rambutan [28].

When performing in vivo antifungal experiments, we were only able to verify the effectiveness of coating treatment with pullulan or with pullulan combined with different concentrations of PM (1 g/L, 5 g/L, 10 g/L). Kiwifruits, following inoculation with four pathogens, showed typical decay symptoms. Meanwhile, the fruit lesion diameters of coating treatments with pullulan combined with different concentrations of PM were less than those of the CK group. However, the coating treatment with pullulan alone promoted the growth of kiwifruit lesions in in vivo experiments (Figure 1) and increased the decay rate during storage (Figure 2), which indicated that pullulan alone had no inhibitory effect on the soft rot of kiwifruit. Therefore, we used pullulan combined with different concentrations of PM coating treatment, and sterile water was applied as a CK during storage experiments.

During postharvest low-temperature storage, kiwifruit often exhibits pathogen infection and a series of physiological and biochemical reactions such as sugar metabolism, as well as respiratory metabolic pathways, resulting in weight loss, decay and physiological disorders [29]. In the experiments in this study, with the extension of storage time, the fruits had a series of quality declines, such as a decrease in firmness, an increase in weight loss and an increase in SSC. The food additive and plant extract were used as coating formulations to effectively preserve postharvest fruits, which has been reported before. Treatment with pullulan, calcium chloride and chitosan might alter sucrose metabolism during fruit ripening by retaining firmness, surface color and total soluble solids and by regulating the metabolic pathways of *Annona squamosa* L. [30]. This study investigated the effects of the pullulan + PM coating treatments on physiological properties. The results showed that the coating reduced ripening changes in kiwifruit during storage, such as by reducing the peak value of respiratory peak and delaying the decrease in hardness and increase in SSC (Figure 4). In addition, the phenomenon observed in pullulan + PM groups, whereby the weight loss was lower and the peel shrinkage was comparatively reduced, could be the consequence of a barrier created to prevent the transpiration of fruits and gas exchange by pullulan (Figures 1 and 4).

Moreover, when kiwifruits are stored for a long time, the off-flavor caused by ethanol significantly reduces consumers' acceptance [26]. In this study, the ethanol content of the coating treatment groups was significantly higher than that of the CK group after 60 days of cold storage (Figure 4E). It was possible that the coating treatment inhibited gas exchange on the fruit surface, resulting in anaerobic respiration. In addition, the ethanol content in the CK group increased rapidly when kiwifruits were transferred from refrigerated conditions to room temperature. In the three coating treatment groups, ethanol content did not increase or decrease with temperature fluctuations, and the ethanol content in the late shelf life was significantly lower than that in the CK group. This was mainly because PU + PM treatment inhibited the infection of pathogens and reduced the rate of fruit decay. Additionally, LF-NMR was used to determine the fruit's state. The distribution curves of transverse relaxation times could reflect the distribution of water in fruits. Observing from left to right, the three primary peaks were classified as bound water, immobilized water and free water. Tylewicz et al. [31] found that the water loss of fruit mainly depended on reductions in free water. The coating treatment significantly inhibited the decrease in free water content and was also related to PM concentration (Figure 4F). This was consistent with the results of weight loss (Figure 4B) and with the fruit's appearance (Figure 2). These results showed that the coating treatment could reduce water loss by limiting the respiration and transpiration of fruit.

Total phenolics and flavonoids are among the secondary products produced by the phenylpropanoid pathway. These are very closely related to disease resistance mechanisms in plants. The contents of total phenolics and flavonoids were gradually increased during fruit ripening and decreased gradually with fruit senescence [28]. The present data showed that the pullulan + PM coating treatment enhanced the disease resistance in kiwifruit during storage (Figure 5). In similar studies, increases in the content of total phenolics and flavonoids were reported as being transmitted into litchi by coatings with pomegranate peel extract [32], and reportedly added to 'XuXiang' kiwifruit by potassium sorbate treatments [33].

A series of reactive oxygen species (ROS) are produced when fruit is subjected to biotic or abiotic stresses. SOD, POD and CAT can induce disease resistance in fruit by scavenging ROS and alleviating ROS toxicity [34]. In addition, PAL improves the disease resistance of fruit by stimulating phenylpropanoid metabolism [35]. In our experiments, POD and CAT activity exhibited an increasing and then decreasing trend during cold storage and shelf life (Figure 6B,C). The initial increase in POD and CAT activity might be the result of excessive fungal attack, which led to the increase in fruit defense against pathogens. The decline in POD and CAT activity with shelf life could be associated with fruit senescence and decay [36]. After 60 days of cold storage, POD and PAL activity increased significantly and fluctuated throughout the shelf life (Figure 6A,D). In addition, pullulan + PM coating treatments significantly increased SOD, POD, CAT and PAL activities during the entire storage period in a concentration-dependent manner (Figure 6). These results demonstrated that pullulan + PM coating treatment could be used to improve fruit disease resistance to reduce soft rot by upregulating antioxidant defense systems. This was in agreement with the study by Apel et al. [34] who found that potassium sorbate treatment reduced pathogen infection in kiwifruit by increasing the activities of POD and SOD. In this study, we also evaluated the expression of antioxidant enzyme genes. The results showed that related gene expression had a similar trend compared to antioxidant enzyme activity. The expression of all related genes of the antioxidant enzymes was upregulated after 60 days of cold storage, and pullulan + PM coating treatments induced the gene expression of SOD, POD, CAT and PPO during shelf life (Figure 7). This suggested that the mechanisms for molecular defense were triggered by pathogens and that coating treatments induced the disease resistance mechanism of fruit. Furthermore, it was also proved that pullulan coating combined with PM treatment had a positive effect on enhancing the metabolism of the antioxidant system during the whole storage period.

There is a correlation between energy metabolism and membrane damage. ATP deficiency causes lipid peroxidation, which damages the fruit's cell structural integrity and makes the fruit lose disease resistance [37]. In our experiments, the ATP content and EC of the coating-treated fruit were higher than those of CK (Figure 8). In addition, the stress response of postharvest fruits was activated in the face of pathogen stress and prevented pathogen infection by producing resistant substances. This reaction requires a lot of energy [38]. In a word, pullulan + PM coating could increase the ATP content and maintain a high energy state of kiwifruit, which is beneficial for enhancing the disease resistance of kiwifruit. Similar results had also been reported for asparagus [37] and longan [39]: higher energy metabolism helps to improve plant disease resistance.

5. Conclusions

Four pathogens were isolated from 'Hongyang' kiwifruit and identified as *Diaporthe* sp., *Botryosphaeria dothidea*, *Phomopsis* sp. and *Alternaria* sp. They had strong pathogenicity to kiwifruit and were the primary pathogens causing soft rot. The results of in vitro experiments showed PM to be the most effective fungicide among 18 food additives. In addition, the pullulan coating combined with PM treatment could significantly inhibit the growth of four pathogens in kiwifruit infected artificially as compared to CK and PU treatments. Our experiments demonstrated that pullulan coating combined with 10 g/LPM treatment could maintain the fruit quality based on visual appearance, firmness, SSC, weight loss rate, respiration rate, ethanol content and water mobility. Moreover, pullulan + PM coating treatment could improve the disease resistance of kiwifruits, mainly by enhancing the energy metabolism, increasing the content of resistant substances (total phenolics and flavonoids) and inducing the activity of disease-resistance-related enzymes (POD, SOD, CAT, PAL) and their gene expression levels. These results confirmed that pullulan-based active coating incorporating PM treatment is an excellent alternative to conventional fungicides for controlling the soft rot of postharvest kiwifruit.

Supplementary Materials: The following supporting information can be downloaded at: https://www.mdpi.com/article/10.3390/foods12173197/s1, Figure S1: Colony morphology, morphology of conidiophore and phylogenetic tree of isolates from kiwifruit; Figure S2: Inhibition rate of 18 food additives on four pathogens.

Author Contributions: Y.T. and L.L. contributed equally. Conceptualization, R.W.; methodology, Y.T., L.L. and R.W.; investigation, Y.T., L.L., N.J., X.Z., C.M. and R.W.; formal analysis, Y.T., L.L., W.G. and J.L.; data curation, Y.T. and L.L.; writing—original draft preparation, Y.T. and L.L.; writing—review and editing, Y.T. and R.W. All authors have read and agreed to the published version of the manuscript.

Funding: This research was funded by the Guizhou Province Key Technology Research and Development and Application of Innovation Base for Agricultural Products Primary Processing (Qi Ke Zhong Yin Di (2020) 4018).

Data Availability Statement: The data used to support the findings of this study can be made available by the corresponding author upon request.

Conflicts of Interest: The authors declare no conflict of interest.

References

1. Luo, A.; Bai, J.; Li, R.; Liu, Z.; Fang, Y.; Wang, D.; Huang, T.; Zhang, L.; Liang, J.; Kou, L. Difference of resistance to postharvest blue mold between Hongyang and Qihong kiwifruits. *Food Chem.* **2019**, *285*, 389–396. [CrossRef] [PubMed]
2. Pan, L.; Zhao, X.; Chen, M.; Fu, Y.; Xiang, M.; Chen, J. Effect of exogenous methyl jasmonate treatment on disease resistance of postharvest kiwifruit. *Food Chem.* **2020**, *305*, 125483. [CrossRef] [PubMed]
3. Li, L.; Pan, H.; Liu, W.; Chen, M.Y.; Zhong, C.H. First report of Alternaria alternata causing postharvest rot of kiwifruit in China. *Plant Dis.* **2017**, *101*, 1046. [CrossRef]
4. Peng, J.; Zhu, S.; Lin, X.; Wan, X.; Zhang, Q.; Njie, A.; Luo, D.; Long, Y.; Fan, R.; Dong, X. Evaluation of Preharvest Melatonin on Soft Rot and Quality of Kiwifruit Based on Principal Component Analysis. *Foods* **2023**, *12*, 1414. [CrossRef] [PubMed]
5. Martinez-Blay, V.; Taberner, V.; Perez-Gago, M.B.; Palou, L. Control of major citrus postharvest diseases by sulfur-containing food additives. *Int. J. Food Microblol.* **2020**, *330*, 108713. [CrossRef] [PubMed]
6. Shah, S.; Hashmi, M.S.; Qazi, I.M.; Durrani, Y.; Sarkhosh, A.; Hussain, I.; Brecht, J.K. Pre-storage chitosan-thyme oil coating control anthracnose in mango fruit. *Sci. Hortic.* **2021**, *284*, 110139. [CrossRef]
7. Tang, X.; Ouyang, Q.; Jing, G.; Shao, X.; Tao, N. Antifungal mechanism of sodium dehydroacetate against Geotrichum citri-aurantii. *World. J. Microb Biot.* **2018**, *34*, 29. [CrossRef]
8. Hua, C.; Li, Y.; Wang, X.; Kai, K.; Su, M.; Shi, W.; Zhang, D.; Liu, Y. The effect of low and high molecular weight chitosan on the control of gray mold (*Botrytis cinerea*) on kiwifruit and host response. *Sci. Hortic.* **2019**, *246*, 700–709. [CrossRef]
9. Chu, Y.; Gao, C.; Liu, X.; Zhang, N.; Xu, T.; Feng, X.; Yang, Y.; Shen, X.; Tang, X. Improvement of storage quality of strawberries by pullulan coatings incorporated with cinnamon essential oil nanoemulsion. *Lwt-Food Sci. Technol.* **2020**, *122*, 109054. [CrossRef]
10. Synowiec, A.; Gniewosz, M.; Krasniewska, K.; Chlebowska-Smigiel, A.; Przybył, J.L.; Bączek, K.; Węglarz, Z. Effect of meadowsweet flower extract-pullulan coatings on rhizopus rot development and postharvest quality of cold-stored red peppers. *Molecules* **2014**, *19*, 12925–12939. [CrossRef]
11. Pobiega, K.; Przybył, J.L.; Żubernik, J.; Gniewosz, M. Prolonging the shelf life of cherry tomatoes by pullulan coating with ethanol extract of propolis during refrigerated storage. *Food Bioprocess Technol.* **2020**, *13*, 1447–1461. [CrossRef]
12. Zhang, Q.; Shi, W.; Zhou, B.; Du, H.; Xi, L.; Zou, M.; Zou, H.; Xin, L.; Gao, Z.; Chen, L. Variable characteristics of microbial communities on the surface of sweet cherries under different storage conditions. *Postharvest Biol. Technol.* **2020**, *173*, 111408. [CrossRef]
13. Duan, X.; Ou Yang, Q.; Jing, G.; Tao, N. Effect of sodium dehydroacetate on the development of sour rot on Satsuma mandarin. *Food Control.* **2016**, *65*, 8–13. [CrossRef]
14. Kou, J.; Wei, C.; Zhao, Z.; Guan, J.; Wang, W. Effects of ethylene and 1-methylcyclopropene treatments on physiological changes and ripening-related gene expression of 'Mopan' persimmon fruit during storage. *Postharvest Biol. Technol.* **2020**, *166*, 111185. [CrossRef]
15. Burdon, J.; Pidakala, P.; Martin, P.; Billing, D. Softening of 'Hayward' kiwifruit on the vine and in storage: The effects of temperature. *Sci. Hortic.* **2017**, *220*, 176–182. [CrossRef]
16. Ali, M.; Raza, M.A.; Li, S.; Zhou, L.; Huan, C.; Shuling, S.; Zheng, X. 1-MCP regulates ethanol fermentation and GABA shunt pathway involved in kiwifruit quality during postharvest storage. *Hortic. Plant J.* **2021**, *7*, 23–30. [CrossRef]
17. Liu, Y.; Zeng, Y.; Hu, R.; Sun, X. Effect of contact ultrasonic power on moisture migration during far-infrared radiation drying of kiwifruit. *J. Food Process. Eng.* **2019**, *42*, e13235. [CrossRef]
18. Liu, J.; Kennedy, J.F.; Zhang, X.; Heng, C.; Chen, W.; Chen, Z.; Wu, X.; Wu, X. Preparation of alginate oligosaccharide and its effects on decay control and quality maintenance of harvested kiwifruit. *Carbohydr. Polym.* **2020**, *242*, 116462. [CrossRef]
19. Wang, Y.S.; Tian, S.P.; Xu, Y.; Qin, G.Z.; Yao, H. Changes in the activities of pro-and anti-oxidant enzymes in peach fruit inoculated with Cryptococcus laurentii or Penicillium expansum at 0 or 20 °C. *Postharvest Biol. Technol.* **2004**, *34*, 21–28. [CrossRef]
20. Lin, Y.; Lin, H.; Fan, Z.; Wang, H.; Lin, M.; Chen, Y.; Lin, Y. Inhibitory effect of propyl gallate on pulp breakdown of longan fruit and its relationship with ROS metabolism. *Postharvest Biol. Technol.* **2020**, *168*, 111272. [CrossRef]
21. Petriccione, M.; Mastrobuoni, F.; Zampella, L.; Scortichini, M. Reference gene selection for normalization of RT-qPCR gene expression data from *Actinidia deliciosa* leaves infected with *Pseudomonas syringae* pv. *actinidiae*. *Sci. Rep.-UK* **2015**, *5*, 16961. [CrossRef] [PubMed]
22. Wu, B.; Guo, Q.; Li, Q.; Ha, Y.; Li, X. Impact of postharvest nitric oxide treatment on antioxidant enzymes and related genes in banana fruit in response to chilling tolerance. *Postharvest Biol. Technol.* **2014**, *92*, 157–163. [CrossRef]
23. Wurms, K.V.; Gould, E.; Chee, A.A.; Taylor, J.; Curran, B.; Reglinski, T. Elicitor induction of defence genes and reduction of bacterial canker in kiwifruit. *N. Z. Plant Prot.* **2017**, *70*, 272–284. [CrossRef]

24. Liu, F.; Yang, N.; Zhang, L.; Cui, B.; Jin, Y.; Jin, Z. Magnetic field delays the senescence of strawberries by maintaining energy state and regulating respiratory metabolism. *Postharvest Biol. Technol.* **2023**, *199*, 112282. [CrossRef]
25. Meena, N.K.; Baghel, M.; Jain, S.K.; Asrey, R. Postharvest biology and technology of kiwifruit. In *Postharvest Biology and Technology of Temperate Fruits*; Springer: Cham, Switzerland, 2018; pp. 299–329. [CrossRef]
26. Zhu, Y.; Yu, J.; Brecht, J.K.; Jiang, T.; Zheng, X. Pre-harvest application of oxalic acid increases quality and resistance to Penicillium expansum in kiwifruit during postharvest storage. *Food Chem.* **2016**, *190*, 537–543. [CrossRef] [PubMed]
27. Hashmi, M.S.; East, A.R.; Palmer, J.S.; Heyes, J.A. Hypobaric treatment stimulates defence-related enzymes in strawberry. *Postharvest Biol. Technol.* **2013**, *85*, 77–82. [CrossRef]
28. Sivakumar, D.; Wijeratnam, R.S.; Wijesundera, R.L.C.; Abeyesekere, M. Control of postharvest diseases of rambutan using controlled atmosphere storage and potassium metabisulphite or Trichoderma harzianum. *Phytoparasitica* **2002**, *30*, 403–409. [CrossRef]
29. Riva, S.C.; Opara, U.O.; Fawole, O.A. Recent developments on postharvest application of edible coatings on stone fruit: A review. *Sci. Hortic.* **2020**, *262*, 109074. [CrossRef]
30. Qiu, J.Y.; Liu, C.Y.; Li, Y.Q.; Xie, B.Y.; Zhu, Z.Y. Effects of postharvest treatment with pullulan, calcium chloride, and chitosan on quality and sugar metabolism of Annona squamosa during storage. *J. Food Process Preserv.* **2022**, *46*, e16615. [CrossRef]
31. Tylewicz, U.; Panarese, V.; Laghi, L.; Rocculi, P.; Nowacka, M.; Placucci, G.; Dalla-Rosa, M. NMR and DSC water study during osmotic dehydration of Actinidia deliciosa and Actinidia chinensis kiwifruit. *Food Biophys.* **2011**, *6*, 327–333. [CrossRef]
32. Kumar, N.; Neeraj; Pratibha; Singla, M. Enhancement of storage life and quality maintenance of litchi (Litchi chinensis Sonn.) fruit using chitosan: Pullulan blend antimicrobial edible coating. *Int. J. Fruit Sci.* **2020**, *20*, 1662–1680. [CrossRef]
33. Ge, M.; Zhang, L.; Ai, J.; Ji, R.; He, L.; Liu, C.H. Effect of heat shock and potassium sorbate treatments on gray mold and postharvest quality of 'XuXiang' kiwifruit. *Food Chem.* **2020**, *324*, 126891. [CrossRef] [PubMed]
34. Apel, K.; Hirt, H. Reactive oxygen species: Metabolism, oxidative stress, and signal transduction. *Annu. Rev. Plant Biol.* **2004**, *55*, 373–399. [CrossRef] [PubMed]
35. Xu, F.; Liu, S.; Liu, Y.; Xu, J.; Liu, T.; Dong, S. Effectiveness of lysozyme coatings and 1-MCP treatments on storage and preservation of kiwifruit. *Food Chem.* **2019**, *288*, 201–207. [CrossRef]
36. Chen, Y.; Hung, Y.C.; Chen, M.; Lin, M.; Lin, H. Enhanced storability of blueberries by acidic electrolyzed oxidizing water application may be mediated by regulating ROS metabolism. *Food Chem.* **2019**, *270*, 229–235. [CrossRef] [PubMed]
37. Zhang, S.; Lin, H.; Lin, Y.; Lin, Y.; Hung, Y.C.; Chen, Y.; Wang, H.; Shi, J. Energy status regulates disease development and respiratory metabolism of Lasiodiplodia theobromae (Pat.) Griff. & Maubl.-infected longan fruit. *Food Chem.* **2017**, *231*, 238–246. [CrossRef] [PubMed]
38. Hu, W.; Zhang, X.; Godana, E.A.; Gu, X.; Zhao, L.; Zhang, H. Yarrowia lipolytica reduces the disease incidence of asparagus infected by Fusarium proliferatum by affecting respiratory metabolism and energy status. *Biol. Control.* **2021**, *159*, 104625. [CrossRef]
39. Chen, Y.; Lin, H.; Zhang, S.; Sun, J.; Lin, Y.; Wang, H.; Shi, J. Phomopsis longanae Chi-induced disease development and pericarp browning of harvested longan fruit in association with energy metabolism. *Front. Microbiol.* **2018**, *9*, 1454. [CrossRef]

foods

MDPI

Article

Use of HPLC-MS to Determine the Loss of Metabolites in Apple Juices under Different Storage Conditions

Jan Juhart, Aljaz Medic *, Jerneja Jakopic, Robert Veberic, Metka Hudina and Franci Stampar

Department of Agronomy, Biotechnical Faculty, University of Ljubljana, 1000 Ljubljana, Slovenia; juhartj@gmail.com (J.J.); jerneja.jakopic@bf.uni-lj.si (J.J.); robert.veberic@bf.uni-lj.si (R.V.); metka.hudina@bf.uni-lj.si (M.H.); franci.stampar@bf.uni-lj.si (F.S.)

* Correspondence: aljaz.medic@bf.uni-lj.si

Abstract: The focus of this experiment was to compare the color and metabolic profile of apple juice from the red-fleshed cultivar 'Baya Marisa' with the white-fleshed cultivar 'Golden Delicious'. The changes in the phenolic compounds, organic acids, and sugar content during high-temperature short-time pasteurization and after one year of storage under different storage conditions were analyzed. A total of 26 individual phenolic compounds were identified and quantified. The total analyzed phenolics content (TAPC) decreased after pasteurization of the juices of both cultivars. The TAPC of fresh 'Baya Marisa' juice after pasteurization increased or remained the same compared to one-year stored 'Baya Marisa' juice, depending on the storage method. The sucrose content of the apple juice of both cultivars remained the same after pasteurization; interestingly, it decreased significantly after one year of storage, while the fructose and glucose content remained the same after pasteurization and increased significantly after one year of storage for both cultivars.

Keywords: *Malus domestica* Borkh.; organic acids; sugars; phenolic compounds; pasteurized apple juice; storage; color; processing; HPLC-MS identification

1. Introduction

The phytochemical composition of apples, which includes the content of organic acids, sugars, and phenolic compounds, is known to depend on the growing season, cultivars, climatic conditions, and storage methods [1]. Apple fruits (*Malus domestica* Borkh.) are renowned for their abundant phenolic compound content, which has been associated with various health benefits for humans [2]. Six main groups of phenolic compounds are found in fresh apple fruit: hydroxycinnamic acids, dihydrochalcones, flavanols, flavonols, anthocyanins [3], and hydroxybenzoic acids [4].

Apples are highly popular among consumers because of their flavor and wide availability throughout the year, whether consumed fresh or in processed forms. A large share of apples is consumed fresh. However, apple fruits are also processed into products such as apple juice (cloudy and clear), dehydrated apple slices, apple chips—dry apple slices, vinegar, and fermented cider [4]. The content of the phenolic compounds during processing is determined using various processing techniques [4]. Fruit is processed into juices in several steps: pressing, filtering, and heating during pasteurization to minimize microbial spoilage and increase the shelf life of fruit juices [5,6].

Usually, the phenolic composition of fruits is influenced by environmental conditions and genetic factors, while storage and processing can modify the phenolic composition due to oxidative reactions [7].

The bioavailability of the phenolic compounds in apples, which are subjected to different processing techniques, may have a positive (release) or negative (degradation) effect on the bioactive compounds in apples [8].

The color of beverages is a crucial attribute because it is the first factor that consumers perceive on store shelves, and it affects the attractiveness of products [9]. The demand for red beverages is increasing because they are associated with healthy properties due to pigments such as anthocyanins. However, anthocyanins are an unstable group of phenolic compounds that are degraded during processing and storage, resulting in a brownish color [9].

The activity of two groups of enzymes, polyphenol oxidase (PPO) and peroxidase (PDO), is important for the color of fruit juices [10]. Cloudy apple juices may experience enzymatic browning, which is a problem because it affects the nutritional and sensory properties of fruit juices [10]. Browning results from the degradation of phenolic compounds to colored quinones (substrate for further reactions leading to pigment formation) as a result of oxidation-reduction enzymes [10].

Consumer interest in foods with preserved functional and nutritional properties is increasing [11]. However, to preserve apple juice and extend its shelf life, apples must undergo thermal (pasteurization or sterilization) or non-thermal (high pressure processes) treatments. Thermal treatments are the most widely used technique for preserving juices. The temperatures of thermal treatments are generally 70–90 °C for pasteurization, 110–120 °C for sterilization, and 140–160 °C for ultrahigh-temperature treatment [12]. Pasteurization is a highly effective technique for prolonging the lifespan of fruit juices, as it mitigates the detrimental impact of enzymes on the juice throughout the storage process. Processing and storage affect the brightness of cloudy apple juices, as fruit juices have been reported to darken during storage, possibly due to browning caused by the enzymes from two groups, polyphenol oxidase and peroxidase [10].

The objective of this research was to conduct a comprehensive comparison of the metabolite composition (including the phenolic compounds, sugars, and organic acids) between the red-fleshed apple cultivar 'Baya Marisa' and the traditional white-fleshed apple cultivar 'Golden Delicious' before and after pasteurization. In addition, the metabolites were also measured after one year of storage of the pasteurized apple juice in a refrigerator, with samples that were either unexposed to light or exposed to light. This research focuses on the analysis of the phenolic compounds, sugars, organic acids, and color in both fresh unpasteurized and fresh pasteurized apple juice. Additionally, it examines the characteristics of pasteurized apple juice stored for one year from 'Baya Marisa' and 'Golden Delicious' apple cultivars. These findings provide valuable insight for future studies exploring the metabolic profile of processed apples and the impact of thermal treatment on apple juice. The results are also important for consumers, as they provide information on which storage method is best in terms of metabolite and color loss and how different storage methods affect apple juice quality (metabolites, color).

2. Materials and Methods

2.1. Plant Material and Sample Preperation

The intensive orchard where both apple cultivars, 'Baya Marisa' and 'Golden Delicious', were harvested is located in the southeastern area of Slovenia, in Zdole (45°98′02″ N; 15°52′07″ E; 307 m.a.s.l.). In the orchard, both cultivars of apple trees were grafted onto M9 rootstocks and spaced at intervals of 3.2 m × 0.9 m and were in their fifth growing season. The rows in the orchard were positioned in a north–south direction. The apple trees were subjected to standard integrated pest management practices. The harvesting of 'Golden Delicious' took place on 8 September 2022, followed by the harvest of 'Baya Marisa' on 16 September 2022. Subsequently, the apples were transported to a laboratory for analysis and subsequent processing.

The analyses were performed in the laboratory of the Department of Agronomy, Biotechnical Faculty, University of Ljubljana (Slovenia), where 30 apples, 15 per cultivar, were randomly selected as replicates and further processed (juice color, organic acids, sugars, and phenolic compounds of unpasteurized and pasteurized apple juice were the subjects of this study). The cores of the apples were removed, and the apples were then pressed in a VitaExtract Slow Juicer (Bosch, Gerlingen, Germany) and filtered through MN615 Ø 150 mm filter paper (Macherey-Nagel). The yield of each apple cultivar was about 60%, 1.8 L per cultivar. There were three replicates per treatment, with each bottle representing a biological replicate (Figure 1). Three bottles of 100 mL juice of both apple cultivars were analyzed immediately after pressing. The remaining bottles were pasteurized. Three bottles per cultivar were analyzed immediately after pasteurization, while the rest were stored under standard household conditions (in a refrigerator, at room temperature unexposed to light, and at room temperature exposed to light). Pasteurization was performed by the widely known and used conventional VHTST (very high-temperature short-time) method, 85 °C for 30 s [13]. The apple juice of both cultivars was cloudy because it was not treated with the enzyme pectinase, which causes depectinization, the main substance responsible for apple juice cloudiness [14]. The unpasteurized juice was not stored and measured because it would be spoiled by microbial activity after one year of storage.

Figure 1. Apple juice in 100 mL glass bottles of cultivars 'Baya Marisa' (**left**) and 'Golden Delicious' (**right**).

2.2. Apple Juice Color Measurements

The color of the apple juice (chromatic parameters) was measured by spectrophotometric determination (Perkin-Elmer, UV-vis Lambda Bio 20 spectrophotometer, Waltham, MA, USA) for the immediately pressed unpasteurized, pasteurized apple juice, and stored pasteurized apple juice of both cultivars (refrigerated, unexposed to light, and exposed to light). There were three replicates per treatment, with 100 mL representing one biological replicate. The tonality (To) and color intensity (CI) were measured according to a previous method [15]. The color intensity was calculated using the sum of the absorbance values at 420, 520, and 620 nm in a quartz cuvette with a 10 mm optical path. The tonality indicates the ratio between yellow–orange (420 nm) and red (520 nm) pigments. The CI, To, % yellow, % red, and % blue were calculated using the following formulas:

$$CI = A420 + A520 + A620 \quad (1)$$

$$TO = A420/A520 \quad (2)$$

$$\%\ yellow = (A420/CI) \times 100 \quad (3)$$

$$\%\ red = (A520/CI) \times 100 \quad (4)$$

$$\%\ blue = (A620/CI) \times 100 \quad (5)$$

2.3. *Analysis of Sugars, Organic Acids, and Phenolic Compounds*

For the analysis of the sugars, organic acids, and phenolic compounds, the method described in previous protocols was used [16–18]. All the treatments, i.e., the various juices, were filtered using 0.2 μm cellulose filters (Chromafil Xtra MV-20/25; Macherey-Nagel, Düren, Germany) into vials for the subsequent analysis of the sugars and organic acids. For the analysis of the phenolic compounds, 0.2 μm polyamide filters (Chromafil AO-20/25; Macherey-Nagel, Düren, Germany) were used for filtration. The vials containing the filtered samples were stored at −20 °C until further analysis of the organic acids, sugars, and phenolic compounds. The ascorbic acid was extracted in a 1:1 ratio, whereby 1 mL of apple juice was mixed with 1 mL of 2% metaphosphoric acid.

2.4. *HPLC Analysis of Organic Acids and Sugars*

Commercial standards and Vanquish, a UHPLC system from Thermo Scientific (San Jose, CA, USA), were used to identify the individual organic acids and sugars. To separate the individual sugars, a Rezex RCM monosaccharide column from Phenomenex (Torrence, CA, USA) was used and operated at a temperature of 85 °C and a flow rate of 0.6 mL/min. The eluted carbohydrates were measured using a refractive index detector (Refractomax 520, Idex health and science KK 5-8-6, Nishiaoki Kawaguchi, Japan) in accordance with the methodology described by Medic et al. [19]. For the identification of the organic acids, a Rezex ROA column from Phenomenex operated at 65 °C and a UV detector set at 210 nm were used, following the same UHPLC system and methodology previously described by Medic et al. [20]. The ascorbic acid analysis was conducted at 20 °C with a UV detector at 210 nm using the same UHPLC system, based on the approach outlined by Bizjak et al. [21]. The mobile phase for the sugar analysis consisted of double-distilled water, while the mobile phase for the organic acids analysis comprised 4 mM of sulfuric acid in double-distilled water. The content of the organic acids and sugars was expressed in milligrams per liter (mg/L) of juice.

2.5. *HPLC-MS Analysis of Phenolic Compounds*

The analysis of the phenolic compounds was conducted using a UHPLC system (Thermo Scientific; San Jose, CA, USA). The diode array detector was configured at specific wavelengths: 350 nm for flavonols, 530 nm for anthocyanins, and 280 nm for other phenolic compounds. The experimental conditions adhered to the established protocols [18], and a 20 μL injection volume was utilized. The spectra were recorded within the range of 200 to 600 nm. A C18 column (Gemini; 150 × 4.60 mm, 3 μm; Phenomenex, Torrance, CA, USA) was employed for compound separation, operating at a temperature of 25 °C. The phenolic compounds were identified through tandem mass spectrometry (MS/MS; LCQ Deca XP MAX; Thermo Scientific, Waltham, MA, USA), utilizing heated electrospray ionization in positive ion mode for detecting the anthocyanins and negative ion mode for detecting the hydroxycinnamic acids, dihydrochalcones, flavanols, and flavonols. For analysis, MS scans ranging from 50 to 2000 m/z were acquired. The data acquisition was facilitated by Xcalibur 2.2 software (Thermo Fischer Scientific Institute, Waltham, MA, USA). Whenever available, external standards were employed for both identification and quantification. In cases where standards were not available, literature data and fragmentation patterns were utilized to identify the unknown compounds.

The total analyzed phenolic content (TAPC) represents the cumulative amount of all the identified phenolic compounds. Both the TAPC and individual phenolic compounds are expressed in milligrams per liter (mg/L) of juice.

2.6. *Chemicals*

The Milli-Q water purification system (Millipore, Bedford, MA, USA) was used to purify bi-distilled water for sample preparation and analysis. Metaphosphoric acid was purchased from Sigma-Aldrich (Steinheim, Germany).

The standards employed for the analysis included the following compounds: caffeic acid, citric acid, cyanidin-3-O-galactoside, (-) epicatechin, ferulic acid, fructose, fumaric acid, glucose, malic acid, sorbitol, sucrose, phloridzin, p-coumaric acid (obtained from Fluka Chemie GmbH, Buchs, Switzerland), chlorogenic acid, cryptochlorogenic acid, protocatechuic acid (purchased from Sigma-Aldrich Chemie GmbH, Steinheim, Germany); cyanidin-3-O-arabinoside and peonidin-3-galactoside, malvidin-3-glucoside, and malvidin-3-pentoside (sourced from Apin Chemicals, Abingdon, UK).

2.7. Statistical Analysis

The data collection was conducted using Microsoft Excel 2016. The statistical analysis was carried out using R commander. Significant differences among the data were determined through one-way analysis of variance (ANOVA) with Tukey tests. To establish significance ($p < 0.05$), the statistical means were calculated with a 95% confidence level. All the data are presented as means ± standard errors (SE).

3. Results and Discussion

3.1. Juice Color

One of the most important visual characteristics of apple juice is the appealing color, which is represented by the composition of the percentage of red, yellow, and blue that contribute to the overall color appearance, while the total amount of color is represented by the color intensity (Table 1).

Table 1. The color parameters (represented as mean ± standard error) of apple juice derived from the red-fleshed 'Baya Marisa' and white-fleshed 'Golden Delicious' apple cultivars were assessed both at the time of harvest and following one year of storage.

Cultivar	Color Intensity	% Yellow	% Red	% Blue	Tonality
'Baya Marisa' non-pasteurized	5.54 ± 0.01 a	43.17 ± 0.02 i	41.27 ± 0.04 b	15.57 ± 0.00 g	1.05 ± 0.00 h
'Baya Marisa' pasteurized	5.43 ± 0.01 b	43.12 ± 0.01 i	41.52 ± 0.01 a	15.63 ± 0.00 f	1.04 ± 0.00 h
'Baya Marisa' refrigerated	2.82 ± 0.00 g	44.37 ± 0.03 h	40.12 ± 0.03 c	15.52 ± 0.00 g	1.11 ± 0.00 g
'Baya Marisa' unexposed to light	3.88 ± 0.02 d	51.49 ± 0.09 e	33.87 ± 0.04 e	14.64 ± 0.12 h	1.52 ± 0.00 e
'Baya Marisa' exposed to light	5.33 ± 0.01 c	47.48 ± 0.01 g	34.67 ± 0.01 d	17.85 ± 0.00 b	1.37 ± 0.00 f
'Golden Delicious' non pasteurized	3.14 ± 0.00 e	55.72 ± 0.02 a	28.18 ± 0.01 j	16.10 ± 0.01 e	1.98 ± 0.00 a
'Golden Delicious' pasteurized	3.11 ± 0.01 e	55.61 ± 0.01 b	28.26 ± 0.01 i	16.13 ± 0.01 e	1.97 ± 0.00 a
'Golden Delicious' refrigerated	3.06 ± 0.00 f	51.22 ± 0.01 f	29.74 ± 0.00 f	19.04 ± 0.01 a	1.72 ± 0.00 d
'Golden Delicious' unexposed to light	2.22 ± 0.01 i	55.44 ± 0.02 c	28.41 ± 0.01 h	16.15 ± 0.01 d	1.95 ± 0.00 b
'Golden Delicious' exposed to light	2.44 ± 0.00 h	53.36 ± 0.01 d	29.02 ± 0.00 g	17.62 ± 0.01 c	1.84 ± 0.00 c

Statistically significant differences are presented with different letters after the mean values within the same columns. Same letters after the mean values in the same columns indicates no statistically significant differences.

The color intensity of the red-fleshed 'Baya Marisa' decreased after pasteurization, while the color intensity of the white-fleshed 'Golden Delicious' did not change after pasteurization. After one year of storage in a refrigerator, unexposed to light, and exposed to light, the color intensity of both apple juices decreased. 'Baya Marisa', which was stored under exposure to light, showed a slightly decreased color intensity compared to the pasteurized and unpasteurized apple juice of the same cultivar, but it was significantly higher than the color intensity of the apple juice stored in the refrigerator and unexposed to light. The highest color intensity of the stored samples of 'Golden Delicious' juice was in the refrigerator, and the lowest was in the samples unexposed to light. The percentage of yellow color in 'Golden Delicious' decreased after pasteurization and after one year of storage in the refrigerated samples, the samples unexposed to light, and samples that were exposed to light. It should be noted that the percentage of yellow of the stored apple juice of the 'Golden Delicious' cultivar was highest in the unexposed to light samples, but still slightly lower than the fresh pasteurized and unpasteurized apple juice of the same cultivar. The percentage of yellow in the red-fleshed 'Baya Marisa' apple juice increased after one year of storage, with the percentage of the samples unexposed to light being the highest. The percentage of red color increased slightly after pasteurization but decreased after one year of storage, with the lowest percentage of red occurring in the apple juice of 'Baya Marisa' in closed storage. The percentage of red color increased in the 'Golden Delicious' apple juice after pasteurization and continued to increase after one year of storage, with the highest percentage occurring in 'Golden Delicious' apple juice stored in a refrigerator. The percentage of blue color of the apple juice of 'Baya Marisa' increased slightly after pasteurization and was the highest in apple juice exposed to light, while it was lowest in apple juice of the same cultivar unexposed to light. In 'Golden Delicious' apple juice, the blue content did not change after pasteurization, but increased after one year of storage, with the highest blue content in the apple juice of 'Golden Delicious' stored in a cool place. The tonality was highest in 'Baya Marisa' apple juice stored in the refrigerator, while in 'Golden Delicious', the unpasteurized and freshly pasteurized apple juice had the highest tonality, i.e., the highest proportion of yellow color and the lowest proportion of red color.

Previous studies have reported comparable findings, as well as consistent results from Petrovic et al. [22], where an increase in the percentage of yellow color and a loss of red color were observed in apple liqueur after the evaluated storage time. The decrease of red color in the red-fleshed 'Baya Marisa' could be due to the precipitation of the anthocyanin pigments or the degradation of anthocyanins by oxidation [23]. On the other hand, the observed decline in the red color of both apple juices following storage may be attributed to a potential process involving the oxidation of chlorogenic acid in the presence of oxygen, leading to the formation of the corresponding quinone. This quinone then interacts with anthocyanins, resulting in the production of brown condensation products [24]. In our study, the HTST technique was used for pasteurization; however, some studies report that the loss of red color (degradation of anthocyanins) can be prevented by lower temperatures combined with a short processing time, since anthocyanins are more stable at lower temperatures [24]. Since the red color in the apple juice was more intense in the pasteurized apple juices from 'Baya Marisa' stored at lower temperatures (in a refrigerator), this suggests a more rapid degradation of anthocyanins at higher temperatures [24].

3.2. Content of Sugars and Organic Acids

The sugar and organic acid composition, as well as their ratio, play a crucial role in determining the quality and taste of fruit juices [25]. In the apple juice of these two cultivars, four different sugars and three different organic acids were identified. Table 2 presents the content of the sugars and organic acids in freshly pressed unpasteurized apple juice, freshly pasteurized apple juice, and pasteurized juice stored for one year in the refrigerator, both unexposed to light and exposed to light.

The sucrose content of the fresh and pasteurized juice of 'Baya Marisa' was similar, while the sucrose content decreased significantly after one year of storage in the refrigerator, unexposed to light, and exposed to light, with the content being the same among the treatments. The sucrose content of the fresh pasteurized and unpasteurized apple juice of 'Golden Delicious' was the same. After one year of storage, the sucrose content continued to decrease, with the highest content occurring in apple juice stored in the refrigerator compared to the other treatments. The sucrose content of the apple juice from 'Golden Delicious' stored unexposed to light and exposed to light was the same. Similar results were reported by Mezey and Meyezová [26], where a loss of 60.5% of the sucrose content in apple juice was observed after 140 days of cold storage. We suggest that the decrease in the sucrose content was partly attributable to the inversion of sucrose into glucose and fructose due to activity of enzyme invertase [27], as the content of glucose and fructose increased after one year of storage in both cultivars. After one year, the pasteurized apple juice from the 'Golden Delicious' cultivar exhibited a higher overall fructose content compared to the 'Baya Marisa' cultivar. The sorbitol content was the same across all the treatments and between the cultivars.

Table 2. Individual sugars and organic acids (mean ± SE, in g/L) in unpasteurized, freshly pasteurized, and pasteurized apple juice stored for one year in a refrigerator, unexposed to light, and exposed to light, of cultivars 'Baya Marisa' and 'Golden Delicious'.

	'Baya Marisa'					'Golden Delicious'				
Compounds	Fresh	Pasteurization	Refrigerator	Unexposed to Light	Exposed to Light	Fresh	Pasteurization	Refrigerator	Unexposed to Light	Exposed to Light
Sugars										
Sucrose	43.11 ± 5.94 a	42.81 ± 4.02 a	5.19 ± 0.30 d	4.38 ± 0.30 e	4.08 ± 0.60 e	41.19 ± 3.84 a	37.71 ± 5.22 a	21.27 ± 0.60 b	8.67 ± 0.06 c	8.58 ± 0.15 c
Glucose	20.97 ± 1.77 b	21.57 ± 1.59 b	25.17 ± 0.66 a	25.53 ± 0.87 a	25.95 ± 0.57 a	23.94 ± 2.82 b	23.58 ± 1.17 b	26.4 ± 0.18 a	28.83 ± 0.11 a	29.79 ± 0.15 a
Fructose	56.31 ± 0.33 c	58.51 ± 0.93 c	67.26 ± 0.33 b	66.93 ± 0.96 b	65.07 ± 0.22 b	68.04 ± 0.36 b	65.85 ± 0.33 b	73.5 ± 0.39 a	71.16 ± 0.60 a	69.54 ± 0.69 a
Sorbitol	6.93 ± 1.11 a	7.08 ± 0.66 a	15.76 ± 0.57 a	5.67 ± 0.54 a	5.76 ± 0.51 a	6.36 ± 2.31 a	6.27 ± 0.57 a	5.70 ± 0.57 a	5.70 ± 0.60 a	5.82 ± 0.57 a
Organic acids										
Citric acid	6.03 ± 0.87 a	5.73 ± 0.36 a	8.16 ± 1.77 a	7.05 ± 0.48 a	6.93 ± 0.36 a	4.74 ± 0.81 b	4.35 ± 0.57 b	5.37 ± 0.36 b	4.77 ± 0.72 b	5.01 ± 1.71 b
Malic acid	37.5 ± 4.98 b	32.94 ± 5.43 b	62.85 ± 3.24 a	58.80 ± 3.36 a	54.96 ± 3.48 a	26.67 ± 3.06 c	24.96 ± 3.27 c	57.87 ± 3.57 a	56.97 ± 3.21 a	56.64 ± 3.18 a
Ascorbic acid	0.018 ± 0.003 a	0.012 ± 0.003 b	0.003 ± 0.000 c	0.003 ± 0.000 c	0.003 ± 0.000 c	0.015 ± 0.003 a	0.009 ± 0.003 b	0.003 ± 0.000 c	0.003 ± 0.000 c	0.003 ± 0.000 c

Different letters between treatments (within the same rows) following mean values are significantly different ($p < 0.05$).

The overall citric acid content was higher in the apple juice of 'Baya Marisa' than in the juice of the 'Golden Delicious', but it did not change significantly in either cultivar after pasteurization and after one year of storage in the refrigerator, unexposed to light, and exposed to light. Similar results were obtained by Farnworth et al. [28], in which no changes in the citric acid content were measured in pasteurized and unpasteurized fruit juices and during storage.

The malic acid content was higher in the fresh pasteurized and unpasteurized apple juice of 'Baya Marisa' than in the same treatments of 'Golden Delicious'; no differences were found after pasteurization. The malic acid content in the apple juices of both cultivars increased after one year of storage in the refrigerator, exposed to light, and unexposed to light. These results are in agreement with the earlier findings of Rehman et al. [29], in which the highest malic acid content was found in apple juice at the end of the 30-day storage period.

The ascorbic acid content decreased after pasteurization of the fresh apple juices of both cultivars. After one year of storage, the ascorbic acid content in the apple juices of both cultivars decreased significantly. The loss of the ascorbic acid could partly be due to the heat treatment and air in the headspace of the 100 mL glass bottles during storage [30]. On the other hand, enzymes such as ascorbic acid oxidase, peroxidase, and cytochrome oxidase are also responsible for the loss of ascorbic acid [30]. The degradation of ascorbic acid by heat treatment and storage was also reported by Tikekar et al. [31].

3.3. Identification of Individual Phenolic Compounds

Based on the available literature and mass spectra, a comprehensive analysis identified a total of 26 distinct phenolic compounds in the unpasteurized juice, pasteurized juice, and pasteurized juice after one year of storage. Among the phenolic compounds, 20 hydroxycinnamic acid derivatives, 1 hydroxybenzoic acid derivative, 1 dihydrochalcone, 1 flavanol, and 3 anthocyanins were identified (Table 3). Fewer individual phenolic compounds were identified compared with our previous study on fresh apples of the same cultivar (Juhart, J. et al., 2022). We hypothesize that the phenolic compounds were degraded by the thermal processing and juice pressing.

Table 3. Tentative identification of the 26 individual phenolic compounds of apple juices of *Malus domestica* Borkh. cultivars 'Baya Marisa' and 'Golden Delicious' and used standards.

Phenolic Compounds	Rt (min)	$[M-H]^-$ (m/z)	M^+ (*m/z*)	MS^2 (*m/z*)	MS^3 (*m/z*)	Expressed as
Hydroxycinnamic acids						
p-coumaric acid hexoside derivative 1	10.2	371		325,163		*p*-coumaric acid
p-coumaric acid hexoside derivative 2	12.3	371		325,163		*p*-coumaric acid
p-coumaric acid hexoside 1	14.2	325		265,235,163		*p*-coumaric acid
Caffeic acid derivative	16.9	295		251,189,179	179,135	caffeic acid
Caffeic acid derivative 2	12.9	311		179		caffeic acid
Caffeic acid hexoside 1	10.6	341		179,135		caffeic acid
Dicaffeic acid derivative 1	10.8	457		277,185,179	179,135	caffeic acid
Dicaffeic acid derivative 2	19.3	403		233,179	179,135	caffeic acid
Dicaffeic acid derivative 3	22.6	429		249,205,179		caffeic acid
Dicaffeic acid derivative 4	24.3	473		429,249,179	249,205,179	caffeic acid
Dihydrodicaffeic acid derivative	20.5	405		225,181		caffeic acid
4-*O*-*p*-Coumaroylquinic acid	16.2	337		173,163		chlorogenic acid
5-*O*-*p*-Coumaroylquinic acid	17.1	337		191,163,119		chlorogenic acid
Chlorogenic acid (5-caffeoylquinic acid)	13.4	353		191,179		chlorogenic acid
Caffeoylferuoylquinic acid	13.9	563		517,385,321,179	205,191,103	chlorogenic acid
Ferouoylquinnic acid gallate	10.9	658		466,385,273	193,191	chlorogenic acid
Feruloylquinic acid derivative 1	12.7	431		385,331	193,191	chlorogenic acid
Feruoylquinnic acid hexoside	20.9	547		385,235	193,191	chlorogenic acid
Ferulic acid hexoside derivative	11.6	401		355	193	ferulic acid

Table 3. *Cont.*

Phenolic Compounds	Rt (min)	[M − H]⁻ (m/z)	M⁺ (m/z)	MS² (m/z)	MS³ (m/z)	Expressed as
Cryptochlorogenic acid (4-caffeoylquinic acid)	15.3	353		191,179		cryptochlorogenic acid
Hydroxybenzoic acids						
Protocatechuic acid derivative	14.2	481		327,153	153,109	protocatechuic acid
Dihydrochalcones						
Phloretin-2-*O*-xyloside	22.2	567		273,167		phloridzin
Flavanols						
(epi)catechin derivative	22.1	579		289,271,245,205		(-) epicatechin
Anthocyanins						
Cyanidin-3-*O*-galactoside	8.9		449	287		cyanidin-3-*O*-galactoside
Cyanidin-3-*O*-arabinoside	11.9		419	287		cyanidin-3-*O*-arabinoside
Peonidin-3-*O*-galactoside	11.0		463	301		peonidin-3-*O*-galactoside

Rt, retention time; $[M - H]^-$ pseudomolecular ion identified in a negative ion mode; M^+, pseudomolecular ion identified in a positive ion mode.

The information regarding all the identified phenolic compounds in the apple juice of 'Baya Marisa' and 'Golden Delicious' cultivars are presented in Table 2.

The phenolic compounds were identified by analyzing the mass-to-charge (m/z) ratio of specific molecular fragmentation ions. In cases where standard samples for phenolic compounds were not accessible, the initial identification relied on pseudomolecular ions, specifically $[M - H]^-$ and M^+ ions. Whenever possible, comparisons were made with authentic standards (Table 3).

3.4. Quantification of Individual Phenolic Compounds and Total Analyzed Phenolic Content

As shown in Table 4, the phenolic compounds were identified in all the treatments for the apple juice of both cultivars, red-fleshed 'Baya Marisa' and white-fleshed 'Golden Delicious'.

Among the identified phenolic groups in the apple juice, hydroxycinnamic acids were found to be the predominant group in both the red-fleshed 'Baya Marisa' and white-fleshed 'Golden Delicious' cultivars. Chlorogenic acid was identified as the most abundant among these hydroxycinnamic acids. However, it was observed that the content of the hydroxycinnamic acids decreased following the pasteurization process in both the 'Baya Marisa' and 'Golden Delicious' apple juices. We hypothesize that the decrease in the total hydroxycinnamic acid after pasteurization was due to oxidative reactions during heating [32]. Similar results were obtained by Tian et al. [33], who found a decrease in chlorogenic acid of up to 30% after pasteurization. On the other hand, an increase in the total hydroxycinnamic acid content after pasteurization was reported, which may be due to the inactivation of enzymes from the polyphenol oxidase group [34]. However, in the samples after one year of storage in the refrigerator, at room temperature unexposed to light, and at room temperature exposed to light, the total hydroxycinnamic acid content in the pasteurized 'Golden Delicious' apple juice increased and was higher than that of the unpasteurized 'Golden Delicious' apple juice. Similar results were obtained for the apple juice of the 'Baya Marisa' cultivar, in which the total hydroxycinnamic acid content in the refrigerated apple juice increased and was higher than in the unpasteurized fresh apple juice. The total hydroxycinnamic acid content in the 'Baya Marisa' apple juice stored unexposed to light and exposed to light also increased after pasteurization, although not as much as in the 'Golden Delicious' juice. The cumulative content of the hydroxycinnamic acids in 'Baya Marisa' apple juice stored unexposed to light increased to the same level as before pasteurization, while the apple juice stored exposed to light increased to a slightly higher level than the freshly pasteurized juice. The increase in the total hydroxycinnamic acid content is mainly due to the increase in the chlorogenic acid. We suggest that our results may be due to the release of chlorogenic acid from the cell walls into the food matrix [35], since cloudy apple juice contains particles from the pulp of apple fruits (sediments), which are known to contain the highest amounts of hydroxycinnamic acids [17]. Our results are partly in agreement with those of Deng et al. [10], in which the HTST treatment did not decrease the content of the phenolic compounds in apple juice during storage. The total hydroxycinnamic acid content was higher in the samples stored in the refrigerator, which is consistent with the results of Zhao et al. [36], in which the chlorogenic acid content was lower in the samples stored at higher temperatures.

The only hydroxybenzoic acid identified was the protocatechuic acid derivative in the apple juice of red-fleshed 'Baya Marisa'. No hydroxybenzoic acid was found in the apple juice of 'Golden Delicious'. The total hydroxybenzoic acid content decreased after pasteurization and further decreased after one year of storage unexposed to light and exposed to light. However, the total hydroxybenzoic acid content increased significantly after one year of storage in the refrigerator. As reported by Patras et al. [24], anthocyanins can be hydrolyzed by acids during processing and storage and, moreover, can be cleaved into hydroxybenzoic acids. Hydroxybenzoic acids are known to be degradation products of anthocyanins with antimicrobial activity in processed foods [37]. We suspect that the total hydroxybenzoic acid content is partly due to the processing and degradation of the anthocyanins, as it was not detected in our previous study on fresh apples of the same cultivar [38]. The higher content of the hydroxybenzoic acid in the refrigerator-stored apple juice from 'Baya Marisa' suggests that the anthocyanins were cleaved into hydroxybenzoic acid. However, these results do not agree with the color measurements, as the percent red of the stored samples was highest in the refrigerator-stored samples of 'Baya Marisa'.

Phloretin-2-*O*-xyloside was the only dihydrochalcone detected. The total dihydrochalcone content did not change significantly after pasteurization in either the 'Baya Marisa' and 'Golden Delicious' apple juice. The total dihydrochalcone content increased in all the storage methods for the pasteurized 'Golden Delicious' apple juice and for the 'Baya Marisa' apple juice stored in the refrigerator and unexposed to light, while the total dihydrochalcone content did not change in the 'Baya Marisa' juice exposed to light. We suspect that the higher content of phloretin-2-*O*-xyloside in the stored pasteurized apple juices of 'Golden Delicious' and in the refrigerator-stored apple juices of 'Baya Marisa' could be due to the release of this phenolic compound from small peel and pulp particles in the cloudy apple juice [6], from which phenols were released. As reported by Salazar-Orbea et al. [35], heat treatments and comminution increase the release of phenols from the peel and pulp. Our results are in partial agreement with those of Salazar-Orbea et al. [35], who reported that in some apple juices of different cultivars, the total dihydrochalcone content increased from 8% to 767% during processing and storage.

The only flavanol detected in the apple juices of 'Baya Marisa' and 'Golden Delicious' was the epicatechin derivative. The content of the epicatechin derivative decreased after pasteurization of both the apple juices of the selected cultivars. However, the content of the epicatechin derivative increased after one year of storage in the apple juices of 'Baya Marisa' stored in the refrigerator and unexposed to light, while the content of epicatechin derivatives in the apple juice of 'Baya Marisa' stored exposed to light did not change after pasteurization. Similar results were obtained for the pasteurized apple juice of 'Golden Delicious', in which the content of the epicatechin derivative did not change in the samples exposed to light, while it slightly increased in the samples stored in the refrigerator and unexposed to light. The low level of flavanols could be due to the destruction of the integrity of the cells during processing (pressing, mixing), which increases the exposure of flavanols to environmental factors, such as oxygen, light, enzymes, and interactions with other biomacromolecules, since this phenolic group is known to have low stability under various environmental conditions [39]. In addition, procyanidin B2 is converted by intermolecular coupling to procyanidin A2, which can be depolymerized by processing to phenolic acids. The stability of flavanols depends on the type of processing, which leads to thermal degradation, depolymerization, and polymerization [39].

Table 4. Phenolic compounds in unpasteurized and pasteurized apple juice of both apple (*Malus domestica* Borkh.) cultivars 'Baya Marisa' and 'Golden Delicious' (mean ± SE, in mg/L juice) immediately after pressing and after one year of storage in a refrigerator, at room temperature unexposed to light (RMU), and at room temperature exposed to light (RME).

	'Baya Marisa'					'Golden Delicious'				
Compounds	**Fresh**	**Pasteurization**	**Refrigerator**	**RMU**	**RME**	**Fresh**	**Pasteurization**	**Refrigerator**	**RMU**	**RME**
Hydroxycinnamic acids										
p-coumaric acid hexoside derivative 1	1.82 ± 0.01 a	0.85 ± 0.02 b	1.04 ± 0.34 b	0.25 ± 0.34 c	0.39 ± 0.15 c	nd	nd	nd	nd	nd
p-coumaric acid hexoside derivative 2	1.27 ± 0.03 a	1.00 ± 0.03 b	0.73 ± 0.09	0.24 ± 0.08	0.76 ± 0.14	nd	nd	nd	nd	nd
p-coumaric acid hexoside 1	0.25 ± 0.01 c	0.17 ± 0.04 c	1.02 ± 0.12 a	0.46 ± 0.19 b	0.43 ± 0.18 b	0.48 ± 0.07 b	0.36 ± 0.08 b	0.71 ± 0.14 a	0.74 ± 0.10 a	0.92 ± 0.25 a
Caffeic acid derivative	2.07 ± 0.11 a	1.23 ± 0.21 b	0.91 ± 0.16 c	0.56 ± 0.14 d	0.49 ± 0.09 d	0.99 ± 0.11 c	0.25 ± 0.01 e	1.47 ± 0.20 b	1.40 ± 0.17 b	0.71 ± 0.13 c
Caffeic acid derivative 2	12.1 ± 0.03 a	9.44 ± 0.02 b	nd	nd	nd	0.7 ± 0.03 c	0.7 ± 0.01 c	nd	nd	nd
Dicaffeic acid derivative 1	0.39 ± 0.02 a	0.27 ± 0.03 b	nd	nd	nd	0.22 ± 0.01 a	0.19 ± 0.01 a	nd	nd	nd
Dicaffeic acid derivative 2	0.96 ± 0.07 c	0.85 ± 0.09 c	1.94 ± 0.24 a	1.03 ± 0.11 c	0.35 ± 0.09 d	0.83 ± 0.07 c	0.68 ± 0.09 b	1.02 ± 0.04 b	0.86 ± 0.11 c	0.54 ± 0.09 d
Dicaffeic acid derivative 3	0.48 ± 0.02 b	0.39 ± 0.01 b	1.78 ± 0.25 a	0.89 ± 0.02 b	0.14 ± 0.01 c	0.47 ± 0.09 b	0.31 ± 0.00 b	0.57 ± 0.04 b	1.53 ± 0.07 a	0.29 ± 0.02 c
Dicaffeic acid derivative 4	0.23 ± 0.05 a	0.16 ± 0.03 b	nd	nd	nd	0.37 ± 0.09 a	0.19 ± 0.02 b	0.21 ± 0.02 b	0.15 ± 0.02 c	0.14 ± 0.01 c
Caffeic acid hexoside 1	1.66 ± 0.07 a	0.88 ± 0.03 b	nd	nd	nd	nd	nd	nd	nd	nd
Dihydrodicaffeic acid derivative	0.30 ± 0.02 c	0.28 ± 0.04 c	1.57 ± 0.07 a	0.56 ± 0.05 b	0.34 ± 0.02 c	0.36 ± 0.01 c	0.37 ± 0.00 c	0.47 ± 0.06 c	0.41 ± 0.03 c	0.63 ± 0.03 b
4-*O*-*p*-Coumaroylquinic acid	1.00 ± 0.03 c	0.55 ± 0.03 d	3.39 ± 0.20 a	1.80 ± 0.30 b	1.77 ± 0.27 b	0.41 ± 0.05 d	0.35 ± 0.03 d	2.23 ± 0.25 ab	1.90 ± 0.32 b	1.17 ± 0.20 b
5-*O*-*p*-Coumaroylquinic acid	0.08 ± 0.01 d	0.09 ± 0.01 d	2.47 ± 0.10 c	2.37 ± 0.73 c	1.03 ± 0.08 c	6.29 ± 0.03 b	1.75 ± 0.05 c	14.62 ± 0.65 a	12.51 ± 0.49 a	6.52 ± 0.45 b
Chlorogenic acid (5-caffeoylquinic acid)	37.08 ± 0.18 e	16.00 ± 0.11 f	67.56 ± 2.68 d	54.70 ± 1.77 e	36.00 ±1.97 e	55.47 ± 0.04 d	13.99 ± 0.08 f	124.49 ± 6.58 a	109.59 ± 4.87 b	94.29 ± 4.19 c
Caffeoylferuoylquinic acid	1.40 ± 0.11 b	0.76 ± 0.12 c	2.26 ± 0.35 a	1.83 ± 0.26 b	0.74 ± 0.17 c	1.67 ± 0.01 b	0.56 ± 0.05 c	1.37 ± 0.18 bc	1.66 ± 0.25 b	1.71 ± 0.33 b
Ferouoylquinnic acid gallate	4.84 ± 0.06 b	3.40 ± 0.08 c	2.96 ± 0.49 c	1.26 ± 0.11 d	1.5 ± 0.48 d	8.07 ± 0.02 a	6.46 ± 0.13 b	7.81 ± 0.19 b	5.98 ± 1.01 b	3.80 ± 0.71 c
Feruloylquinnic acid derivative 1	3.06 ± 0.08 a	2.91 ± 0.12 a	3.75 ± 0.79 a	4.15 ± 0.91 a	3.64 ± 0.89 a	0.94 ± 0.01 b	0.85 ± 0.04 b	1.45 ± 0.15 a	1.67 ± 0.24 a	1.73 ± 0.10 a
Ferulic acid hexoside derivative	0.35 ± 0.02 c	0.18 ± 0.03 d	1.57± 0.19 a	0.69 ± 0.18 b	0.27 ± 0.09 c	0.33 ± 0.01 c	nd	nd	nd	nd
Feruoylquinic acid hexoside	1.68 ± 0.02 a	1.29 ± 0.02 b	0.78 ± 0.09 c	0.41 ± 0.10 cd	0.13 ± 0.01 d	0.83 ± 0.07 c	0.52 ± 0.05 c	0.63 ± 0.09 c	0.64 ± 0.07 c	0.70 ± 0.09 c
Cryptochlorogenic acid (4-caffeoylquinic acid)	3.34 ± 0.05 a	2.51 ± 0.04 b	2.49 ± 0.07 b	0.68 ± 0.15 d	0.82 ± 0.04 d	1.47 ± 0.07 c	1.00 ± 0.04 c	1.17 ± 0.14 c	1.01 ± 0.04 c	0.94 ± 0.03 c
Hydroxybenzoic acids										
Protocatechic acid derivative	8.53 ± 0.09 b	5.66 ± 0.08 c	17.01 ± 0.93 a	1.54 ± 0.24 d	1.53 ± 0.24 d	nd	nd	nd	nd	nd
Dihydrochalcones										
Phloretin-2-*O*-xyloside	1.52 ± 0.25 d	1.02 ± 0.09 d	10.70 ± 0.13 a	5.74 ± 0.28 b	2.47 ± 0.37 c	1.01 ± 0.02 d	0.99 ± 0.04 d	5.96 ± 0.68 b	6.50 ± 0.48 b	6.69 ± 0.75 b
Flavanols										
(epi)catechin derivative	2.54 ± 0.09 b	0.52 ± 0.03 d	5.43 ± 0.23 a	2.11 ± 0.24 b	0.55 ± 0.12 d	1.26 ± 0.01 c	0.35 ± 0.03 d	0.78 ± 0.03 dc	0.66 ± 0.18 dc	0.55 ± 0.06 d
Anthocyanins										
Cyanidin-3-arabinoside	0.81 ± 0.13 a	0.58 ± 0.09 a	nd	nd	nd	nd	nd	nd	nd	nd
Cyanidin-3-galactoside	58.17 ± 0.19 a	19.07 ± 0.08 b	4.80 ± 0.25 c	0.50 ± 0.02 d	0.36 ± 0.04 d	nd	nd	nd	nd	nd
Peonidin-3-galactoside	0.90 ± 0.16 a	0.41 ± 0.01 b	nd	nd	nd	nd	nd	nd	nd	nd
Total hydroxycinnamic acids	74.39 ± 0.59 c	43.18 ± 0.58 de	106.90 ± 7.23 b	74.21 ± 1.89 c	51.01 ± 0.67 d	79.89 ± 0.10 c	28.49 ± 0.45 e	162.35 ± 6.75 a	148.26 ± 1.56 a	122.40 ± 2.00 b
Total hydroxybenzoic acids	8.53 ± 0.09 a	5.66 ± 0.08 b	17.01 ± 0.93 a	1.54 ± 0.24 d	1.53 ± 0.24 d	nd	nd	nd	nd	nd
Total dihydrochalcones	1.52 ± 0.25 c	1.02 ± 0.09 c	10.70 ± 0.13 a	5.74 ± 0.28 b	2.47 ± 0.37 c	1.01 ± 0.02 c	0.99 ± 0.04 c	5.96 ± 0.68 b	6.50 ± 0.48 b	6.69 ± 0.75 b
Total flavanols	2.54 ± 0.09 b	0.52 ± 0.03 d	5.43 ± 0.23 a	2.11 ± 0.24 b	0.55 ± 0.12 d	1.26 ± 0.01 c	0.35 ± 0.00 d	0.78 ± 0.03 dc	0.66 ± 0.18 dc	0.55 ± 0.06 d
Total anthocyanins	59.88 ± 0.38 a	20.07 ± 0.13 b	4.80 ± 0.10 c	0.50 ± 0.04 d	0.35 ± 0.01 d	nd	nd	nd	nd	nd
TAPC	146.86 ± 0.93 bc	70.45 ± 0.43 de	146.65 ± 8.20 bc	84.46 ± 1.78 d	56.18 ± 1.04 e	82.16 ± 0.08 d	29.83 ± 0.42 f	169.08 ± 6.08 a	155.42 ± 1.45 ab	129.64 ± 2.62 c

The data presented in the table are expressed as means ± standard error. Different letters are used to indicate significant differences ($p < 0.05$) among the means within rows for unpasteurized, pasteurized, and stored pasteurized apple juice kept in the refrigerator, at room temperature unexposed to light, and at room temperature exposed to light. The abbreviation "nd" indicates that the component was not detected.

As expected, anthocyanins were detected only in the red-fleshed and red-peeled apple juice of the 'Baya Marisa'. A comparison of the anthocyanins in the fresh unpasteurized juice, fresh pasteurized juice, and pasteurized juice stored for one year showed that this group of phenolic compounds was the most unstable, as the total anthocyanin content decreased significantly after pasteurization (heat treatment) and after one year of storage in the refrigerator, unexposed to light, and exposed to light. Similar results were reported by Knebel et al. [40], in which the content of anthocyanins in depectinized apple juice of the red-fleshed cultivar 'Maggy' decreased significantly after processing. Our results suggest that anthocyanin degradation is associated with a higher storage temperature, as the apple juice of 'Baya Marisa' stored in the refrigerator had a higher anthocyanin content than the apple juice stored unexposed to light and exposed to light, indicating the higher stability of anthocyanins in a colder environment.

The anthocyanins detected in the fresh, unpasteurized, and pasteurized apple juice were cyanidin-3-arabinoside, cyanidin-3-galactoside, and peonidin-3-galactoside. Only cyanidin-3-galactoside was found in the stored apple juice from 'Baya Marisa' with a significantly lower content. The interaction between the fourth ring (C4) of anthocyanins and the ascorbic acid could lead to color loss in different fruit juices, since the presence of ascorbic acid in anthocyanin-containing solutions can accelerate color loss by degradation. The result of this process is the absence of the original color expression of the parental pigments [41].

The most abundant group of phenolic compounds were hydroxycinnamic acids, as shown in Figure 2.

The TAPC value was higher in the fresh, unpasteurized juice of 'Baya Marisa' than in 'Golden Delicious'. The TAPC value of the apple juice for both 'Baya Marisa' and 'Golden Delicious' decreased significantly after pasteurization. However, the TAPC value of 'Golden Delicious' apple juice increased after one year of storage in the refrigerator, unexposed to light, and exposed to light, and was higher than that of the fresh apple juice before pasteurization. For the 'Baya Marisa' apple juice, the TAPC value also increased in the refrigerator and slightly increased for the apple juice unexposed to light, while the apple juice exposed to light showed the smallest decrease after one year of storage compared to the fresh juice after pasteurization. The TAPC value of the pasteurized 'Baya Marisa' apple juice stored in the refrigerator increased to the same TAPC value as before pasteurization. The 'Baya Marisa' apple juice stored in the refrigerator showed a minimal increase in the TAPC value, which was slightly higher than the fresh apple juice after pasteurization.

Overall, the TAPC value of the fresh, unpasteurized, pasteurized, and stored pasteurized apple juice of both apple cultivars was significantly lower compared to the fresh apples of these cultivars [38].

Our results show that apple juice has significantly lower levels of hydroxycinnamic acids, dihydrochalcones, flavanols, flavonols, and anthocyanins in comparison to fresh apples of the same cultivar, which were evaluated in a previously conducted study. However, hydroxybenzoic acid was detected in the apple juice, but was not found in the fresh apples of these cultivars. Higher or consistent levels of some phenolic compounds in the apple juices after one year of storage could be due to the pectin present in cloudy apple juice, as it has a protective effect by forming colloidal suspensions that limit the degradation of phenolic compounds [35].

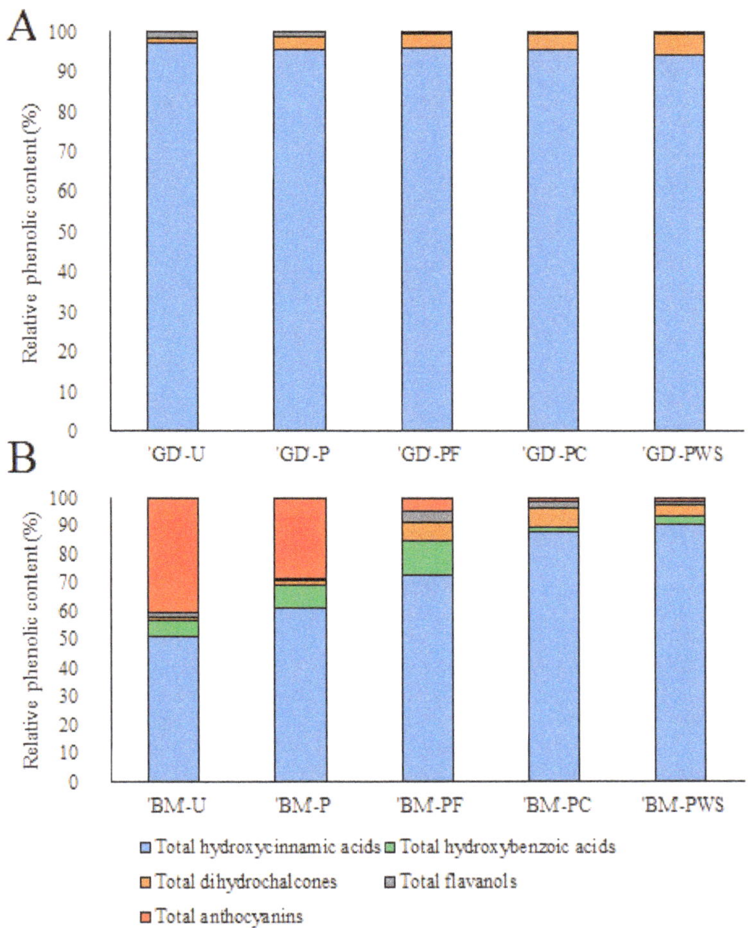

Figure 2. Identified phenolic compounds presented in relative content in fresh unpasteurized, pasteurized, and pasteurized and stored for one year apple (*Malus domestica* Borkh.) juice of 'Golden Delicious' (**A**) and 'Baya Marisa' (**B**). 'BM'-U ('Baya Marisa' unpasteurized), 'GD'-U ('Golden Delicious' unpasteurized), 'BM'-P ('Baya Marisa' pasteurized), 'GD'-P ('Golden Delicious' pasteurized), 'BM'-PF ('Baya Marisa' pasteurized, refrigerator-stored), 'GD'-PF ('Golden Delicious' pasteurized, refrigerator-stored), 'BM'-PC ('Baya Marisa' pasteurized, unexposed to light), 'GD'-PC ('Golden Delicious' pasteurized, unexposed to light), 'BM'-PWS ('Baya Marisa' pasteurized, exposed to light), 'GD'-PWS ('Golden Delicious' pasteurized, exposed to light).

4. Conclusions

The TAPC and sugar and organic acid content were found to be lower overall in the apple juices of both cultivars than in fresh apples of the same cultivars. For consumers, this means that they ingest fewer phenolic compounds overall with juice pressed from the fresh apple.

The total hydroxycinnamic acid content decreased after pasteurization of the juices of both cultivars and increased after one year of storage in all the treatments, except in the RME samples of cultivar 'Baya Marisa', compared to freshly pasteurized apple juice.

Hydroxybenzoic acids were detected only in the cultivar 'Baya Marisa'; however, the total hydroxybenzoic acid content decreased after pasteurization, then decreased even further after one year of storage, except for the refrigerator-stored samples, in which the total hydroxybenzoic acid content increased compared to the freshly pasteurized apple juice.

The total dihydrochalcone content remained stable after pasteurization in the apple juices of both cultivars and increased after a year of storage compared to the freshly pasteurized apple juice of both cultivars, except in the RME samples of the cultivar 'Baya Marisa'.

The total flavanol content decreased after pasteurization in both apple juices and remained the same after a year of storage, except in the refrigerated samples and the RMU samples of the 'Baya Marisa' apple juice, in which the total flavanol content increased compared to the freshly pasteurized samples.

The total anthocyanin content decreased after pasteurization, then decreased even further after one year of storage, which indicates the high instability of this group of phenolic compounds.

Different storage methods had different effects on the color parameters, suggesting the influence of the storage technique on the preservation of the original color, which is one of the most important factors in the food industry.

The results of this study show that the best storage location for apple juice is the refrigerator, as TAPC content is highest after one year of storage. The lowest levels were measured in samples of both apple juices exposed to light at room temperature. Our results also suggest that phenolic compound preservation is cultivar-dependent. In addition, studies on enzyme activity and air exposure should be carried out, as they affect the degradation, polymerization, and depolymerization of secondary metabolites and, thus, the interactions of primary and secondary metabolites.

Author Contributions: Conceptualization, J.J. (Jan Juhart) and A.M.; software, J.J. (Jan Juhart); validation, A.M. and F.S.; formal analysis, J.J. (Jan Juhart); investigation, J.J. (Jan Juhart), A.M. and F.S.; resources, J.J. (Jan Juhart), F.S., M.H. and A.M.; data curation, J.J. (Jan Juhart) and A.M.; writing—original draft preparation, J.J. (Jan Juhart); writing—review and editing, F.S., A.M., M.H., R.V. and J.J. (Jerneja Jakopic); visualization, J.J. (Jan Juhart) and A.M.; supervision, F.S. and A.M.; funding acquisition, J.J. (Jan Juhart), A.M., F.S., M.H. and R.V. All authors have read and agreed to the published version of the manuscript.

Funding: This study is a part of the program P4-0013-0481, which is funded by the Slovenian Research and Innovation Agency (ARIS).

Institutional Review Board Statement: Not applicable.

Informed Consent Statement: Not applicable.

Data Availability Statement: The data presented in this study are available on request from the corresponding author.

Conflicts of Interest: The authors declare no conflict of interest.

References

1. Faramarzi, S.; Pacifico, S.; Yadollahi, A.; Lettieri, A.; Nocera, P.; Piccolella, S. Red-fleshed apples: Old autochthonous fruits as a novel source of anthocyanin antioxidants. *Plant Foods Hum. Nutr.* **2015**, *70*, 324–330. [CrossRef]
2. Khanizadeh, S.; Tsao, R.; Rekika, D.; Yang, R.; DeEll, J. Phenolic composition and antioxidant activity of selected apple genotypes. *J. Food Agric. Environ.* **2007**, *5*, 61–66.
3. Vrhovsek, U.; Rigo, A.; Tonon, D.; Mattivi, F. Quantitation of polyphenols in different apple varieties. *J. Agric. Food Chem.* **2004**, *52*, 6532–6538. [CrossRef]
4. Feng, S.; Yi, J.; Li, X.; Wu, X.; Zhao, Y.; Ma, Y.; Bi, J. Systematic review of phenolic compounds in apple fruits: Compositions, distribution, absorption, metabolism, and processing stability. *J. Agric. Food Chem.* **2021**, *69*, 7–27. [CrossRef] [PubMed]
5. Toydemir, G.; Capanoglu, E.; Gomez Roldan, M.V.; De Vos, R.C.H.; Boyacioglu, D.; Hall, R.D.; Beekwilder, J. Industrial processing effects on phenolic compounds in sour cherry (*Prunus cerasus* L.) fruit. *Food Res. Int.* **2013**, *53*, 218–225. [CrossRef]
6. Oszmiański, J.; Wolniak, M.; Wojdyło, A.; Wawer, I. Influence of apple purée preparation and storage on polyphenol contents and antioxidant activity. *Food Chem.* **2007**, *107*, 1473–1484. [CrossRef]

7. Kim, A.N.; Lee, K.Y.; Rahman, M.S.; Kim, H.J.; Kerr, W.L.; Choi, S.G. Thermal treatment of apple puree under oxygen-free condition: Effect on phenolic compounds, ascorbic acid, antioxidant activities, color, and enzyme activities. *Food Biosci.* **2021**, *39*, 100802. [CrossRef]
8. Salazar-Orbea, G.L.; García-Villalba, R.; Bernal, M.J.; Hernández, A.; Tomás-Barberán, F.A.; Sánchez-Siles, L.M. Stability of phenolic compounds in apple and strawberry: Effect of different processing techniques in industrial set up. *Food Chem.* **2023**, *401*, 134099. [CrossRef]
9. Vegara, S.; Martí, N.; Mena, P.; Saura, D.; Valero, M. Effect of pasteurization process and storage on color and shelf-life of pomegranate juices. *LWT Food Sci. Technol.* **2013**, *54*, 592–596. [CrossRef]
10. Deng, H.; Zhao, P.T.; Yang, T.G.; Meng, Y.H. A comparative study of the cloudy apple juice sterilized by high-temperature short-time or high hydrostatic pressure processing: Shelf-life, phytochemical and microbial view. *Food Sci. Technol.* **2022**, *42*, e63620. [CrossRef]
11. Román, S.; Sánchez-Siles, L.M.; Siegrist, M. The importance of food naturalness for consumers: Results of a systematic review. *Trends Food Sci. Technol.* **2017**, *67*, 44–57. [CrossRef]
12. Misra, N.N.; Koubaa, M.; Roohinejad, S.; Juliano, P.; Alpas, H.; Inácio, R.S.; Saraiva, J.A.; Barba, F.J. Landmarks in the historical development of twenty first century food processing technologies. *Food Res. Int.* **2017**, *97*, 318–339. [CrossRef] [PubMed]
13. Aguilar-Rosas, S.F.; Ballinas-Casarrubias, M.L.; Nevarez-Moorillon, G.V.; Martin-Belloso, O.; Ortega-Rivas, E. Thermal and pulsed electric fields pasteurization of apple juice: Effects on physicochemical properties and flavour compounds. *J. Food Eng.* **2007**, *83*, 41–46. [CrossRef]
14. Dey, T.B.; Banerjee, R. Application of decolourized and partially purified polygalacturonase and α-amylase in apple juice clarification. *Braz. J. Microbiol.* **2014**, *45*, 97–104. [CrossRef]
15. Veberic, R.; Stampar, F.; Schmitzer, V.; Cunja, V.; Zupan, A.; Koron, D.; Mikulic-Petkovsek, M. Changes in the contents of anthocyanins and other compounds in blackberry fruits due to freezing and long-term frozen storage. *J. Agric. Food Chem.* **2014**, *62*, 6926–6935. [CrossRef]
16. Mikulic-Petkovsek, M.; Koron, D.; Rusjan, D. The impact of food processing on the phenolic content in products made from juneberry (*Amelanchier lamarckii*) fruits. *J. Food Sci.* **2020**, *85*, 386–393. [CrossRef]
17. Persic, M.; Mikulic-Petkovsek, M.; Slatnar, A.; Veberic, R. Chemical composition of apple fruit, juice and pomace and the correlation between phenolic content, enzymatic activity and browning. *LWT Food Sci. Technol.* **2017**, *82*, 23–31. [CrossRef]
18. Medic, A.; Zamljen, T.; Hudina, M.; Veberic, R. Time-dependent degradation of naphthoquinones and phenolic compounds in walnut husks. *Biology* **2022**, *11*, 342. [CrossRef]
19. Medic, A.; Jakopic, J.; Hudina, M.; Solar, A.; Veberic, R. Identification and quantification of the major phenolic constituents in Juglans regia L. peeled kernels and pellicles, using HPLC–MS/MS. *Food Chem.* **2021**, *352*, 129404. [CrossRef]
20. Medic, A.; Hudina, M.; Veberic, R. The effect of cane vigour on the kiwifruit (*Actinidia chinensis*) and kiwiberry (*Actinidia arguta*) quality. *Sci. Rep.* **2021**, *11*, 12749. [CrossRef]
21. Bizjak, J.; Mikulic-Petkovsek, M.; Stampar, F.; Veberic, R. Changes in primary metabolites and polyphenols in the peel of "Braeburn" apples (*Malus domestica* Borkh.) during advanced maturation. *J. Agric. Food Chem.* **2013**, *61*, 10283–10292. [CrossRef]
22. Petrović, M.; Veljović, S.; Tomić, N.; Zlatanović, S.; Tosti, T.; Vukosavljević, P.; Gorjanović, S. Formulation of novel liqueurs from juice industry waste: Consumer acceptance, phenolic profile and preliminary monitoring of antioxidant activity and colour changes during storage. *Food Technol. Biotechnol.* **2021**, *59*, 282–294. [CrossRef]
23. Liu, S.X.; Yang, H.Y.; Li, S.Y.; Zhang, J.Y.; Li, T.; Zhu, B.Q.; Zhang, B.L. Polyphenolic compositions and chromatic characteristics of bog bilberry syrup wines. *Molecules* **2015**, *20*, 19865–19877. [CrossRef]
24. Patras, A.; Brunton, N.P.; O'Donnell, C.; Tiwari, B.K. Effect of thermal processing on anthocyanin stability in foods, mechanisms and kinetics of degradation. *Trends Food Sci. Technol.* **2010**, *21*, 3–11. [CrossRef]
25. Begic-Akagić, A.; Spaho, N.; Gasi, F.; Drkenda, P.; Vranac, A.; Meland, M. Sugar and organic acid profiles of the traditional and international apple cultivars for processing. *J. Hyg. Eng. Des.* **2014**, *7*, 190–196.
26. Mezey, J.; Mezeyová, I. Changes in the levels of selected organic acids and sugars in apple juice after cold storage. *Czech J. Food Sci.* **2018**, *36*, 175–180. [CrossRef]
27. Singh, S.K.; Sharma, M. Review on biochemical changes associated with storage of fruit juice. *Int. J. Curr. Microbiol. Appl. Sci.* **2017**, *6*, 236–245. [CrossRef]
28. Farnworth, E.R.; Lagacé, M.; Couture, R.; Yaylayan, V.; Stewart, B. Thermal processing, storage conditions, and the composition and physical properties of orange juice. *Food Res. Int.* **2001**, *34*, 25–30. [CrossRef]
29. Rehman, M.A.; Khan, M.R.; Sharif, M.K.; Ahmad, S.; Shah, F.-U.-H. Study on the storage stability of fruit juice concentrates. Pakistan. *J. Food Sci.* **2014**, *24*, 101–107.
30. Hameed, F.; Kumar, A.; Hamid, N. Effect of thermal treatment on the quality of cloudy apple juice. *J. Pharmacogn. Phytochem.* **2019**, *8*, 1976–1979.
31. Tikekar, R.V.; Anantheswaran, R.C.; LaBorde, L.F. Ascorbic acid degradation in a model apple juice system and in apple juice during ultraviolet processing and storage. *J. Food Sci.* **2011**, *76*, H62–H71. [CrossRef] [PubMed]
32. Kim, J.K.; Kim, E.H.; Lee, O.K.; Park, S.Y.; Lee, B.; Kim, S.H.; Park, I.; Chung, I.M. Variation and correlation analysis of phenolic compounds in mungbean (*Vigna radiata* L.) varieties. *Food Chem.* **2013**, *141*, 2988–2997. [CrossRef] [PubMed]

33. Tian, Y.; Sun, L.; Yang, Y.; Gou, X.; Niu, P.; Guo, Y. Changes in the physicochemical properties, aromas and polyphenols of not from concentrate (NFC) apple juice during production. *CyTA J. Food* **2018**, *16*, 755–764. [CrossRef]
34. Le Bourvellec, C.; Bouzerzour, K.; Ginies, C.; Regis, S.; Plé, Y.; Renard, C.M.G.C. Phenolic and polysaccharidic composition of applesauce is close to that of apple flesh. *J. Food Compos. Anal.* **2011**, *24*, 537–547. [CrossRef]
35. Salazar-Orbea, G.L.; García-Villalba, R.; Tomás-Barberán, F.A.; Sánchez-Siles, L.M. High–pressure processing vs. thermal treatment: Effect on the stability of polyphenols in strawberry and apple products. *Foods* **2021**, *10*, 2919. [CrossRef] [PubMed]
36. Zhao, G.Y.; Zong, W.; An, G.J. Effect of storage on cloud stability of cloudy apple juice. *Food Sci. Technol. Int.* **2008**, *14*, 105–113. [CrossRef]
37. Sadilova, E.; Stintzing, F.C.; Carle, R. Thermal degradation of acylated and nonacylated anthocyanins. *J. Food Sci.* **2006**, *71*, C504–C512. [CrossRef]
38. Juhart, J.; Medic, A.; Veberic, R.; Hudina, M.; Jakopic, J.; Stampar, F. Phytochemical composition of red-fleshed apple cultivar 'Baya Marisa' compared to traditional, white-fleshed apple cultivar 'Golden Delicious'. *Horticulturae* **2022**, *8*, 811. [CrossRef]
39. Liu, X.; Le Bourvellec, C.; Guyot, S.; Renard, C.M.G.C. Reactivity of flavanols: Their fate in physical food processing and recent advances in their analysis by depolymerization. *Compr. Rev. Food Sci. Food Saf.* **2021**, *20*, 4841–4880. [CrossRef] [PubMed]
40. Knebel, T.; Braun, P.; Dietrich, H. Degradation kinetics of anthocyanins and polyphenols during storage of red apple juice produced from red-fleshed apples. *Eur. Food Res. Technol.* **2018**, *244*, 1741–1750. [CrossRef]
41. Farr, J.E.; Monica Giusti, M. Investigating the interaction of ascorbic acid with anthocyanins and pyranoanthocyanins. *Molecules* **2018**, *23*, 744. [CrossRef] [PubMed]

foods

MDPI

Article

Effect of Different Walnut and Hazelnut Leaf Compost Treatments on Yield and Phenolic Composition of *Lactuca sativa* L.

Aljaz Medic *, Anita Solar, Metka Hudina, Robert Veberic and Tilen Zamljen

Department of Agronomy, Biotechnical Faculty, University of Ljubljana, SI 1000 Ljubljana, Slovenia; anita.solar@bf.uni-lj.si (A.S.); metka.hudina@bf.uni-lj.si (M.H.); tilen.zamljen@bf.uni-lj.si (T.Z.)

* Correspondence: aljaz.medic@bf.uni-lj.si

Abstract: The use of compost made from the leaves of *Juglans regia* has long been controversial because of its inhibitory effect due to the presence of juglone. Therefore, the aim of our study was to replicate the typical habits of farmers and gardeners, where the dried leaves are collected at the end of the season and placed in a composter. Then, the effects of the different treatments on the yield of the plant (lettuce), secondary metabolism, and possible toxicity of the compost of the grown plant were evaluated. The lowest yield of lettuce was obtained in soil with composted walnut and hazelnut leaves, while the highest yield was recorded in in soil with compost control, soil with composted walnut leaves and grass with the addition of composting agent and soil with composted walnut leaves with addition of composting agent. Some allelochemicals were still present in the compost but at such low levels that they did not affect yield. We suggest that dry walnut leaves and cut grass can be used for composting, while dry hazelnut leaves still contain some allelochemicals after two years that significantly inhibit plant growth and thus yield, so we would not recommend their use for composting.

Keywords: allelopathy; lettuce; naphthoquinones; toxic residue; HPLC; mass spectrometry; LC-MS

1. Introduction

Composting is a technical process in which organic waste is microbiologically degraded to humic substances. Composting is suitable for the treatment of many solid wastes, including farm manure, sewage sludge, and especially crop residues. The goal of composting is to reduce the amount of undesirable substances such as plant and animal pathogens and organic compounds that are considered hazardous to the environment, as well as to reduce the carbon concentration and increase the concentration of plant nutrients. Composting essentially converts materials that are not suitable for land application into materials that can be used for land application [1]. In addition, the application of composted organic waste is considered efficient in terms of energy savings and solid waste disposal costs. Currently, composting is an important industry involving cooperatives and companies as well as farms and organic gardeners [2].

During the growing season, plants produce a variety of agro-waste or plant residues that can release allelochemicals into the soil. Depending on their effect on the growing plant, these allelochemicals are divided into positive ones, which stimulate plant growth, and negative ones, which inhibit growth. They can cause malformation and wilting, induce chlorosis, reduce plant vigor, hinder plant development and growth, and slow or prevent plant germination. All these may eventually lead to the complete collapse of a plant [3]. Moreover, allelochemicals can not only alter the yield of plants but also the metabolism of the plant, especially the secondary metabolism and phenolic profile [4,5]. Since phenolic compounds are considered to be the most important factor in the health benefits of fruits and vegetables besides vitamins [6], this effect is very important but still poorly studied.

Phenolic compounds, in addition to their health-promoting effects (e.g., antimicrobial, antiallergenic, anti-inflammatory, and antioxidant activity) [7–10], also play an important role in the quality of fruits and vegetables by influencing their flavor, appearance, and stability [11].

The use of compost made from the leaves of *Juglans regia* L. has long been controversial, as it has been associated with its germination- and growth-inhibiting effects, due to juglone presence which was previously reported to affect germination, growth, yield, and secondary metabolism of the grown plants [4,5,12]. Previously, young compost (less than six months) from fresh walnut leaves was reported to have a negative effect on germination and growth of test plants, while older compost (nine to ten months) from fresh walnut leaves did not inhibit germination or plant growth. In addition, the nutrients of walnut leaf compost lead to higher green mass of the grown plants compared to the control [13]. However, there are very few studies on this topic, and the only study [13] used fresh leaves for compost, which is not so relevant because the leaves falling from walnut trees are usually not fresh but already dry. Since the dry leaves are usually collected at the end of the growing season and placed in a composter, the objective of our study was to replicate the typical habits of farmers and gardeners where the dried leaves are collected at the end of the season and placed in a composter. Since other plant waste is usually added to the compost but also to ensure homogeneity, the compost was mixed with various plant residues and additives to speed up composting. Then, the effects of the different treatments on the yield of the plant (lettuce), secondary metabolism, and possible toxicity of the grown plant were evaluated. These results will further improve the knowledge of compost from walnut residues and answer the question of how the compost affects the yield and phenolic profile and thus the health benefits of the plants grown in this compost.

2. Materials and Methods

2.1. Compost Preparation

Composting was performed at the Experimental Field for Nut Crops in Maribor (Slovenia; 46°34′01″ N; 15°37′51″ E; 275 m a.s.l.). All plant material was also obtained at this site. Four compost bins were used for the experiment, three of which were divided into two halves by a wooden barrier, as one half of the compost bin was used for the treatments without the composting agent and the other half was used for the treatments with the composting agent (see Figure 1). The composting agent selected was used because it is commonly used by gardeners and farmers for faster composting to see the effects on the yield of the crops grown. The composting agent used was COMPO BIO (COMPO GmbH, Münster, Germany).

For the experiment, 7 compost treatments and a control group were used. One compost bin was used to compost various plant residues commonly used by an allotment gardener or farmer (excluding grass, hazelnut, and walnut leaves), which was a positive control. The remaining three bins were divided into two parts (one with composting agent and one without) as previously described and treated as follows: walnut leaves, walnut leaves with the addition of the composting agent, walnut and hazelnut leaves (ratio 1:1), walnut and hazelnut leaves (ratio 1:1) with the addition of the composting agent, freshly cut grass and walnut and hazelnut leaves (ratio 1:1:1), and freshly cut grass and walnut and hazelnut leaves (ratio 1:1:1) with the addition of the composting agent. Composting was carried out for 2 years, as is common practice among gardeners.

After two years, the compost was carefully emptied and packed into bags and transported to the experimental field of the Department of Agronomy at the Biotechnical Faculty of the University of Ljubljana, Slovenia. There, the compost was mixed separately with soil in a 9:1 ratio for each treatment. After mixing, the mixture of soil and compost was labeled as further described in the text: K2, soil with compost control; O, soil with composted walnut leaves; O + D, soil with composted walnut leaves with addition of composting agent; O + L, soil with composted walnut and hazelnut leaves; O + L + D, soil with composted walnut and hazelnut leaves with addition of composting agent; O + T, soil with composted walnut leaves and grass; and O + T + D, soil with composted walnut leaves and grass with addition of composting agent. In addition, a control treatment with pure soil without compost was added, which is a negative control and is referred to as K1 in the text.

Figure 1. Compost bins used and divided by the wooden barriers.

2.2. Plant Material and Growing Conditions

The plants treated were lettuce (*Lactuca sativa*), as it grows quickly and is a common garden vegetable to which compost is usually added before planting. Plants were grown in a greenhouse to better control growing conditions and watered evenly every other day. Seeds were obtained from Austrosaat AG, Wien, Austria, and the cultivar chosen was 'Tourbillon'. Plants were germinated from seeds in the greenhouse and sown in pots with standard peat substrate and repotted into soil with different compost treatments after the appearance of the third leaf (29 March 2022). After technological maturity (60 days after the seeds were planted), the plants were sampled (28 May 2022). There were 10 replicates per treatment. A total of 80 lettuce plants were used. Of these 10 plants per treatment, all 10 were weighed to determine their yield; then, 10 plants (2 plants were used per replicate) were used for metabolite analysis (Figure 2), giving 5 replicates per treatment for metabolite analysis.

2.3. Chemicals

The following standards were used for the identification and quantification of phenolic compounds: *p*-coumaric acid, caffeic acid, gallic acid, sinapic acid, chlorogenic acid (3-caffeoylquinic acid), cryptochlorogenic acid (4-caffeoylquinic acid), sinapic acid (Sigma-Aldrich Chemie GmbH, Buchs, Switzerland), quercetin-3-*O*-glucoside, and kaempferol-3-*O*-glucoside (Fluka Chemie GmbH, Buchs, Switzerland).

A Milli-Q water purification system from Millipore (Bedford, MA, USA) was used for the bi-distillation and purification of the water used. The acetonitrile and formic acid used as mobile phases for MS analysis were of HPLC-MS quality (Fluka Chemie GmbH). The methanol used for the extraction of the phenolic compounds was of the quality HPLC-MS (Sigma-Aldrich Chemie GmbH).

Figure 2. Visual representation of the experiment: Different treatments affecting the yield of the *L. sativa*.

2.4. Sampling of the Plants and Extraction of the Phenolic Compounds

After sampling, whole plants were immersed in liquid nitrogen and lyophilized, then ground to powder using a mill (A11 basic analytical mill, IKA®-Werke GmbH &Co., KG, Staufen, Germany), and stored until further analysis.

The extraction was performed according to the protocol described by Medic et al. [5]. Briefly, 0.2 g of the previously lyophilized sample was extracted at a tissue-to-solution ratio of 1:30 (w/v) using an extraction medium of 80% methanol and 3% formic acid in water.

2.5. HPLC—Mass Spectrometry Analysis of Individual Phenolic Compounds

A tandem mass spectrometer (LTQ XL, Thermo Scientific, Waltham, MA, USA) coupled to a UHPLC (Vanquish, Thermo Scientific, Waltham, MA, USA) operated in negative ion mode and heated electrospray ionization was used for the identification of phenolic compounds. Parameters were as described by Medic et al. [5]. Data acquisition was performed using Xcalibur 2.2 software (Thermo Fischer Scientific Institute, Waltham, MA, USA).

Quantification was performed on a Vanquish UHPLC system with a diode detector set at 350 nm for flavanols and 280 nm for the other phenolic compounds. A Gemini 150 mm × 4.60 mm, 3 µm C18 column (Phenomenex, Torrance, CA, USA) was used for the separation of phenolic compounds. The solvents, column temperature, elution flow gradient, gradient, and other parameters were as described by Medic et al. [5].

Commercial standards were used wherever possible. Where these were not available, identification was based on literature data and the specific fragmentation pattern of the compounds, and quantification was performed using a similar standard as shown in Table 1.

Table 1. Tentative identification of the 18 phenolic compounds in *L. sativa* and the standards they are expressed to.

Compound	Rt (min)	[M−H]⁻ (*m*/*z*)	MS² (*m*/*z*)	MS³ (*m*/*z*)	MS⁴ (*m*/*z*)	Expressed As
3,5-Dicaffeoylquinic acid	8.99	515	353 (100), 191 (74), 179 (2)			Chlorogenic acid
Dihydroxybenzoic acid hexoside 1	9.67	315	153 (100), 152 (30), 165 (14)	109 (100)		Galic acid
Dihydroxybenzoic acid hexoside 2	10.53	315	153 (100), 109 (6)	109 (100)		Galic acid
Esculetin glucoside	10.74	339	177 (100), 241 (14)	133 (100), 177 (12)		Galic acid
Caffeoylquinic acid dehydrodimer	11.71	705	513 (100), 339 (10)	339 (100), 321 (59), 469 (28), 229 (26)	229 (100), 295 (76), 293 (12)	Chlorogenic acid
Chlorogenic acid (3-caffeoylquinic acid)	13.64	353	191 (100), 179 (5)			Chlorogenic acid
Sinapoyl hexoside derivative	14.01	431	385 (100), 299 (44)			Sinapic acid
Caffeoyltartaric acid	14.30	311	149 (100), 179 (51), 135 (3)			Chlorogenic acid
Quercetin-3-*O*-(6″-malonyl)-glucoside-7-*O*-glucoside	15.24	711	667 (100)	505 (100), 463 (35), 301 (29), 462 (20)	301 (100), 300 (69)	Quercetin-3-*O*-glucoside
Cryptochlorogenic acid (4-caffeoylquinic acid)	15.49	353	191 (100), 179 (5)			Cryptochlorogenic acid
Quercetin-3-*O*-(6″-malonyl)-glucoside-7-*O*-glucuronide	16.36	725	681 (100), 505 (17)	505 (100)	301 (100), 300 (69)	Quercetin-3-*O*-glucoside
cis-5-*O*-*p*-coumaroylquinic acid	17.24	337	191 (100), 163 (7), 173 (1)			*p*-Coumaric acid
trans-5-*O*-*p*-coumaroylquinic acid	18.68	337	191 (100), 163 (7), 173 (1)			*p*-Coumaric acid
Caffeoyl malate	19.28	295	179 (100), 133 (58), 135 (5)	135 (100)		Caffeic acid
Caffeic acid derivative	21.11	415	179 (100), 225 (20), 161 (11), 143 (11)	135 (100)		Caffeic acid
Quercetin-3-*O*-glucoside	21.66	463	301 (100), 300 (17), 179 (1)			Quercetin-3-*O*-glucoside
Kaempferol-3-*O*-glucuronoside	22.47	463	287 (100), 288 (6), 151 (5)			Kaempferol-3-*O*-glucoside
Quercetin-3-*O*-(6″-malonyl)-glucoside	23.52	549	505 (100)	301 (100), 300 (69)		Quercetin-3-*O*-glucoside

Rt, retention time; [M−H]⁻, pseudo-molecular ion identified in negative ion mode; (), relative abundance of fragment ions.

2.6. Statistical Analysis

Data were compiled using R commander (package Rcmdr) version 2.7.1 (Team, R.D.C., 2008, Stanford, CA, USA) and Microsoft Excel (MS Office 2016). Five replicates (for each treatment) were performed. Data are presented as means ± standard error (SE). One-way analysis of variance (ANOVA) with Tukey tests was used to determine differences between treatments. Statistical means were calculated at a 95% confidence level to determine the significance of differences.

3. Results and Discussion

3.1. Identification of Individual Phenolic Compounds in L. sativa

A total of 18 phenolic compounds were identified in *L. sativa* based on the available literature, pseudomolecular ion $[M-H]^-$, and their specific fragmentation patterns (MS^2, MS^3, and MS^4). Among these 18 compounds were 10 hydroxycinnamic acids, 2 hydroxybenzoic acids, 5 flavanols, and 1 coumarin. The identified phenolic compounds, their fragmentation, and the relative standards to which they relate are listed in Table 1.

Of the 10 identified hydroxycinnamic acids, chlorogenic acid, sinapoyl hexoside derivative, cryptochlorogenic acid, *cis*-5-*O*-*p*-coumaroylquinic acid, *trans*-5-*O*-*p*-coumaroylquinic acid, caffeoyl malate, and caffeic acid derivative were already identified/determined in our previous study on *L. sativa* [5], based on their specific fragmentation patterns and the use of standards. 3,5-Diccaffeoylquinic acid was identified based on the specific fragmentation pattern of MS^2 m/z 353, 191, and 179, and caffeoyltartaric acid was identified based on the specific fragmentation pattern of MS^2 m/z 149, 179, and 135, as previously reported in *L. sativa* [14]. The caffeic acid derivative was identified based on the fragmentation pattern of MS^n ions m/z 179 and 135, as previously reported by Liu et al. [15], but was also detected for the first time in *L. sativa*. The caffeoylquinic acid dehydrodimer previously identified in *Berberis thunbergii* leaves followed its specific fragmentation pattern of MS^n m/z 513, 339, and 295, as previously reported by Fernández-Poyatos at al. [16].

Both hydroxybenzoic acids and one coumarin were identified in *L. sativa* in our previous study [5]. Of the five flavanols identified, kaempferol-3-*O*-glucuronoside, quercetin-3-*O*-glucoside, and quercetin-3-*O*-(6″-malonyl)-glucoside were identified by their specific fragmentation patterns and the use of standards, as previously reported by Medic et al. [5]. Quercetin-3-*O*-(6″-malonyl)-glucoside-7-*O*-glucoside and quercetin-3-*O*-(6″-malonyl)-glucoside-7-*O*-glucuronide were identified by their fragmentation pattern as reported by Abu-Reidah et al. [17] for *L. sativa*.

3.2. Effects of Different Composting Treatments on L. sativa Yield and Phenolic Composition

Looking at Table 2, we can see that the lowest yield for lettuce was for O + L, followed by K1 and O + L + D. The highest yield was reported for K2, O + T + D, and O + D. Interestingly, lettuce yield was higher when composting agent was added, regardless of compost type. In addition, the composting agent used also contained minerals (NPK(MgO): 4-3-1 (8)), which most likely contribute to growth and yield of treated plants. Therefore, in terms of yield, the composting agent was found to be effective in increasing the overall yield. In addition, it has been described that fresh walnut leaves still contain some allelochemicals that inhibit plant growth [13]. This is not entirely the case with the dry walnut leaves composted for 2 years, as the yield of the plants is indeed lower than with K2 (positive control) but higher than K1 (negative control), while the dry walnut leaves with the addition of composting agent have a comparable yield to K2 and higher than K1. Therefore, we suggest that dry walnut leaves and cut grass can be used for composting. Dry hazelnut leaves, on the other hand, still contain some unknown allelochemicals after 2 years that significantly and strongly inhibit plant growth and thus yield, so we would not recommend their use for composting. However, future research is needed to explain and address this problem. The higher yield of the plants with compost addition compared to K1 is related to the process of composting, as it reduces the amount of organic compounds considered harmful to the environment and reduces the carbon concentration and increases the concentration of plant nutrients, thus promoting their growth [1].

Table 2. *L. sativa* yield measurements.

	K1	K2	O	O + D	O + L	O + L + D	O + T	O + T + D
Whole plant (g)	29.25 ± 1.18 b	81.50 ± 5.91 e	57.00 ± 6.36 d	76.00 ± 6.26 de	18.00 ± 1.58 a	37.25 ± 2.32 c	65.00 ± 4.51 d	89.75 ± 6.09 e

Data are means ± standard error. Means followed by different letters across the treatments (within rows) are significantly different ($p < 0.05$). (g), grams; K1: soil without compost control, K2: soil with compost control, O: soil with composted walnut leaves, O + D: soil with composted walnut leaves with addition of composting agent, O + L: soil with composted walnut and hazelnut leaves, O + L + D: soil with composted walnut and hazelnut leaves with the addition of composting agent, O + T: soil with composted walnut leaves and grass, O + T + D: soil with composted walnut leaves and grass with the addition of composting agent.

In Table 3, we can see the individual phenolic compounds found in *L. sativa*. It can be seen that the most important phenolic compound, regardless of treatment, is chlorogenic acid, which accounts for up to 80% of the total hydroxycinnamic acids in *L. sativa*. Chlorogenic acid was followed by two flavanols (quercetin-3-*O*-(6″-malonyl)-glucoside-7-*O*-glucoside and quercetin-3-*O*-(6″-malonyl)-glucoside-7-*O*-glucuronide) and caffeoyl malate. The content of the different phenolic groups is shown in Figure 3.

Table 3. Comparison of the content of individual phenolic compounds and phenolic groups in *L. sativa*, expressed in mg/kg dry weight.

Phenolic Compound	Quantification According to Cultivar (mg/kg Dry Weight)							
	K1	K2	O	O + D	O + L	O + L + D	O + T	O + T + D
3,5-Dicaffeoylquinic acid	0.11 ± 0.02 b	0.12 ± 0.03 b	0.18 ± 0.02 b	0.20 ± 0.00 b	0.01 ± 0.01 a	0.14 ± 0.01 b	0.11 ± 0.01 b	0.12 ± 0.03 b
Dihydroxybenzoic acid hexoside 1	0.04 ± 0.01 a	0.03 ± 0.01 a	0.03 ± 0.01 a	0.03 ± 0.02 a	0.02 ± 0.01 a	0.03 ± 0.02 a	0.03 ± 0.01 a	0.02 ± 0.00 a
Dihydroxybenzoic acid hexoside 2	0.05 ± 0.01 a	0.06 ± 0.01 a	0.30 ± 0.07 c	0.27 ± 0.04 c	0.05 ± 0.00 a	0.25 ± 0.03 bc	0.10 ± 0.03 ab	0.23 ± 0.03 bc
Esculetin glucoside	0.02 ± 0.01 a	0.02 ± 0.00 a	0.04 ± 0.01 a	0.04 ± 0.01 a	0.04 ± 0.00 a	0.08 ± 0.02 a	0.05 ± 0.02 a	0.08 ± 0.04 a
Caffeoylquinic acid dehydrodimer	0.09 ± 0.01 a	0.10 ± 0.02 a	0.11 ± 0.01 ab	0.09 ± 0.01 a	0.07 ± 0.00 a	0.10 ± 0.02 ab	0.16 ± 0.03 ab	0.19 ± 0.03 b
Chlorogenic acid (3-caffeoylquinic acid)	3.07 ± 0.11 ab	2.95 ± 0.37 a	5.40 ± 0.48 c	6.54 ± 0.62 c	2.64 ± 0.15 a	6.23 ± 0.06 c	4.87 ± 0.57 bc	2.02 ± 0.41 a
Sinapoyl hexoside derivative	0.04 ± 0.01 a	0.05 ± 0.02 a	0.02 ± 0.00 a	0.05 ± 0.01 a	0.04 ± 0.00 a	0.05 ± 0.01 a	0.03 ± 0.00 a	0.01 ± 0.00 a
Caffeoyltartaric acid	0.07 ± 0.03 a	0.07 ± 0.01 a	0.09 ± 0.01 a	0.09 ± 0.02 a	0.08 ± 0.01 a	0.11 ± 0.02 a	0.07 ± 0.01 a	0.12 ± 0.06 a
Quercetin-3-*O*-(6″-malonyl)-glucoside-7-*O*-glucoside	0.12 ± 0.05 a	0.16 ± 0.07 a	0.26 ± 0.05 a	0.34 ± 0.05 a	0.15 ± 0.02 a	0.25 ± 0.04 a	0.27 ± 0.07 a	0.15 ± 0.02 a
Cryptochlorogenic acid (4-caffeoylquinic acid)	0.13 ± 0.06 ab	0.18 ± 0.03 ac	0.25 ± 0.03 bc	0.25 ± 0.03 bc	0.09 ± 0.01 ab	0.31 ± 0.05 c	0.13 ± 0.03 ab	0.07 ± 0.04 a
Quercetin-3-*O*-(6″-malonyl)-glucoside-7-*O*-glucuronide	0.22 ± 0.02 ab	0.20 ± 0.01 a	0.35 ± 0.02 bc	0.30 ± 0.05 ac	0.22 ± 0.01 ab	0.36 ± 0.02 c	0.24 ± 0.04 ac	0.21 ± 0.02 a
cis-5-*O*-*p*-coumaroylquinic acid	0.17 ± 0.03 ab	0.16 ± 0.02 ab	0.26 ± 0.01 b	0.23 ± 0.04 ab	0.16 ± 0.01 ab	0.26 ± 0.02 b	0.18 ± 0.03 ab	0.13 ± 0.01 a
trans-5-*O*-*p*-coumaroylquinic acid	0.08 ± 0.01 a	0.09 ± 0.01 a	0.11 ± 0.01 a	0.11 ± 0.03 a	0.09 ± 0.01 a	0.12 ± 0.02 a	0.10 ± 0.02 a	0.08 ± 0.01 a
Caffeoyl malate	0.23 ± 0.04 ab	0.26 ± 0.04 ac	0.38 ± 0.02 bcd	0.44 ± 0.02 d	0.18 ± 0.03 a	0.40 ± 0.04 cd	0.31 ± 0.01 ad	0.22 ± 0.04 a
Caffeic acid derivative	0.02 ± 0.01 a	0.02 ± 0.01 a	0.02 ± 0.01 a	0.02 ± 0.01 a	0.01 ± 0.00 a	0.03 ± 0.01 a	0.02 ± 0.01 a	0.02 ± 0.01 a
Quercetin-3-*O*-glucoside	0.05 ± 0.01 a	0.04 ± 0.01 a	0.10 ± 0.03 a	0.07 ± 0.01 a	0.04 ± 0.00 a	0.08 ± 0.00 a	0.04 ± 0.01 a	0.06 ± 0.02 a
Kaempferol-3-*O*-glucuronoside	0.02 ± 0.01 a	0.02 ± 0.01 a	0.05 ± 0.01 ab	0.02 ± 0.01 a	0.02 ± 0.01 a	0.06 ± 0.01 b	0.03 ± 0.00 ab	0.02 ± 0.01 a
Quercetin-3-*O*-(6″-malonyl)-glucoside	0.02 ± 0.01 a	0.04 ± 0.04 a	0.07 ± 0.02 ab	0.03 ± 0.00 a	0.03 ± 0.01 a	0.14 ± 0.02 b	0.04 ± 0.02 a	0.02 ± 0.01 a
Total hydroxycinnamic acids	4.01 ± 0.18 a	3.99 ± 0.51 a	6.84 ± 0.50 c	8.02 ± 0.65 c	3.38 ± 0.17 a	7.74 ± 0.07 c	5.98 ± 0.60 bc	2.97 ± 0.44 a
Total hydroxybenzoic acids	0.11 ± 0.03 a	0.11 ± 0.02 a	0.37 ± 0.06 b	0.34 ± 0.05 b	0.11 ± 0.01 a	0.35 ± 0.06 b	0.18 ± 0.06 ab	0.33 ± 0.04 b
Total Flavanols	0.42 ± 0.08 a	0.46 ± 0.13 a	0.82 ± 0.08 bc	0.77 ± 0.09 b	0.46 ± 0.02 a	0.89 ± 0.04 c	0.62 ± 0.11 ab	0.46 ± 0.02 a
TAPC	4.54 ± 0.28 a	4.56 ± 0.63 a	8.02 ± 0.52 cd	9.13 ± 0.68 d	3.95 ± 0.15 a	8.99 ± 0.15 cd	6.78 ± 0.70 bc	3.76 ± 0.44 a

Data are means ± standard error. Means followed by different letters across the treatments (within rows) are significantly different ($p < 0.05$); K1: soil without compost control, K2: soil with compost control, O: soil with composted walnut leaves, O + D: soil with composted walnut leaves with addition of composting agent, O + L: soil with composted walnut and hazelnut leaves, O + L + D: soil with composted walnut and hazelnut leaves with the addition of composting agent, O + T: soil with composted walnut leaves and grass, O + T + D: soil with composted walnut leaves and grass with the addition of composting agent.

Hydroxycinnamic acids were the major phenolic group in lettuce, accounting for up to 85% of TAPC (total analyzed phenolic content) (Figure 3). They were followed by flavanols and hydroxybenzoic acids, which is consistent with previous studies on phenolic compounds in lettuce [5,14,17]. Interestingly, the highest TAPC was reported for plants grown in O, O + D, and O + L compost, suggesting that some allelochemicals were still present in the compost of these treatments after two years, but in such a small amount that they did not affect yield, as the highest yield was reported for O + D, O + T + D, and K2. However, it did affect plant metabolism by stimulating the synthesis of phenolic compounds. The synthesis of phenolic compounds is related to the reaction of plants to stress conditions and is a defense mechanism of plants [18,19]. Phenolic content of plants grown in O + T + D did not differ from that of controls, indicating that the compost mixture was sufficiently composted not to interfere with plant metabolism or induce stress. Interestingly, the phenolic content of plants grown in O + L compost also did not differ from that of the controls, but the yield of these plants was the lowest, suggesting that O + L could either inhibited the growth so much that the plants metabolic synthesis was obscured as well and that higher levels of compost in the soil might even cause plant collapse. There were no differences in plant metabolism in the control treatments, suggesting that the dry walnut leaves, while affecting plant yield, also have an effect on plant metabolism by stimulating the synthesis of phenolic compounds. As previously reported, allelochemicals alter the phenolic profile and stimulate secondary metabolism in plants [5]. Since some allelochemicals can be taken up by plants and are thus toxic to consumers [20] plants were also screened for known toxic allelochemicals in walnut, i.e., juglone. However, juglone was not found in the treated plants, nor was it found in our previous study of juglone toxicity, in which we concluded that this was likely due to the fact that juglone and other naphthoquinones do not dissolve well in polar solvents and are thus also difficult to transport through the plant and might only prove problematic if the plants grown were tubers or root crops that come into direct contact with juglone [4].

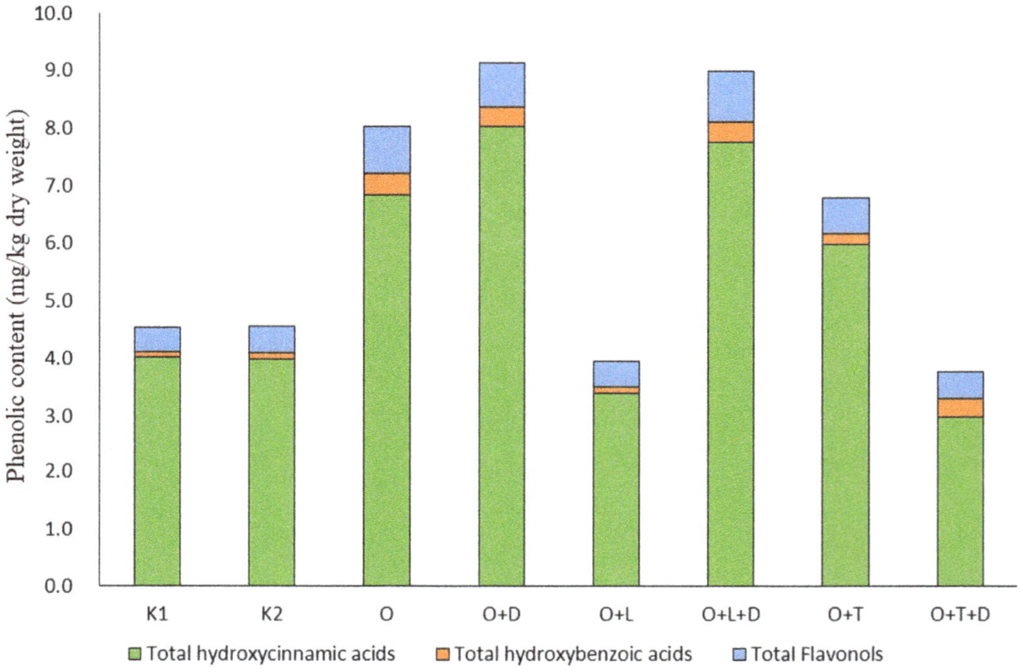

Figure 3. Contents of the total phenolic groups identified in *L. sativa* relative to dry weight. K1; soil without compost control, K2; soil with compost control, O; soil with composted walnut leaves, O + D; soil with composted walnut leaves with addition of composting agent, O + L; soil with composted walnut and hazelnut leaves, O + L + D; soil with composted walnut and hazelnut leaves with the addition of composting agent, O + T; soil with composted walnut leaves and grass, O + T + D; soil with composted walnut leaves and grass with the addition of composting agent.

Overall, walnut leaf compost does not negatively affect plant yield, but rather stimulates their growth; however, it does affect plant metabolism by stimulating the synthesis of phenolic compounds. Since phenolic compounds are considered beneficial to human health [21,22], adding walnut leaf compost to cultivated plants soil is recommended to some extent. However, as little is known about how easily these compounds leach from the soil with rainfall events, the application of walnut leaf compost should be further studied, since allelochemicals could bond with the soil and continued application could exceed the toxicity limit of the treated plant, as was the case with O + L compost, affecting not only plant metabolism but also crop yield, especially since it is known that plant phenols play an important role in plant defense against abiotic and biotic stresses [19,23]. However, to produce these valuable compounds, the plant must use nutrients and energy intended for growth and other primary functions needed to produce these defensive compounds [24,25]. The higher the content of phenolic compounds, the lower the growth of the plants, and consequently the lower the yield [5]. However, the use of hazelnut leaves for compost should be avoided until further studies are conducted to explain this phenomenon.

4. Conclusions

A total of 18 phenolic compounds were identified in *L. sativa*, including 10 hydroxycinnamic acids, 2 hydroxybenzoic acids, 5 flavanols, and 1 coumarin. Lettuce yield was higher when composting agent was added, regardless of compost type, indicating that composting agents are also effective. The major phenolic compound found in lettuce was chlorogenic acid, which accounted for up to 80% of the total hydroxycinnamic acids. Chlorogenic acid was followed by two flavanols (quercetin-3-*O*-(6″-malonyl)-glucoside-7-*O*-glucoside and quercetin-3-*O*-(6″-malonyl)-glucoside-7-*O*-glucuronide) and caffeoyl malate. Hydroxycinnamic acids were the major phenolic group in lettuce, accounting for up to 85% of the total analyzed phenolic content. The highest total analyzed phenolic content was found for plants grown in compost containing walnut leaves, indicating that some allelochemicals were still present in the compost of these treatments after two years, but at such a low level that they did not affect yield. Phenolic content of plants grown in compost from walnut and hazelnut leaves was also no different from controls, but yield of these plants was the lowest, suggesting that compost from walnut and hazelnut leaves could inhibit growth enough to affect plant metabolic synthesis.

We suggest that dry walnut leaves and cut grass can be used for composting, while dry hazelnut leaves, on the other hand, still contain some unknown allelochemicals after two years, which significantly and strongly inhibit plant growth and thus yield, so we would not recommend their use for composting. However, future research is needed to explain and solve this problem.

Author Contributions: Conceptualization, A.M., T.Z. and A.S.; data curation, A.M. and T.Z.; formal analysis, A.M.; funding acquisition, A.M., M.H., A.S. and R.V.; investigation, A.M. and T.Z.; methodology, A.M.; project administration, R.V.; resources, A.M., M.H., A.S. and R.V.; software, A.M.; supervision, A.S. and R.V.; validation, A.S. and R.V.; visualization, A.M., T.Z. and A.S.; writing—original draft preparation, A.M.; writing—review and editing, T.Z., A.S., M.H. and R.V. All authors have read and agreed to the published version of the manuscript.

Funding: This study is a part of program P4-0013-0481, which is funded by the Slovenian Research and Innovation Agency (ARIS).

Data Availability Statement: The remaining data presented in this study are available on request from the corresponding author. The remaining data are not publicly available due to privacy.

Conflicts of Interest: The authors declare no conflict of interest.

References

1. Barker, A.V. Composition and uses of compost. In *Agricultural Uses of by-Products and Wastes*, 1st ed.; Rechcigl, J.E., MacKinnon, H., Eds.; ACS Symposium Series; American Chemical Society: Orlando, FL, USA, 1997; Volume 668, pp. 140–162.
2. Mathur, S.P.; Owen, G.; Dinel, H.; Schnitzer, M. Determination of compost biomaturity. I. literature review. *Biol. Agric. Hortic.* **1993**, *10*, 65–85. [CrossRef]
3. Islam, A.K.M.M.; Widhalm, J.R. Agricultural uses of juglone: Opportunities and challenges. *Agronomy* **2020**, *10*, 1500. [CrossRef]
4. Medic, A.; Zamljen, T.; Slatnar, A.; Hudina, M.; Grohar, M.C.; Veberic, R. Effect of juglone and other allelochemicals in walnut leaves on yield, quality and metabolites of snack cucumber (*Cucumis sativus* L.). *Foods* **2023**, *12*, 371. [CrossRef]
5. Medic, A.; Zamljen, T.; Grohar, M.C.; Slatnar, A.; Hudina, M.; Veberic, R. Using HPLC–MS/MS to assess the quality of beet, mizuna, lettuce and corn salad after juglone and walnut leaf extract treatments. *Agronomy* **2022**, *12*, 347. [CrossRef]
6. Liu, R.H. Health-promoting components of fruits and vegetables in the diet. *Adv. Nutr.* **2013**, *4*, 384S–392S. [CrossRef]
7. Martins, S.; Mussatto, S.I.; Martínez-Avila, G.; Montañez-Saenz, J.; Aguilar, C.N.; Teixeira, J.A. Bioactive phenolic compounds: Production and extraction by solid-state fermentation. A review. *Biotechnol. Adv.* **2011**, *29*, 365–373. [CrossRef] [PubMed]
8. Santos, M.J.; Pinto, T.; Vilela, A. Sweet Chestnut (*Castanea sativa* Mill.) Nutritional and Phenolic Composition Interactions with Chestnut Flavor Physiology. *Foods* **2022**, *11*, 4052. [CrossRef] [PubMed]
9. Rana, A.; Samtiya, M.; Dhewa, T.; Mishra, V.; Aluko, R.E. Health benefits of polyphenols: A concise review. *J. Food Biochem.* **2022**, *46*, e14264. [CrossRef] [PubMed]

10. Liu, H.; Wan, Y.; Wang, Y.; Zhao, Y.; Zhang, Y.; Zhang, A.; Weng, Q.; Xu, M. Walnut polyphenol extract protects against fenitrothion-induced immunotoxicity in murine splenic lymphocytes. *Nutrients* **2018**, *10*, 1838. [CrossRef]
11. Di Noia, J. Defining powerhouse fruits and vegetables: A nutrient density approach. *Prev. Chronic Dis.* **2014**, *11*, E95. [CrossRef]
12. Medic, A.; Zamljen, T.; Slatnar, A.; Hudina, M.; Veberic, R. Is juglone the only naphthoquinone in *Juglans regia* L. with allelopathic effects? *Agriculture* **2021**, *11*, 784. [CrossRef]
13. Tirczka, I.; Hayes, M.; Prokaj, E. Evaluation of walnut (*Juglans regia* L.) leaf compost as growing media. *Hung. Agric. Eng.* **2015**, *24*, 27–30.
14. Ribas-Agustí, A.; Gratacós-Cubarsí, M.; Sárraga, C.; García-Regueiro, J.A.; Castellari, M. Analysis of eleven phenolic compounds including novel *p*-coumaroyl derivatives in lettuce (*Lactuca sativa* L.) by ultra-high-performance liquid chromatography with photodiode array and mass spectrometry detection. *Phytochem. Anal.* **2011**, *22*, 555–563. [CrossRef] [PubMed]
15. Liu, P.; Li, L.; Song, L.; Sun, X.; Yan, S.; Huang, W. Characterisation of phenolics in fruit septum of *Juglans regia* Linn. by ultra performance liquid chromatography coupled with orbitrap mass spectrometer. *Food Chem.* **2019**, *286*, 669–677. [CrossRef] [PubMed]
16. Fernández-Poyatos, M.d.P.; Ruiz-Medina, A.; Zengin, G.; Llorent-Martínez, E.J. Phenolic characterization, antioxidant activity, and enzyme inhibitory properties of *Berberis thunbergii* DC. leaves: A valuable source of phenolic acids. *Molecules* **2019**, *24*, 4171. [CrossRef]
17. Abu-Reidah, I.M.; Contreras, M.M.; Arráez-Román, D.; Segura-Carretero, A.; Fernández-Gutiérrez, A. Reversed-phase ultra-high-performance liquid chromatography coupled to electrospray ionization-quadrupole-time-of-flight mass spectrometry as a powerful tool for metabolic profiling of vegetables: *Lactuca sativa* as an example of its application. *J. Chromatogr. A* **2013**, *1313*, 212–227. [CrossRef]
18. Treutter, D. Significance of flavonoids in plant resistance and enhancement of their biosynthesis. *Plant Biol.* **2005**, *7*, 581–591. [CrossRef] [PubMed]
19. Medic, A.; Solar, A.; Hudina, M.; Veberic, R. Phenolic response to walnut anthracnose (*Ophiognomonia leptostyla*) infection in different parts of *Juglans regia* husks, using HPLC-MS/MS. *Agriculture* **2021**, *11*, 659. [CrossRef]
20. Yahyazadeh, M.; Nowak, M.; Kima, H.; Selmar, D. Horizontal natural product transfer: A potential source of alkaloidal contaminants in phytopharmaceuticals. *Phytomedicine* **2017**, *34*, 21–25. [CrossRef]
21. Singh, A.; Holvoet, S.; Mercenier, A. Dietary polyphenols in the prevention and treatment of allergic diseases. *Clin. Exp. Allergy* **2011**, *41*, 1346–1359. [CrossRef] [PubMed]
22. Schroeter, H.; Heiss, C.; Spencer, J.P.E.; Keen, C.L.; Lupton, J.R.; Schmitz, H.H. Recommending flavanols and procyanidins for cardiovascular health: Current knowledge and future needs. *Mol. Aspects Med.* **2010**, *31*, 546–557. [CrossRef] [PubMed]
23. Ahmed, S.; Stepp, J.R. Beyond yields: Climate change effects on specialty crop quality and agroecological management. *Elementa Sci. Anthr.* **2016**, *4*, 000092. [CrossRef]
24. Li, Z.-H.; Wang, Q.; Ruan, X.; Pan, C.-D.; Jiang, D.-A. Phenolics and plant allelopathy. *Molecules* **2010**, *15*, 8933–8952. [CrossRef] [PubMed]
25. Zubay, P.; Kunzelmann, J.; Ittzés, A.; Németh Zámboriné, É.; Szabó, K. Allelopathic effects of leachates of *Juglans regia* L., *Populus tremula* L. and juglone on germination of temperate zone cultivated medicinal and aromatic plants. *Agrofor. Syst.* **2021**, *95*, 431–442. [CrossRef]

foods

MDPI

Article

Cuticular Wax Triterpenes Maintain Storage Quality of Blueberries by Reducing Water Loss

Qi Kong [1], Ruiling Liu [2], Weijie Wu [2], Xiangjun Fang [2], Hangjun Chen [2], Yanchao Han [2,*] and Jianye Chen [1,*]

1 State Key Laboratory for Conservation and Utilization of Subtropical Agro-Bioresources/Guangdong, Provincial Key Laboratory of Postharvest Science of Fruits and Vegetables/Engineering Research Center of Southern Horticultural Products Preservation, Ministry of Education, College of Horticulture, South China Agricultural University, Guangzhou 510642, China; 18363900697@163.com

2 Key Laboratory of Post-Harvest Handing of Fruits, Ministry of Agriculture and Rural Affairs, Key Laboratory of Fruits and Vegetables Postharvest and Processing Technology Research of Zhejiang Province, Key Laboratory of Postharvest Preservation and Processing of Fruits and Vegetables, China National Light Industry, Food Science Institute, Zhejiang Academy of Agricultural Sciences, Hangzhou 310021, China; ruilingliu2005@163.com (R.L.); wuweijie87@163.com (W.W.); fangxiangjun2004@163.com (X.F.); hang_junchen@163.com (H.C.)

* Correspondence: hanyc@zaas.ac.cn (Y.H.); chenjianye@scau.edu.cn (J.C.)

Abstract: Cuticular wax contributes to maintaining postharvest storage quality against fruit water loss and softening. Triterpenoids, such as oleanolic acid (OA) and ursolic acid (UA), are the main components in blueberry cuticular wax, but their role in water migration during the storage of blueberries remains to be determined. Here, we examined the relationship between the content of OA and UA and the storage quality of blueberry fruit (25 °C). The results revealed that the UA content during eight-day postharvest storage ranged from 58 to 77 $\mu g\ cm^{-2}$, which was negatively related to weight loss. Additionally, we investigated the effect of exogenous OA and UA on water migration in the blueberry fruit during storage at room temperature; the weight loss was significantly lower (by 22%) with UA treatment than in the control fruit. Our findings indicate that OA and UA effectively affect water migration in blueberry fruit during postharvest storage, which could contribute to improving postharvest preservation techniques.

Keywords: blueberry; cuticular wax; triterpenoids; storage quality; hydrophobicity; moisture migration

1. Introduction

As the outer barrier between the environment and higher plants, the structure, morphology and function of the cuticle have been studied extensively [1]. Esterified fatty acids and cuticular wax make it highly hydrophobic. Cuticular wax is a crucial component of the cuticle and serves as the primary physical barrier dedicated to resisting varying external environmental stresses, such as regulating non-stomatal water loss, reflecting UV-radiation, avoiding extreme temperature damage, and providing a self-cleaning mechanism, as well as preventing pathogens and insect infection [2]. Therefore, a mounting number of researchers have pointed out that cuticular wax is strongly associated with the postharvest quality performance of fruit [3,4].

Blueberries have a powdery wax that coats their bluish-black skin, which can effectively prevent fruit from cracking and maintain texture, sensory, and nutritional quality [5]. Chu et al. [4] found that removing the cuticular wax on blueberries accelerated water loss and softening as well as a decline in apparent and physiological qualities. In addition, it is worth noting that the main defects of blueberries are usually fruit softening and dehydration during storage, seriously affecting consumer acceptance and their market value [6]. Some studies on blueberry fruit softening have concentrated on cell wall metabolism, and other potentially related factors' specificity remain unclear. Interestingly, a handful of recent research has shown that water loss was one of the factors affecting firmness changes in postharvest fruit [7,8], and noticeable differences vary among different species. It is well known that cuticular wax is one of the determinants in water-proofing barrier performance [9,10]. According to several previous reports, the removal of citrus fruit cuticular wax significantly promoted postharvest moisture loss [11], while maintaining a high cuticular wax amount in pepper fruit during storage contributed to the reduction of moisture loss [12]. Additionally, triterpenoids and sterols contained in cuticular wax also significantly influenced the postharvest moisture loss performance of pepper fruit [13]. However, cuticular wax was a kind of complex mixture, the specific components involved in the moisture migration process of blueberry fruit are largely unknown.

Cuticular wax is a complex mixture mainly made up of very-long-chain aliphatics, including fatty acids, alkanes, aldehydes, ketones, and alcohols, as well as alicyclic compounds such as triterpenoids and aromatics [14]. These compositions allow great variations among and within different species and cultivars in response to varying external environments [5,15]. Triterpenoids are the most abundant components in blueberry cuticular wax, and ursolic acid (UA) or oleanolic acid (OA) are authenticated as the major triterpenoids among nine commercial cultivars [16]. These have been shown to be highly hydrophobic and can improve the physical stability of the cuticular wax water barrier. Recently, studies on tomatoes, sweet cherries, and peaches have indicated a significant correlation between water loss performance and the proportion of n-alkanes to triterpenoids [17–19]. Moggia et al. [6] also demonstrated that triterpenoids (UA contents) in the cuticular wax of the highbush blueberry have a potential effect on fruit weight loss and softening performance. Interestingly, the blueberry cuticular wax content (especially the content of triterpenoids) continuously increased under low temperature stress (at 0.5 °C and 95% RH) [20], which contrasted with the blueberry storage at 4 °C and 90% RH [21]. However, how the changes in cuticular wax components UA and OA influence water transport and migration during storage of blueberries at room temperature (25 °C) needs to be elucidated.

Therefore, the work reported herein is intended to investigate the changes in cuticular wax components UA and OA in blueberry fruit during postharvest storage (at 25 °C and 85% RH) and then, to reveal the fruit water migration process after being treated by exogenous UA and OA. These results will help to explain the functional role of cuticular wax components in the postharvest preservation of fruit.

2. Materials and Methods

2.1. Fruit Sampling

Blueberry fruit (*Vaccinium ashei* cv. "Britewell") was harvested from a commercial field located in Anji, China. Samples for wax extraction were carefully collected with scissors and placed directly into a conical flask; each bottle contained ten fruit (a total of 15 bottles were used for subsequent wax extraction). Other samples were placed in plastic clamshells and sealed in a biological ice pack foam box and transported to the laboratory via refrigerated truck. Samples without visible wounds or disease, as well as of uniform size were picked for follow-up experiments. All samples were randomly selected and divided into 125 g plastic clamshells (containing 30 fruit each), stored at 25 °C, 85% relative humidity (RH) for eight days. Then, they were sampled every two days for subsequent index determination and cuticular wax extraction.

In order to conduct OA and UA treatment experiments, fifteen boxes were randomly divided into three groups (CK, OA, and UA treatment groups), three boxes were used only for weight loss measurement, and the other two boxes were randomly sampled (six fruit each) for water migration analysis. The OA and UA (Shanghai Macklin Biochemical technology Co., Ltd., Shanghai, China) treatment method was according to Yang et al. [22] and slightly modified: samples were treated by coating with a morpholine fatty acid salt. The emulsion system (containing morpholine fatty acids, OA, and UA) was prepared by using Tween 80 (Shanghai Aladdin Biochemical technology Co., Ltd., Shanghai, China) as surfactant under the condition of continuous heating and agitation. The sample coated with morpholine fatty acid salt alone was Control (CK), and the morpholine fatty acid salt supplemented with 10% OA and 10% UA (w/w), respectively, was treated as the experimental group. Two coats were applied evenly to each fruit and stored at 25 °C, 85% RH. The weight-loss rate and moisture migration of samples were measured every two days.

2.2. Wax Extraction and Determination

A total of 30 mature samples were taken for total wax extraction based on the method performed by Chu et al. [16]. An extract of ten blueberries was obtained by dipping them twice in 10 mL chloroform for 1 min at room temperature, then filtered by quantitative filter paper and dried under nitrogen flow at 40 °C. The dried extract is the total wax (μg). The experiment was conducted as fully randomized and triplicated.

Fruit surface area after wax extraction was determined according to the method of Ketata et al. [23]. Briefly, the equatorial diameter $[d_1]$ and polar diameter $[d_2]$ of the fruit were determined by digital vernier caliper. The surface area (cm^2) = $4\pi r^2$, $r = (d_1 + d_2)/4$. Wax content was recorded as microgramme per unit (μg cm^{-2}).

High-performance liquid chromatography (HPLC) analysis was performed on Agilent 1200 series, Eclipse XDB-C18 column (250 × 4.6 mm, 5 μm). The dried extract was made up to exactly 5 mL with methanol and filtered through a 0.22 μm membrane filter.

A mixed standard solution was prepared by OA and UA standards with 5 mL methanol. The standard working solutions contained 0.05, 0.1, 0.15, 0.2, 0.25, and 0.3 mg mL^{-1} of OA and 0.1, 0.2, 0.3, 0.4, 0.5, and 0.6 mg mL^{-1} of UA. The mobile phase of methanol: water: acetic acid: triethylamine = 265:35:0.1:0.05 ($v/v/v/v$) was used for sample testing; some test conditions were followed: flow rate (1.0 mL min^{-1}), detection wavelength (210 nm), column temperature (35 °C), and sample quantity (20 μL).

2.3. Quality Assessments

Fruit firmness was measured based on the method performed by Jiang's method [24]. Ten samples were chosen randomly and compressed to pedicle area of 5 mm under speed of 1.0 mm s^{-1}. The peak force (N) during the test represents fruit firmness. A total of 90 samples (three boxes) were selected to determine the water-loss rate. The samples were weighed as W_1 and W_2 before and after storage, and the results were then calculated by the following formula: (%) $= [(w_1 - w_2)/w_1] \times 100$ [25]. The diameter of the sample was measured in the same way as the surface area.

The scanning electron microscope (TM 3000, Hitachi, Tokyo, Japan) was used to observe the cuticular wax shape of blueberry fruit. Blueberry pericarp pieces (3 mm × 3 mm) were excised from the equatorial surface of blueberries with a blade and dried in a vacuum freeze dryer. The dried sample was then sputter coated with gold (20 nm thick) in an ion sputter coater (SBC-12, KYKY, KYKY Technology Co., Ltd., Beijing, China) [16].

2.4. Hydrophobic Properties

The water contact angles of OA and UA were quantified using Data Physics OCA-15EC [26]. The powdered samples of OA and UA were made into a round slice (radius was 1 cm) before contact angle measurements. A volume of 3 μL of deionized water droplet was placed on the circular sample sheet. The angle at which the droplet touched the sample was then recorded. The results were derived from three independent experiments.

The polarity of the two components was reflected by the diffusive position of the thin chromatography silica gel plate. The starting and ending lines were marked 1 cm from each end of the silica gel plate. The solvent system used for developing the TLC plates is petroleum ether: ethyl acetate = 2:1 (v/v). The samples were spotted on the starting line with a capillary pipette. After the samples were expanded by unfolding agent, the color was developed using a 95% sulfuric acid ethanol solution.

2.5. Measurement of Moisture Migration

According to the method performed by Zhu et al. [27], low-field nuclear magnetic resonance (LF-NMR) relaxation measurements of the samples were conducted on a MacroMR (MacroMR12-150V-I, Suzhou Niumag Analytical Instrument Corporation, Suzhou, China) with a neutron resonance frequency of 12 MHz. The standard water samples were placed into the magnet coils for individual samples, and the instrument automatically adjusted the center and RF frequency. Then, the blueberry samples were placed in the center of the magnetic field, and the transverse relaxation time (T_2) was determined by the Carr–Purcell–Meiboom–Gill (CPMG) sequence with τ value of 300 µs. Data from 12,000 echoes were acquired at 32 °C. The repetition delay between two subsequent scans was 3000 ms.

2.6. Data Analysis

Statistical analyses of the data were executed with the SPSS software V23.0. The significance was assessed using Duncan's multiple range test ($p < 0.05$), then mapped using Origin 2017. The experiment was conducted as fully randomized and triplicated, and the data were presented as means ± SD.

3. Results

3.1. Determination of Total Wax Content and Storage Quality Performance

Chloroform extracts were used to obtain total wax as well as two triterpenoid acids (OA and UA). Figure 1A,B showed that the mobile phase of HPLC can accurately separate OA and UA, and the content changes in both were shown in Figure 1C. As can be seen from Figure 1C, the total amount of wax continued to decrease with the passage of storage time, while there was no significant difference in the first four days of storage. The amount of wax present during storage decreased constantly between 512 and 333 µg cm^{-2}; by the end of storage, the wax amount had been reduced to 65% of its initial quantity. To study the effects of cuticular wax and component changes on the storage quality of blueberries, especially on the water loss and softening, we measured the weight loss, hardness, fruit diameter, and wax appearance of blueberry fruit during storage (Figure 2). Obviously, it was expected that the weight-loss rate increased significantly and reached 11% at the end of storage (Figure 2A). The fruit diameter (Figure 2B) and firmness (Figure 2C) decreased to 79% and 48% of their initial storage levels, respectively, along with smoothing out the waxy epidermis (Figure 2D). According to these results, wax content in the fruit epidermis may have a significant impact on the exterior quality during storage. Therefore, we hypothesized that the wax content may have an effect on the process of water migration.

3.2. Correlation between Storage Qualities and Contents of Triterpenoid Substance OA and UA

Triterpenoids are major components in blueberry cuticular wax, mainly OA and UA [16]. According to Figure 1A,B, the peak spectra of HPLC was used to determine the contents of OA and UA. As Figure 1C showed, UA content was higher than OA, and it appears that the changes in OA and UA were not fully consistent with the changes in total wax content, as the content of UA and OA increased during the first four days of storage when the total wax was unchanged. These different evolutions could be attributed to different metabolic processes among complex components of cuticular wax. Interestingly, the content at the end of storage was not significantly different from that at the beginning. A typical UA sample had an initial storage content of about 58 µg cm^{-2}, increased to about 77 µg cm^{-2} after four days of storage, and decreased to 60 µg cm^{-2} after the last day of storage, which amounted to no significant changes compared with the beginning of storage. After four days of storage, the total amount of wax gradually decreased; UA and OA showed the same trend, with a significant decrease on the sixth day. Subsequently, we conducted correlation analysis between fruit storage quality indices (weight loss, hardness, fruit diameter) and changes in total wax and OA and UA contents (Figure 3). The results showed that there was a significant correlation ($p \leq 0.05$) between fruit quality and wax change during storage. A negative relationship was found between persistent water loss during storage and wax amount, UA contents, fruit diameters, and hardness (r < −0.95). Moreover, as the most abundant components of blueberry wax (OA and UA), the changes in UA and OA content were closely related to the wax content, despite some differences in wax content and their changing trends during storage. These results indicated that the contents of total wax and UA may be closely related to the storage quality of blueberries, especially affecting the water loss, shrinkage, and softening of blueberries during storage.

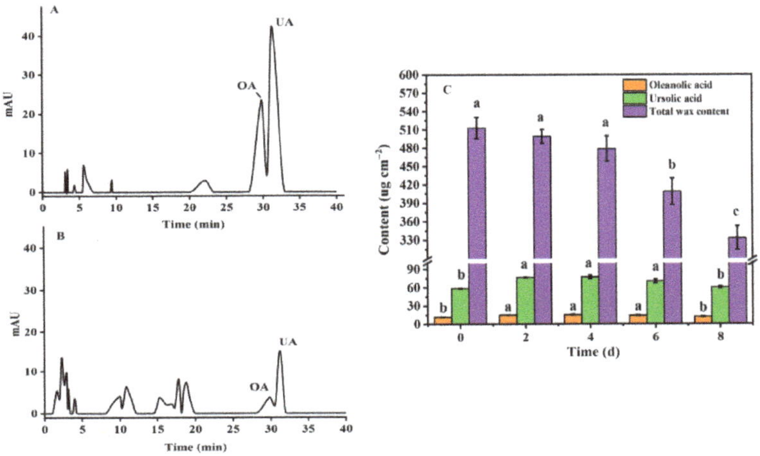

Figure 1. Changes in total wax and triterpenoid composition content during storage at room temperature (25 °C) in blueberry. (**A**) and (**B**) were HPLC peak spectra of the standard and sample, respectively, suggesting that the mobile phase used in the experiment can separate the two components. (**C**) Changes in total wax and triterpenoid composition content. Data are presented as means ± SD. Letters show significance ($p < 0.05$).

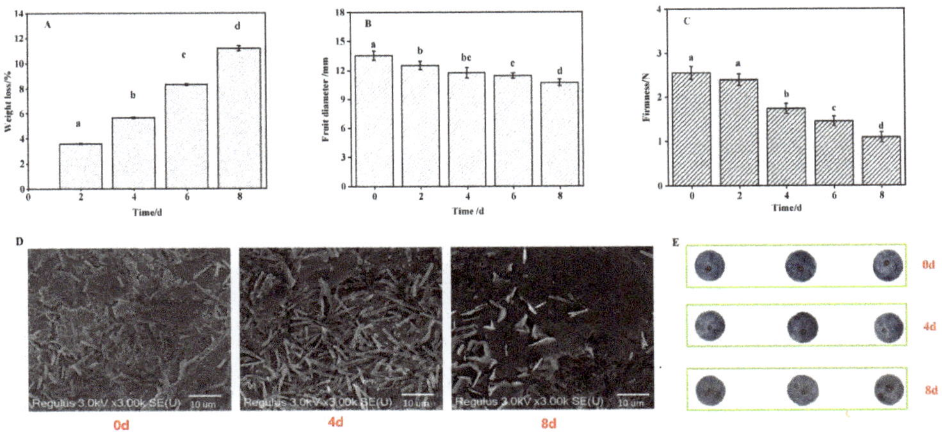

Figure 2. Storage quality performance of fruit: (**A**) weight loss, (**B**) fruit diameter, (**C**) firmness, (**D**) status of cuticular wax, and (**E**) appearance of samples. The image of the status of cuticle wax in blueberry fruits has been magnified 3000 times. In (**A–C**), mean differences between the bars are significant at $p < 0.05$ level for different letters but not significant for the same letters.

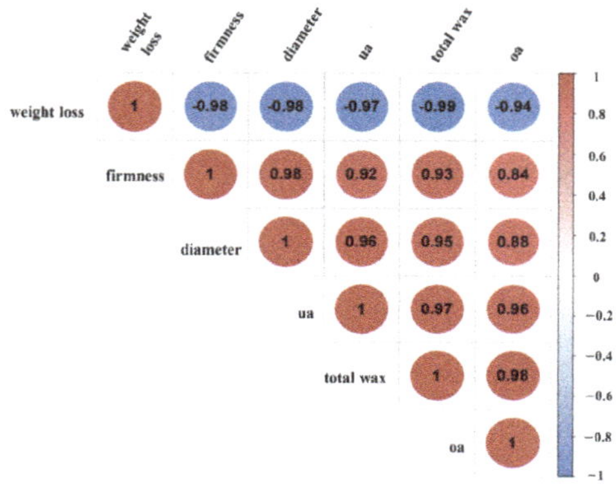

Figure 3. Correlation coefficients between fruit characteristics and wax content. Pearson correlation coefficient calculation method was used to establish correlation ($p < 0.05$). The red zone demonstrates a positive correlation between two variables, the blue suggests a negative correlation between two variables, and the numbers in each cell indicate the correlation coefficient.

3.3. Hydrophobic Characteristics of OA and UA

To explore the hydrophobic characteristics of the wax components OA and UA, we characterized the two components by molecular polarity experiments and water droplet contact angles. As can be seen from Figure 4A, for OA and UA in the same polarity diffusion agent, thin-layer chromatography on silica gel plate diffusion rate and position showed that UA has a lower molecular polarity, indicating that UA has stronger hydrophobic ability; when attached to the blueberry fruit skin, it may be more likely to form a dense hydrophobic protective layer, helping to suppress the water loss of the fruit. Meanwhile, the analysis after measuring the droplet contact angles of the two components also showed that although both of them were larger than 90 and had hydrophobic ability, the droplet contact angle of UA was larger than that of OA, suggesting that UA has a stronger hydrophobic ability (Figure 4B,C). Based on these results, UA was more hydrophobic with lower polarity than OA, despite their both being hydrophobic. Therefore, we hypothesized that blueberry fruit may benefit from the treatment of the wax components (OA and UA) during storage by slowing down water migration and retaining water in storage, extending its shelf life.

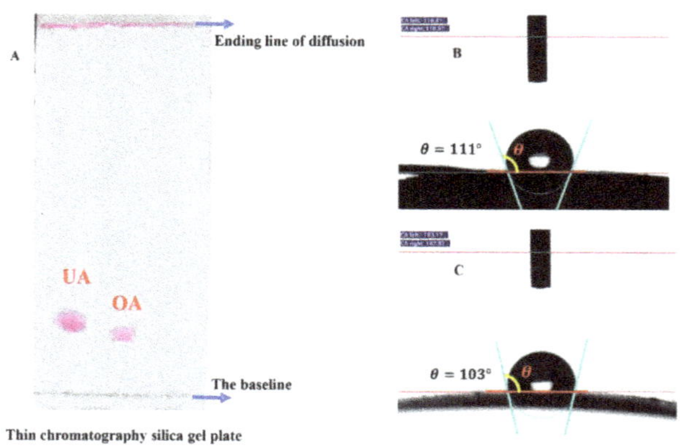

Figure 4. Hydrophobic characterization of OA and UA. (**A**)The polarity of the two components was reflected by the diffusion position of the thin chromatography silica gel plate, the faster the rate of diffusion, the less polar the component. (**B**) and (**C**) were the droplet contact angles of UA and OA, respectively. Water droplet contact angle greater than 90 is hydrophobic material; the larger the angle, the stronger the performance.

3.4. Effects of OA and UA on Fruit Moisture Migration

In light of the above findings, cuticular wax components may be involved in the water retention process due to their hydrophobic properties and the formation of hydrophobic layers on fruit surfaces. To further verify the role of triterpenoid substances OA and UA in fruit water migration during storage, CK, OA, and UA treatment experiments were conducted, and the water migration process of fruit was measured within eight days by LF-NMR. As can be seen from Figures 5 and 6, based on the inversion of lateral relaxation time (T_2), it was determined that the content and composition of water inside the fruit were mostly free water (T_{23}), and the peak area A_{23} represents the changes in the contents. It was not only that the free water of fruit kept losing over time, but it also went through a process of transforming from free to bound water. Since the CK group had the fastest fruit water loss, there was no significant difference in water content between all groups during the first four days storage due to the transition from free to bound water. As the storage period progressed, the water content of the CK group decreased significantly, and the differences with that of the other two treatment groups (OA and UA) became apparent after four days. In turn, this clearly leads to substantial differences in fruit weight-loss rates. In terms of bound water conversion, OA and UA treatment groups performed significantly better than the CK group, which may be a consequence of a hydrophobic layer that prevents free water from dispersing and encourages its conversion into bound water on the surface of the fruit. However, it is worth noting that there was no significant difference between OA and UA treatment groups in fruit water content change or weight loss change during storage, which indicates that OA and UA blueberry wax components may play similar roles in fruit water migration.

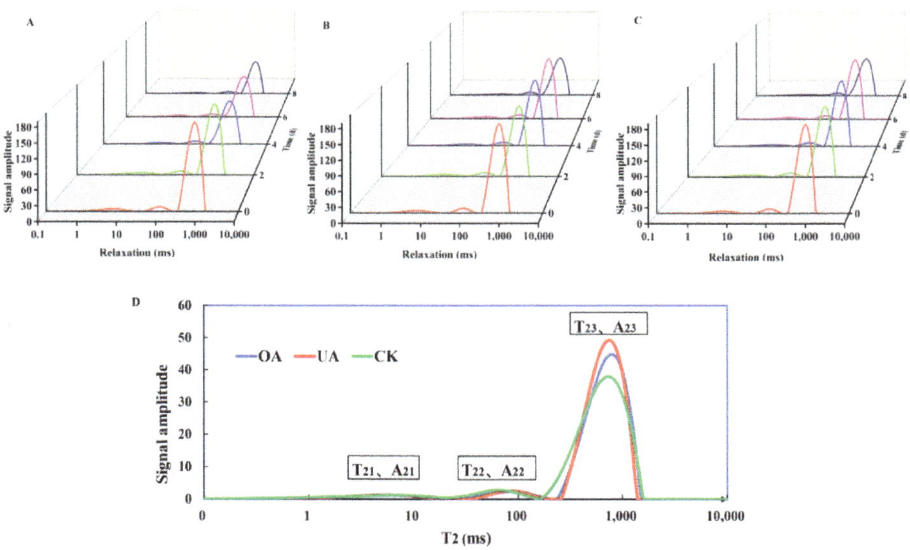

Figure 5. LF-NMR T_2 relaxation time curve of blueberries during storage. (**A–C**) were the changes in signal amplitudes of different water molecules in blueberry fruits treated with CK, OA, and UA, respectively. The signal amplitude was positively correlated with the content. (**D**): T_{21}, T_{22}, and T_{23} refers to the transverse relaxation times of the bound water and free water molecules, respectively. A_{2i} is the peak area of T_2 signal inversion of different component, which can reflect the component content. It can be seen that the peak area of T_{23} of all samples is the largest, indicating that the dominance of the free water proportion. In (**A**), (**B**) and (**C**), the five colors represent the peak spectra at 0, 2, 4, 6 and 8 days, respectively, corresponding to the time axis.

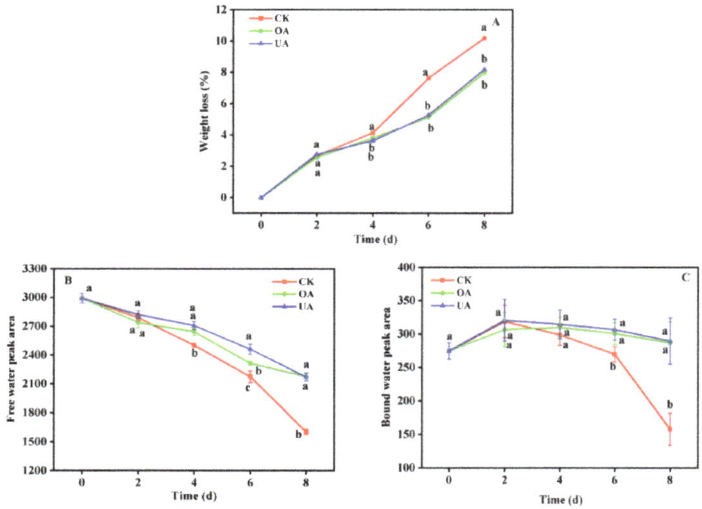

Figure 6. Effects of OA and UA treatments on moisture loss. (**A**) Changes in weight loss during storage. (**B**) and (**C**) were the changes in fruit free water and bound water peak areas under different treatments, respectively, which represent the changes in water molecules status in the fruit. In (**A–C**), mean differences between the bars are significant at $p < 0.05$ level for different letters but not significant for the same letters.

4. Discussion

Many of the quality characteristics of fleshy fruit depend on moisture, which makes up a large percentage of their wet weight and plays a crucial role in metabolic activity and quality, such as protein denaturation, enzyme activity, and hardness [28]. After harvesting, the process of absorbing water from the plant roots is terminated, and the consumption of water by various life activities leads to changes in the structure, texture, and the surface of the fruit. In the present study, fruit softening, shrinking, and other phenomena gradually occur as the weight of the fruit decreases during storage, and there is a strong correlation between them [8]. Water loss is a common phenomenon in postharvest fruit, which is easily affected by fruit surface structure. The cuticle is the outermost structure of the fruit, and its hydrophobicity endows the fruit an effective barrier to prevent water loss [29]. Cuticular wax attached to the cuticle surface has been fully confirmed to affect the water permeability of fruit, inhibit non-stomatal water loss, and contribute to the maintenance of storage quality [30,31]. Cuticular wax thickness and total wax content are important factors affecting water loss [32]. Verifying the correlation between cuticular wax and storage-quality wax has been carried out in blueberry fruit [4]. For instance, remove the cuticular wax from the blueberry significantly enhanced fruit water loss and rot rate, decreased fruit sensory quality, and reduced shelf life [4]. Here, as can be observed from Figures 1 and 3, the total wax content decreased continuously, which was negatively related to the weight loss of fruit during storage. Further evidence by analysis of hardness and fruit diameter as well as by scanning electron microscope observation verified that the content of wax in blueberries was closely related to water loss and quality (Figure 3). However, the wax content during cold storage (0.5 °C) has the opposite change trend [20]; the increase in wax content may be a stress reaction against cold stress. Therefore, the variety of storage environments may contribute to different wax deposition in blueberries.

Moreover, considerable effort was also focused on the effect of the wax composition and structure on the permeability of the epidermis [33]. As reported in previous studies, aliphatic VLCFA derivatives could greatly help to enhance the effect of the epidermal transpiration barrier [34,35]. The existence of other wax components such as triterpenoids, which are highly hydrophobic, and can improve the physical stability of water barrier [22]. Likewise, in the current study, although the triterpenoids OA and UA showed first an increase and then a decrease in their content, the correlation analysis further demonstrated that they were closely related to fruit weight loss as well as the amount of cuticular wax (Figure 3). Therefore, the explanation for the correlation between UA content and fruit quality may be linked to their effect on water loss. Moreover, as the substances with the highest contents of waxy triterpenes in blueberry fruit, the droplet contact angles of OA and UA proved their hydrophobic properties. Further evidence by the thin-plate chromatography experiment showed that UA and OA exhibit similar molecular polarity in the solvent systems, with very little polarity. Therefore, triterpenes may play an important role in limiting water loss during storage.

Recently, LF-NMR has been gradually conducted to monitor changes in the moisture status of fruit and vegetables under a variety of conditions due to the advantages of large amplitude range, sensitive phase response, and short detection time [36,37]. The detected T_2 signal (transverse relaxation time) could display the content and status of water molecules in samples [38]. As mentioned in the results of LF-NMR measurements, water migration and state changes in sweet cherries were highly correlated with water loss and fruit softening [27]. Similarly, in this work, the coating treatments with OA and UA on blueberries demonstrated the potential role of triterpenoids in suppressing water loss. Our study was similar to the previous studies on highbush blueberries, which suggested that the content and proportion of triterpenoids negatively correlated with the water-loss rate [20]. There were no obvious differences between OA and UA treatment in free water and bound water, while they were significantly higher than CK. It can therefore be assumed that the treatment of OA and UA not only suppresses the free water loss but also promotes the process of conversion from free water to bound water. Obviously, this results in lower fruit weight loss. At the end of the storage period, weight loss was markedly lower (by 22%) in the blueberry fruit with UA treatment than in the control fruit. Hence, it is conceivable to suppose that part of the life activities involved in water in fruit may be inhibited, which in turn prolongs the shelf life. This important water migration process that we found was also reported in the blueberries stored at low temperatures [28]. Interestingly, they also pointed out that dehydration was a manifestation of chilling injury.

In addition, there are many factors affecting fruit water loss [39]. Although triterpenoids may have a potential effect on preventing fruit water loss, definitive conclusions cannot be drawn directly. A study by Moggia [6] showed that the samples suffered higher deterioration rates with higher UA content during storage. They found differences in UA and OA content between two different blueberry varieties with different water loss performances, but there was no significant difference in the total amount of triterpenoids. This suggests that triterpenoids may not be the only factor affecting water loss in cuticular wax components. The existence of other cuticular wax components such as alkanes similarly plays an important role in elevating the hydrophobic properties [22,40]. Leide et al. [17] reported that, compared with the wild-type, the cuticular wax of mutant tomato fruit was almost completely devoid of n-alkanes and lost water very easily. Yang et al. [22] reported that water loss in six stored citrus varieties was significantly correlated with cuticle alkanes, and coating with C_{28} alkane could significantly decrease moisture loss. Interestingly, some studies have also shown that in some stored fruit and vegetables, such as tomatoes [17], peppers [13], sweet cherries [18], and peaches [19], the water-loss rate was closely correlated with the ratio of normal alkanes to triterpenes and sterols. However, in this work, the alkanes did not received the necessary attention and research due to their relatively low amount in blueberry cuticular wax. Therefore, factors such as variety differences, cuticle structure and thickness, other cuticular wax components, and even storage environment need to be investigated in further studies.

5. Conclusions

In conclusion, cuticular wax is of great significance for the preservation of blueberries after harvesting. Through HPLC content detection and correlation analysis, the content of the triterpenoid substance UA in blueberry cuticular wax was positively correlated with the total wax content, fruit firmness, and fruit diameter. Moreover, UA and OA have small molecular polarities and stable hydrophobic properties, which play important roles in the formation of surface water barriers in fruits. More importantly, the LF-NMR detection of moisture migration in UA- and OA-treated fruit provided substantial evidence that water loss performance was significantly inhibited, and the free and bound water content of the fruit slowly decreased. Our findings will help to reveal the functional role of cuticular wax components in fruit preservation, and they suggest a new perspective in the development of natural packaging to maintain the postharvest quality of blueberries.

Author Contributions: Conceptualization, Y.H.; methodology, Q.K. and R.L.; formal analysis, W.W.; investigation, X.F.; resources, H.C.; data curation, Q.K.; writing—original draft preparation, Q.K.; writing—review and editing, Y.H. and J.C.; project administration, Y.H. All authors have read and agreed to the published version of the manuscript.

Funding: This research was funded by the National Natural Science Foundation of China (Grant No. 31772042), the Key Research & Development Program of Zhejiang Province (2021C02015).

Data Availability Statement: The data used to support the findings of this study can be made available by the corresponding author upon request.

Conflicts of Interest: The authors declare no conflict of interest.

References

1. Philippe, G.; Sørensen, I.; Jiao, C.; Sun, X.; Fei, Z.; Domozych, D.S.; Rose, J.K. Cutin and suberin: Assembly and origins of specialized lipidic cell wall scaffolds. *Curr. Opin. Plant Biol.* **2020**, *55*, 11–20. [CrossRef] [PubMed]
2. Serrano, M.; Coluccia, F.; Torres, M.; L'Haridon, F.; MeTraux, J.-P. The cuticle and plant defense to pathogens. *Front. Plant Sci.* **2014**, *5*, 274. [CrossRef] [PubMed]
3. Zhu, S.; Huang, S.; Lin, X.; Wan, X.; Zhang, Q.; Peng, J.; Luo, D.; Zhang, Y.; Dong, X. The relationships between waxes and storage quality indexes of fruits of three plum cultivars. *Foods* **2023**, *12*, 1717. [CrossRef] [PubMed]
4. Chu, W.; Gao, H.; Chen, H.; Fang, X.; Zheng, Y. Effects of cuticular wax on the postharvest quality of blueberry fruit. *Food Chem.* **2018**, *239*, 68–74. [CrossRef] [PubMed]
5. Trivedi, P.; Karppinen, K.; Klavins, L.; Kviesis, J.; Sundqvist, P.; Nguyen, N.; Heinonen, E.; Klavins, M.; Jaakola, L.; Väänänen, J.; et al. Compositional and morphological analyses of wax in northern wild berry species. *Food Chem.* **2019**, *295*, 441–448. [CrossRef]
6. Moggia, C.; Graell, J.; Lara, I.; Schmeda-Hirschmann, G.; Thomas-Valdés, S.; Lobos, G.A. Fruit characteristics and cuticle triterpenes as related to postharvest quality of highbush blueberries. *Sci. Hortic.* **2016**, *211*, 449–457. [CrossRef]
7. Saei, A.; Tustin, D.; Zamani, Z.; Talaie, A.; Hall, A. Cropping effects on the loss of apple fruit firmness during storage: The relationship between texture retention and fruit dry matter concentration. *Sci. Hortic.* **2011**, *130*, 256–265. [CrossRef]
8. Paniagua, A.; East, A.; Hindmarsh, J.; Heyes, J. Moisture loss is the major cause of firmness change during postharvest storage of blueberry. *Postharvest Biol. Technol.* **2013**, *79*, 13–19. [CrossRef]
9. Fernández, V.; Khayet, M.; Montero-Prado, P.; Heredia-Guerrero, J.A.; Liakopoulos, G.; Karabourniotis, G.; del Río, V.; Domínguez, E.; Tacchini, I.; Nerín, C.; et al. New insights into the properties of pubescent surfaces: Peach fruit as a model. *Plant Physiol.* **2011**, *156*, 2098–2108. [CrossRef]
10. Lara, I.; Belge, B.; Goulao, L.F. The fruit cuticle as a modulator of postharvest quality. *Postharvest Biol. Technol.* **2014**, *87*, 103–112. [CrossRef]
11. Mukhtar, A.; Damerow, L.; Blanke, M. Non-invasive assessment of glossiness and polishing of the wax bloom of European plum. *Postharvest Biol. Technol.* **2014**, *87*, 144–151. [CrossRef]
12. Kissinger, M.; Tuvia-Alkalai, S.; Shalom, Y.; Fallik, E.; Elkind, Y.; Jenks, M.A.; Goodwin, M.S. Characterization of physiological and biochemical factors associated with postharvest water loss in ripe pepper fruit during storage. *J. Am. Soc. Hortic. Sci.* **2005**, *130*, 735–741. [CrossRef]
13. Parsons, E.P.; Popopvsky, S.; Lohrey, G.T.; Lü, S.; Alkalai-Tuvia, S.; Perzelan, Y.; Paran, I.; Fallik, E.; Jenks, M.A. Fruit cuticle lipid composition and fruit post-harvest water loss in an advanced backcross generation of pepper (*Capsicum* sp.). *Physiol. Plant.* **2012**, *146*, 15–25. [CrossRef]
14. Lara, I.; Belge, B.; Goulao, L.F. A focus on the biosynthesis and composition of cuticle in fruits. *J. Agric. Food Chem.* **2015**, *63*, 4005–4019. [CrossRef]

15. Leandro, O.L.; Bénédicte, Q.T.; Claire, D.; Marie-Noëlle, C.; René, L.; Michel, G.; Jean-Luc, P. Cuticular waxes of nectarines (*Prunus persica* L. Batsch) during fruit development in relation to surface conductance and susceptibility to *Monilinia laxa*. *J. Exp. Bot.* **2020**, *71*, 5521–5537.
16. Chu, W.; Gao, H.; Cao, S.; Fang, X.; Chen, H.; Xiao, S. Composition and morphology of cuticular wax in blueberry (Vaccinium spp.) fruits. *Food Chem.* **2017**, *219*, 436–442. [CrossRef]
17. Leide, J.; Hildebrandt, U.; Vogg, G.; Riederer, M. The positional sterile (ps) mutation affects cuticular transpiration and wax biosynthesis of tomato fruits. *J. Plant Physiol.* **2011**, *168*, 871–877. [CrossRef]
18. Belge, B.; Llovera, M.; Comabella, E.; Gatius, F.; Guillén, P.; Graell, J.; Lara, I. Characterization of cuticle composition after cold storage of Celeste and Somerset sweet cherry fruit. *J. Agric. Food Chem.* **2014**, *62*, 8722–8729. [CrossRef]
19. Belge, B.; Llovera, M.; Comabella, E.; Graell, J.; Lara, I. Fruit cuticle composition of a melting and a nonmelting peach cultivar. *J. Agric. Food Chem.* **2014**, *62*, 3488–3495. [CrossRef]
20. Yan, Y.; Castellarin, S.D. Blueberry water loss is related to both cuticular wax composition and stem scar size. *Postharvest Biol. Technol.* **2022**, *188*, 111907. [CrossRef]
21. Chu, W.; Gao, H.; Chen, H.; Wu, W.-J.; Fang, X. Changes in cuticular wax composition of two blueberry cultivars during fruit ripening and postharvest cold storage. *J. Agric. Food Chem.* **2018**, *66*, 2870–2876. [CrossRef] [PubMed]
22. Yang, H.; Zou, Y.; Li, X.; Zhang, M.; Zhu, Z.; Xu, R.; Xu, J.; Deng, X.; Cheng, Y. QTL analysis reveals the effect of CER1-1 and CER1-3 to reduce fruit water loss by increasing cuticular wax alkanes in citrus fruit. *J. Plant Physiol.* **2022**, *185*, 111771. [CrossRef]
23. Ketata, M.; Desjardins, Y.; Ratti, C. Effect of liquid nitrogen pretreatments on osmotic dehydration of blueberries. *J. Food Eng.* **2013**, *116*, 202–212. [CrossRef]
24. Jiang, B.; Liu, R.; Fang, X.; Tong, C.; Chen, H.; Gao, H. Effects of salicylic acid treatment on fruit quality and wax composition of blueberry (*Vaccinium virgatum* Ait). *Food Chem.* **2022**, *368*, 130757. [CrossRef]
25. Wang, S.; Zhou, Q.; Zhou, X.; Wei, B.; Ji, S. The effect of ethylene absorbent treatment on the softening of blueberry fruit. *Food Chem.* **2018**, *246*, 286–294. [CrossRef]
26. Suhas, D.; Aminabhavi, T.; Raghu, A. para-Toluene sulfonic acid treated clay loaded sodium alginate membranes for enhanced pervaporative dehydration of isopropanol. *Appl. Clay Sci.* **2014**, *101*, 419–429. [CrossRef]
27. Zhu, D.; Liang, J.; Liu, H.; Cao, X.; Ge, Y.; Li, J. Sweet cherry softening accompanied with moisture migration and loss during low-temperature storage. *J. Sci. Food Agric.* **2018**, *98*, 3651–3658. [CrossRef]
28. Wang, Y.; Ji, S.; Dai, H.; Kong, X.; Hao, J.; Wang, S.; Zhou, X.; Zhao, Y.; Wei, B.; Cheng, S.; et al. Changes in membrane lipid metabolism accompany pitting in blueberry during refrigeration and subsequent storage at room temperature. *Front. Plant Sci.* **2019**, *10*, 829. [CrossRef]
29. Yeats, T.H.; Rose, J.K. The formation and function of plant cuticles. *Plant Physiol.* **2013**, *163*, 5–20. [CrossRef]
30. Trivedi, P.; Nguyen, N.; Hykkerud, A.L.; Häggman, H.; Martinussen, I.; Jaakola, L.; Karppinen, K. Developmental and environmental regulation of cuticular wax biosynthesis in fleshy fruits. *Front. Plant Sci.* **2019**, *10*, 431. [CrossRef]
31. Lin, X.; Huang, S.A.; Huber, D.J.; Zhang, Q.; Wan, X.; Peng, J.S.; Luo, D.C.; Dong, X.Q.; Zhu, S.L. Melatonin treatment affects wax composition and maintains storage quality in 'kongxin' plum (*Prunus salicina* L. cv) during postharvest. *Foods* **2022**, *11*, 3972. [CrossRef]
32. Yang, M.; Luo, Z.; Li, D.; Ma, C.; Li, L. Role of epicuticular wax involved in quality maintenance of table grapes: Evidence from transcriptomic data. *Food Chem.* **2023**, *195*, 112155. [CrossRef]
33. Kosma, D.; Parsons, E.P.; Isaacson, T.; Lü, S.; Rose, J.K.C.; Jenks, M.A. Fruit cuticle lipid composition during development in tomato ripening mutants. *Physiol. Plant.* **2010**, *139*, 107–117. [CrossRef]
34. Vogg, G.; Fischer, S.; Leide, J.; Emmanuel, E.; Jetter, R.; Levy, A.A.; Riederer, M. Tomato fruit cuticular waxes and their effects on transpiration barrier properties: Functional characterization of a mutant deficient in a very-long-chain fatty acid β-ketoacyl-CoA synthase. *J. Exp. Bot.* **2004**, *55*, 1401–1410. [CrossRef]
35. Jetter, R.; Riederer, M. Localization of the transpiration barrier in the epi- and intracuticular waxes of eight plant species: Water transport resistances are associated with fatty acyl rather than alicyclic components. *Plant Physiol.* **2016**, *170*, 921–934. [CrossRef]
36. Li, L.; Zhang, M.; Yang, P. Suitability of LF-NMR to analysis water state and predict dielectric properties of Chinese yam during microwave vacuum drying. *LWT-Food Sci. Technol.* **2019**, *105*, 257–264. [CrossRef]
37. Sun, Q.; Zhang, M.; Yang, P. Combination of LF-NMR and BP-ANN to monitor water states of typical fruits and vegetables during microwave vacuum drying. *LWT-Food Sci. Technol.* **2019**, *116*, 108548. [CrossRef]
38. Sánchez-Alonso, I.; Martinez, I.; Sánchez-Valencia, J.; Careche, M. Estimation of freezing storage time and quality changes in hake (*Merluccius merluccius* L.) by low field NMR. *Food Chem.* **2012**, *135*, 1626–1634. [CrossRef]
39. Lufu, R.; Ambaw, A.; Opara, U.L. Water loss of fresh fruit: Influencing pre-harvest, harvest and postharvest factors. *Sci. Hortic.* **2020**, *272*, 109519. [CrossRef]
40. Parsons, E.P.; Popopvsky, S.; Lohrey, G.T.; Alkalai-Tuvia, S.; Perzelan, Y.; Bosland, P.; Bebeli, P.J.; Paran, I.; Fallik, E.; Jenks, M.A. Fruit cuticle lipid composition and water loss in a diverse collection of pepper (*Capsicum*). *Physiol. Plant.* **2013**, *149*, 160–174. [CrossRef]

foods

MDPI

Article

Postharvest Calcium Chloride Treatment Strengthens Cell Wall Structure to Maintain Litchi Fruit Quality

Xiaomeng Guo [1,2,3,†], Qiao Li [1,2,3,†], Tao Luo [1,2,3], Dongmei Han [4], Difa Zhu [1,2,3] and Zhenxian Wu [1,2,3,*]

1 College of Horticulture, South China Agricultural University, Guangzhou 510642, China; guoxm_scau@163.com (X.G.); liqiao@stu.scau.edu.cn (Q.L.); luotao0502@scau.edu.cn (T.L.); dfzhu@stu.scau.edu.cn (D.Z.)
2 Guangdong Provincial Key Laboratory of Postharvest Science of Fruits and Vegetables, Engineering Research Center of Southern Horticultural Products Preservation, Ministry of Education, Guangzhou 510642, China
3 Key Laboratory of Biology and Genetic Improvement of Horticultural Crops (South China), Ministry of Agriculture and Rural Affairs, College of Horticulture, South China Agricultural University, Guangzhou 510642, China
4 Institute of Fruit Tree Research, Guangdong Academy of Agricultural Sciences, Key Laboratory of South Subtropical Fruit Biology and Genetic Resource Utilization, Ministry of Agriculture, Guangzhou 510640, China; handongmei@gdaas.cn
* Correspondence: zhenxwu@scau.edu.cn
† These authors contributed equally to this work.

Abstract: Litchi (*Litchi chinensis* Sonn.) fruit deterioration occurs rapidly after harvest and is characterized by pericarp browning, pulp softening, and decay. In this study, we found that calcium chloride ($CaCl_2$) treatment (5 g L^{-1} $CaCl_2$ solution vacuum infiltration for 5 min) affected the cell wall component contents and cell wall-degrading enzyme activities of litchi fruit during storage at room temperature. $CaCl_2$ treatment significantly increased the contents of Ca^{2+} and cellulose, while it decreased the water-soluble pectin content, and the activities of polygalacturonase, β-galactosidase, and cellulase in the litchi pericarp. Meanwhile, the treatment resulted in significantly increased contents of Ca^{2+}, water-soluble pectin, ionic-soluble pectin, covalent-soluble pectin and hemicellulose, and upregulated activities of pectinesterase and β-galactosidase, while significantly decreasing the activities of polygalacturonase and cellulase in litchi pulp. The above results indicate that $CaCl_2$ treatment strengthened the cell wall structure of litchi fruit. More importantly, the enzymatic browning of the pericarp, softening of the pulp, and disease incidence were delayed. The treatment had a more pronounced effect on the pericarp than on the pulp. We consider $CaCl_2$ treatment to be a safe and effective treatment for maintaining the postharvest quality of litchi fruit.

Keywords: litchi; postharvest; calcium chloride; cell wall; pericarp browning; pulp; decay

1. Introduction

Litchi (*Litchi chinensis* Sonn.) is a fruit indigenous to China [1]. China is the leading litchi producer in the world [2]. Pericarp browning, pulp deterioration, and decay are major postharvest problems that significantly reduce the commodity value of litchi fruit, and limit the development of the industry [3]. In recent years, safe and effective treatments have been developed to meet the increasing demand for maintaining the postharvest quality of litchi fruit. Recently, many substances, such as proanthocyanidin [4], melatonin [5], apple polyphenols [6], fluopyram [7], methyl jasmonate [8], benzothiadiazole [9], and α-lipoic acid [10] have been used to treat postharvest litchi fruit and achieved good effects.

Calcium treatment is a safe treatment that effectively delays fruit ripening and senescence, and reduces the incidence of physiological and pathological diseases. Specifically, Ca^{2+} could maintain or strengthen the structure and function of the cell wall and membrane, reduce the respiration and ethylene synthesis of fruit, and regulate enzyme activity and the senescence-related metabolism [11]. Calcium treatment, especially calcium chloride

($CaCl_2$) treatment, has been shown to have a positive effect on the postharvest preservation of many fruits. For example, $CaCl_2$ treatment effectively maintains qualities as the firmness, color, sugar and acid contents of tomato [12], date [13], and apple [14], and enhances the antioxidation of grape [15], jujube [16,17], pear [18], banana [19], blackberry [20], and papaya [21]. $CaCl_2$ treatment also significantly maintains or strengthens the cell wall structure of grape [22], apple [23], red raspberry [24], papaya [25], melon [26], apricot [27], and sweet cherry [28], and improves the cold resistance of apple [29], peach [30], jujube [16], loquat [31], and banana [32], in addition to reducing the decay of melon [26,33]. However, the application of $CaCl_2$ treatment in the postharvest preservation of litchi fruit has not been reported.

On the one hand, the litchi pericarp is very thin, and susceptible to browning and decay due to mechanical damage, disease infection, and water loss. On the other hand, litchi pulp softens quickly after the harvest due to the degradation of cell wall polysaccharides, especially when stored at room temperature [34]. Considering the widely reported effectiveness of $CaCl_2$ treatment in improving postharvest resistance and maintaining fruit quality, we tested the effect of this treatment on the postharvest storage of litchi fruit and analyzed the related physiological and biochemical indices. We believe that $CaCl_2$ treatment could be a promising method to solve the above-mentioned postharvest problems of litchi fruit. This study aims to provide a $CaCl_2$ treatment method for maintaining the quality of litchi fruit stored at room temperature and to investigate the mechanism of $CaCl_2$ treatment on litchi fruit.

2. Materials and Methods

2.1. Plant Material and Treatments

Commercially mature (bright red pericarp) litchi fruits (cultivar 'Guiwei') were obtained from a commercial orchard in Guangzhou, China. Approximately 600 healthy litchi fruits were divided into two groups. Each of the two groups of fruits was vacuum-infiltrated (−33 kPa) in 20 L of water (control), or 5 g L^{-1} $CaCl_2$ (Damao, China) solution ($CaCl_2$ treatment) for 5 min at 20 °C, and then air-dried. The fruits were stored at room temperature (25 °C ± 2 °C) in polyethylene terephthalate trays (approximately 20 fruits per tray), and packaged with 0.01 mm polyvinyl chloride film. A total of 60 litchi fruits per group were analyzed in triplicate on days 0, 1, 3 and 5. Meanwhile, the pericarp and pulp of 20 litchi fruits were divided into three replicates, frozen in liquid nitrogen, and stored at −80 °C.

2.2. Determination of Pericarp Browning Index and Decay

In total, 60 fruits (at the last point in time, there was a total of 100 fruits) were randomly selected for each treatment to evaluate the pericarp browning index and decay according to the method described by Guo et al. [3]. The details are shown in the Supplementary Materials.

2.3. Determination of Calcium Content

The calcium content was determined according to the method described by Erdogan et al. with modifications [35]. A total of 0.2 g of oven-dried pericarp or 0.4 g of oven-dried pulp samples and 5 mL of concentrated HNO_3 were added to the digestion vessel overnight. Then, 3 mL of 30% H_2O_2 was added for microwave digestion. After cooling the digestion vessel to room temperature, the contents were transferred to a volumetric flask, diluted to 50 mL or 25 mL with deionized water, and filtered with filter paper for determination using a flame atomic absorption spectrometer (Spectr AA 220FS, Varian, Palo Alto, CA, USA).

2.4. Determination of Cell Wall Component Contents

The cell wall components, including water-soluble pectin (WSP), ionic-soluble pectin (ISP), covalent-soluble pectin (CSP), hemicellulose, and cellulose, were extracted and determined based on the method described by Lin et al. [36]. Briefly, 5 g of ground samples of pericarp or pulp were boiled with 80% ethanol, and centrifuged repeatedly to inactivate enzymes and remove sugars. The residues were washed three times with chloroform/methanol (1:1), then dissolved in 90% dimethyl sulfoxide overnight, and centrifuged until all starch was removed. The residues were thoroughly dried at 40 °C as cell wall components after washing three times with acetone.

Cell wall components were immersed in ultrapure water for 6 h to obtain WSP. The water-insoluble residues were immersed in 50mM sodium acetate buffer (pH 6.8) containing 1M 1,2-cyclohexanediaminetetraacetic acid (CDTA) for 6 h to obtain ISP. The CDTA-insoluble residues were immersed in 50 mM sodium carbonate (Na_2CO_3) containing 20 mM sodium borohydride ($NaBH_4$) for 6 h to obtain CSP. The Na_2CO_3-insoluble residues were immersed in 50 mM sodium hydroxide containing 100 mM $NaBH_4$ for 6 h to obtain hemicellulose. The residues were washed with 0.1 M potassium hydroxide (KOH) and 8 mM sodium sulfite three times, and then immersed in 8 M KOH for 6 h. The KOH-insoluble residues were washed three times with ethanol and thoroughly dried at 40 °C to obtain cellulose.

The contents of WSP, ISP and CSP were determined by the carbazole method with galacturonic acid as standard. The contents of cellulose and hemicellulose were determined by the anthrone method with glucose as standard.

2.5. Determination of Cell Wall-Degrading Enzyme Activities

The crude enzyme was extracted according to the method described by Lin et al. with modifications [36]. A total of 3 g samples of pericarp or pulp ground with 15 mL of extraction buffer containing 50 mmol L^{-1} sodium acetate buffer (pH 5.2), 100 mmol L^{-1} NaCl, 5 mmol L^{-1} dithiothreitol, and 5% (*w*/*v*) polyvinylpyrrolidone. The supernatant was used as the crude enzyme.

Pectinesterase (PE) activity was determined according to the method described by Lin et al. with modifications [36]. The reaction mixture contained 1.5 mL of crude enzyme and 5 mL of 1 % (*w*/*v*) pectin. After incubation at 37 °C for 30 min, the reaction was stopped by boiling for 5 min. The pH of the reaction solution was titrated to 7.4 with 0.01 mol L^{-1} NaOH. The amount of enzyme consumed 1 μmol of NaOH per hour was defined as one unit of PE activity. Polygalacturonase (PG) activity was determined according to the method described by Chen et al. with modifications [37]. The reaction mixture contained 1.3 mL of 50 mmol L^{-1} sodium acetate buffer (pH 4.5), 0.5 mL of 10 g L^{-1} polygalacturonic acid, and 0.2 mL of crude enzyme. The mixture was incubated at 37 °C for 1 h, and 1.5 mL of 3,5-dinitro salicylic acid reagent was added and boiled for 5 min to stop the reaction. After cooling on ice, the absorbance was measured at 540 nm. The amount of enzyme produced 1 μg of galacturonic acid per hour was defined as one unit of PG activity. Cellulase (Cx) activity was determined according to the method described by Chen et al. with modifications [37]. The reaction mixture contained 1.8 mL of 10 g L^{-1} carboxymethyl cellulose and 0.2 mL of crude enzyme. Cx activity was determined to be the same as PG activity. The amount of enzyme produced 1 mg of glucose per hour was defined as one unit of cellulose activity. The β-galactosidase (β-Gal) activity was determined according to the method described by Lin et al. with modifications [36]. The reaction mixture contained 1 mL of crude enzyme and 1 mL of 5 mmol L^{-1} *p*-nitrophenyl-β-D-galactopyranoside. After incubating at 37 °C for 30 min, the reaction was stopped by adding 4 mL of 1 mol L^{-1} Na_2CO_3. Absorbance was measured at 400 nm. The amount of enzyme produced 1 μmol of *p*-nitrophenol per hour was defined as one unit of β-Gal activity. The protein content was determined according to the method established by Bradford [38]. The activity of each enzyme was expressed as U mg^{-1} protein.

3. Results

3.1. Effects of $CaCl_2$ Treatment on Pericarp Browning and Decay of Litchi Fruit

The pericarp browning index and decay of litchi fruit showed an increasing trend with storage. The pericarp browning index of $CaCl_2$-treated fruit was significantly lower than that of the control on days 3–5 (Figure 1a,c). The decay increased rapidly during days 1–3, and the decay of $CaCl_2$-treated fruit was significantly lower than that of the control on day 5 (Figure 1b).

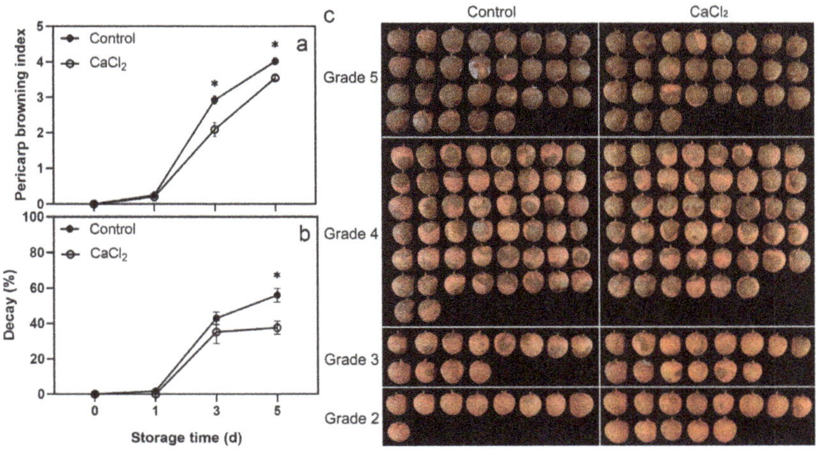

Figure 1. Effects of $CaCl_2$ treatment on pericarp browning index (**a**) and decay (**b**) of litchi fruit. Detailed appearance of control and $CaCl_2$-treated fruits according to their grades of pericarp browning on day 5 (**c**). Asterisks indicate significant differences between $CaCl_2$-treated and control fruits (* $p < 0.05$). The data are presented as the mean ± SEM (standard error of the mean) of three replicates.

3.2. *Effects of $CaCl_2$-Treated on Cell Wall Component Contents and Cell Wall-Degrading Enzyme Activities in Litchi Pericarp*

The Ca^{2+} content in both $CaCl_2$-treated and control pericarp decreased during storage. The Ca^{2+} content in the $CaCl_2$-treated pericarp was significantly higher than that in the control on days 1 and 5 (Figure 2a).

The WSP content in the control pericarp increased rapidly on the first day, then showed a slight decrease from day 1 to day 5. In contrast, the WSP content in the $CaCl_2$-treated pericarp slightly decreased on the first day of storage and then increased slightly. The WSP content in the $CaCl_2$-treated pericarp was significantly lower than that in the control on day 1 (Figure 2b). A quick decline in the ISP content was observed on the first day, whereas later, it increased continuously. A significantly lower ISP content was observed in $CaCl_2$-treated pericarp than in the control on day 3. However, a significantly higher ISP content was observed in $CaCl_2$-treated pericarp than in the control on day 5 (Figure 2c). The cellulose content increased slightly during storage. Compared to the control, the cellulose content in the $CaCl_2$-treated pericarp was significantly higher on day 3 (Figure 2e). $CaCl_2$ treatment did not affect the contents of CSP and hemicellulose in the pericarp (Figure 2d,f).

The $CaCl_2$ treatment significantly inhibited the activities of PG, β-Gal, and Cx in the pericarp. Lower activities of PG, β-Gal, and Cx were observed in the $CaCl_2$-treated pericarp compared to the control pericarp throughout storage (Figure 3b–d). However, $CaCl_2$ treatment did not affect the PE activity in the pericarp (Figure 3a).

The PG activity in the control pericarp increased rapidly from day 1 to day 3, and then decreased. The PG activity in the $CaCl_2$-treated pericarp was always significantly lower than that in the control (Figure 3b). The β-Gal activity in the control pericarp rose on the first day, and then declined, whereas the β-Gal activity in the $CaCl_2$-treated pericarp showed a downward trend and was lower than that in the control pericarp over the storage period (Figure 3c). A sharp increase in the Cx activity was observed in the control pericarp from day 1 to day 3. The Cx activity in the $CaCl_2$-treated pericarp rose continuously during days 1–5 and was significantly lower than that in the control on day 3 (Figure 3d).

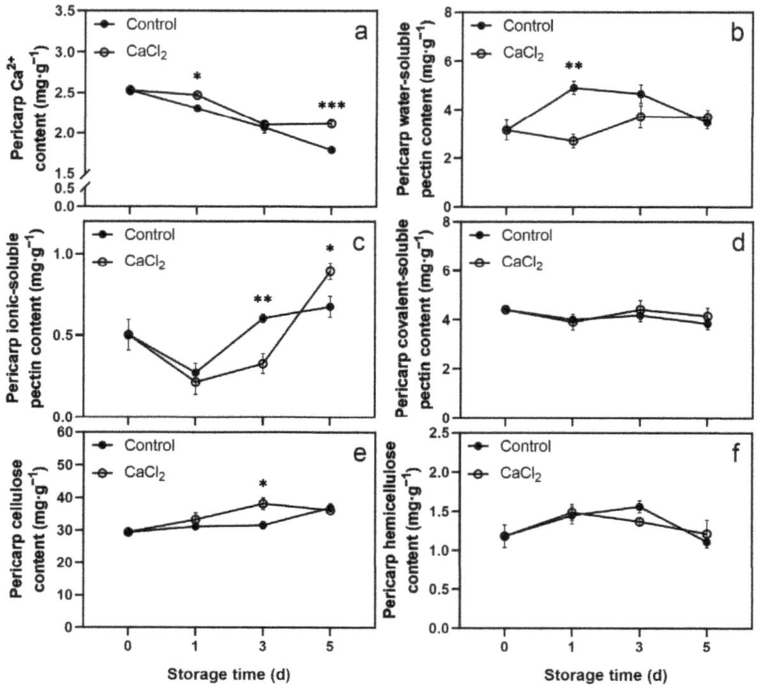

Figure 2. Effects of $CaCl_2$ treatment on the contents of Ca^{2+} (**a**), water-soluble pectin (**b**), ionic-soluble pectin (**c**), covalent-soluble pectin (**d**), cellulose (**e**), and hemicellulose (**f**) in litchi pericarp. Asterisks indicate significant differences between $CaCl_2$-treated and control fruits (* $p < 0.05$, ** $p < 0.01$, *** $p < 0.001$). The data are presented as the mean ± SEM of three replicates.

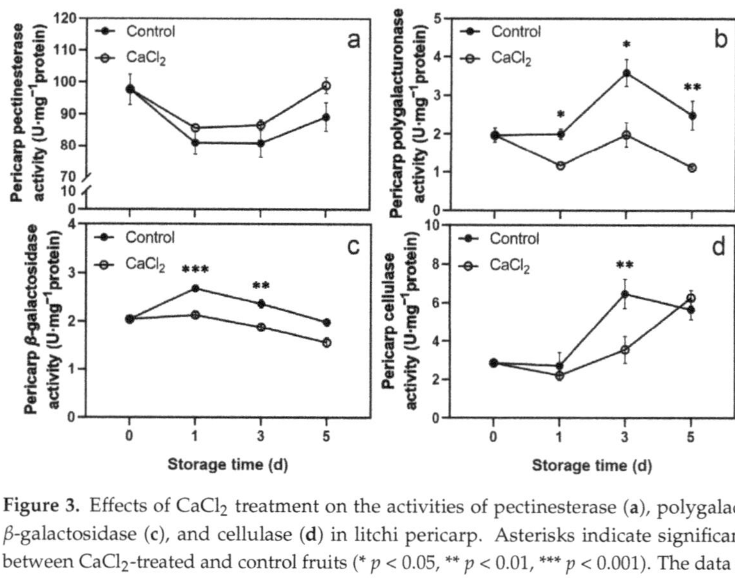

Figure 3. Effects of $CaCl_2$ treatment on the activities of pectinesterase (**a**), polygalacturonase (**b**), β-galactosidase (**c**), and cellulase (**d**) in litchi pericarp. Asterisks indicate significant differences between $CaCl_2$-treated and control fruits (* $p < 0.05$, ** $p < 0.01$, *** $p < 0.001$). The data are presented as the mean ± SEM of three replicates.

3.3. Effects of $CaCl_2$-Treated on Cell Wall Component Contents and Cell Wall-Degrading Enzyme Activities in Litchi Pulp

The Ca^{2+} content in the $CaCl_2$-treated pulp increased sharply on the first day, followed by a decline from day 1 to day 5. The Ca^{2+} content in the control pulp rose continuously during storage. Notably, the Ca^{2+} content in the $CaCl_2$-treated pulp was substantially higher than that in the control pulp on days 1 and 3, whereas it was significantly lower than that in the control pulp on day 5 (Figure 4a).

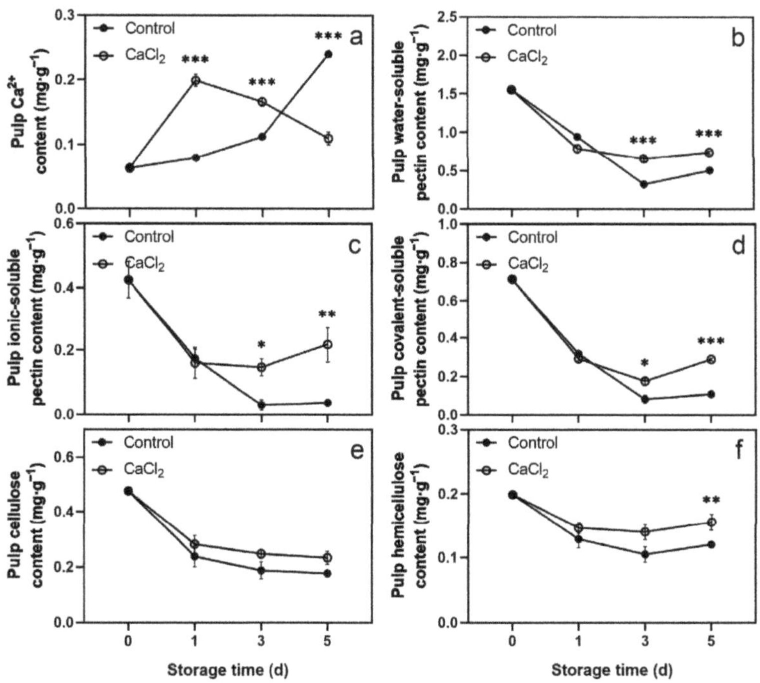

Figure 4. Effects of $CaCl_2$ treatment on the contents of Ca^{2+} (**a**), water-soluble pectin (**b**), ionic-soluble pectin (**c**), covalent-soluble pectin (**d**), cellulose (**e**), and hemicellulose (**f**) in litchi pulp. Asterisks indicate significant differences between $CaCl_2$-treated and control fruits (* $p < 0.05$, ** $p < 0.01$, *** $p < 0.001$). The data are presented as the mean ± SEM of three replicates.

The contents of WSP, ISP, CSP, and hemicellulose in both $CaCl_2$-treated and control pulps showed a downtrend with increasing storage time, and the values in the $CaCl_2$-treated pulp were consistently higher than those in the control pulp. The WSP, ISP, and CSP contents in the $CaCl_2$-treated pulp declined sharply before day 1, while the contents in the control pulp declined sharply before day 3. Furthermore, the WSP, ISP, and CSP in the $CaCl_2$-treated pulp were significantly higher than those in the control pulp on the days 3 and 5. A higher level of hemicellulose content was detected in the $CaCl_2$-treated pulp, with prominent differences observed on day 5. $CaCl_2$ treatment did not affect the cellulose content in the pulp (Figure 4b–f).

The PE activity in both $CaCl_2$-treated and control pulps displayed a downtrend during storage. The activity decreased on day 1 and increased during days 1–3. The difference was that the PE activity in the $CaCl_2$-treated pulp continued to rise from day 3 to day 5, whereas the activity in the control pulp decreased. Therefore, the PE activity in the $CaCl_2$-treated pulp was significantly higher than that in the control pulp on day 5 (Figure 5a). The PG activity in the $CaCl_2$-treated pulp decreased slightly at on day 1 and then increased. The activity in the control fruit increased gradually during storage. Compared to the control pulp, a lower level of PG activity was observed in the $CaCl_2$-treated pulp on day 1 (Figure 5b). The β-Gal activity in the control pulp increased slightly on the first day, followed by a rapid reduction on days 1–3. Compared to the control pulp, the $CaCl_2$-treated pulp showed a lower β-Gal activity on day 1, and a higher β-Gal activity on day 3 (Figure 5c). The Cx activity reduced quickly in both the $CaCl_2$-treated and control pulps. The activity in the $CaCl_2$-treated pulp was significantly lower than that in the control pulp on day 5 (Figure 5d).

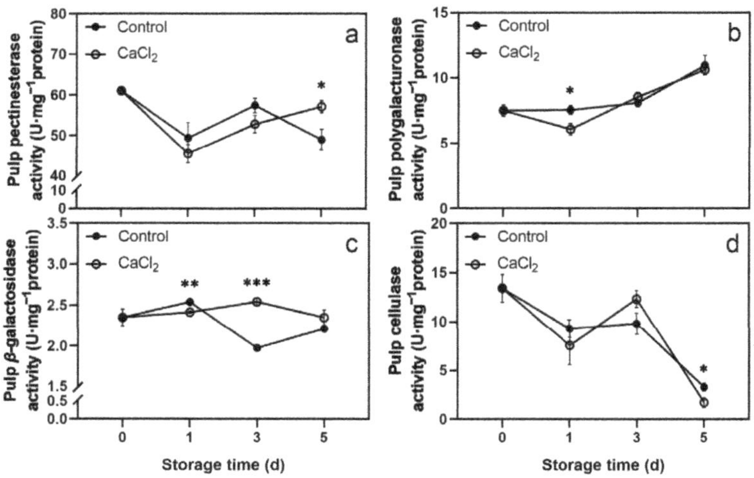

Figure 5. Effects of $CaCl_2$ treatment on the activities of pectinesterase (**a**), polygalacturonase (**b**), β-galactosidase (**c**), and cellulase (**d**) in litchi pulp. Asterisks indicate significant differences between $CaCl_2$-treated and control fruits (* $p < 0.05$, ** $p < 0.01$, *** $p < 0.001$). The data are presented as the mean ± SEM of three replicates.

4. Discussion

4.1. $CaCl_2$ Treatment Strengthened the Cell Wall Structure of Litchi Fruit

The methods involved in postharvest $CaCl_2$ treatment include immersion treatment, vacuum infiltration treatment, and spray treatment. Since the pericarp and pulp of litchi completely separated in the structure, we used the vacuum infiltration method to enhance the effect of $CaCl_2$ treatment. Figure S1 shows that the contents of titratable acid and ascorbic acid, respiration rate, and weight loss were not significantly different between $CaCl_2$-treated and control fruits. Although $CaCl_2$ treatment significantly reduced the total soluble solids content of litchi fruit on storage day 1, it had no significant impact on days 3 and 5. These results indicate that $CaCl_2$ treatment did not cause apparent damage to the quality of litchi fruit. It is worth noting that $CaCl_2$ treatment effectively delayed pericarp browning and decay (Figure 1), and the cell wall component contents and cell wall-degrading enzyme activities of litchi pericarp and pulp were changed (Figures 2–5).

Ca^{2+} can cross-link unesterified homogalacturonan molecules via ionic calcium bridges to form a calcium–pectate gel, thereby increasing cell wall stiffness. Additionally, calcium–pectate cross-links in the middle lamella interlink most cells in ripe fruit [39]. Cell wall components include pectin, cellulose, and hemicellulose. WSP, ISP, and CSP are different forms of pectin. During the ripening and senescence process, ISP and CSP are converted to WSP, and the contents of cellulose and hemicellulose are decreased [36,40]. The cell wall components and structure are gradually degraded under the action of cell wall-degrading enzymes such as PE, PG, β-Gal, and Cx. Among them, PE, PG, and β-Gal act on pectin, and Cx acts on cellulose [36]. Many studies have confirmed the effect of postharvest $CaCl_2$ treatment on the contents of cell wall components and the activities of cell wall-degrading enzymes in fruit. Yu et al. reported that the immersion of grape clusters in a 5 g L^{-1}

$CaCl_2{\cdot}2H_2O$ solution containing 0.01% Tween 20 for 2 min inhibited WSP production, and delayed the degradation of protopectin (non-water-soluble pectin, including ISP and CSP) [22]. Liu et al. found that apricots treated with a 1% $CaCl_2$ solution for 2 min reduced the degradation of WSP, CSP, and hemicellulose [27]. Chen et al. showed that strawberries treated with a 1% $CaCl_2$ solution for 15 min slowed the breakdown of ISP and CSP [41]. Mohebbi et al. declared that dipping apples in a 4% $CaCl_2$ solution for 5 min increased the contents of ISP and CSP, and decreased the activities of β-Gal, α-L-arabinofuranosidase, pectate lyase (PL) and PG [23]. Lv et al. reported that a 0.2 mol L^{-1} $CaCl_2$ solution treatment maintained higher levels of protopectin and cellulose, lower levels of WSP, and decreased the activities of PG, pectin methylesterase (PME), Cx and β-Gal of red raspberry [24]. Gao et al. stated that papaya immersed in a 2.5% $CaCl_2$ solution for 15 min repressed the activities of Cx, β-glucosidase, PG, and PME, and the expression of cell wall softening-related genes [25]. Zhang et al. explained that full exposure to a 0.18 mol L^{-1} $CaCl_2$ solution for 25 min decreased the activities and gene expression of PG, PME and PL of New Queen melon [26]. These studies have illustrated that $CaCl_2$ treatment maintains or strengthens the cell wall structure by inhibiting the production of WSP, the degradation of ISP, CSP, cellulose, and hemicellulose, as well as the activities of cell wall-degrading enzymes, thereby preventing fruit cracking and delaying fruit softening.

In our study, vacuum infiltration enhanced the effect of $CaCl_2$ treatment, resulting in changes in the contents of cell wall components and in the activities of cell wall-degrading enzymes in both the pericarp and the pulp of litchi. Similarly to previous studies, in the pericarp, $CaCl_2$ treatment significantly increased the contents of Ca^{2+} and cellulose while reducing the content of WSP and the activities of PG, β-Gal, and Cx. However, in the pulp, $CaCl_2$ treatment significantly increased the contents of Ca^{2+}, WSP, ISP, CSP, and hemicellulose, and the activities of PE and β-Gal, in addition to reducing the activities of PG and Cx. Thus, this study illustrates that the changes between the pericarp and pulp were different, and the $CaCl_2$ treatment had a more pronounced effect on the pericarp. We believe this was related to the limitation of the treatment effect by the fruit structure, and the difference in the content of each cell wall component between the pericarp and the pulp. The content of each cell wall component in the pericarp was significantly higher than that in the pulp. Overall, $CaCl_2$ treatment strengthened the cell wall structure of litchi fruit, greatly strengthened the firmness of the pericarp, and delayed the softening of the pulp to a certain extent.

4.2. $CaCl_2$ Treatment Delayed the Pericarp Browning of Litchi

Enzymatic browning is one of the most frequent causes of litchi pericarp browning. Under normal circumstances, oxidative enzymes and their substrates exist in distinct subcellular compartments. The compartmentation limits enzymatic browning in the healthy pericarp [42]. We found that $CaCl_2$ treatment maintained or strengthened the structure and function of the cell wall (Figure 2), and the firmness of the pericarp was strengthened. Therefore, the integrity of subcellular compartments remained and the litchi pericarp enzymatic browning was delayed. In addition, enzymatic browning was associated with oxidation resistance. $CaCl_2$ treatment can decrease reactive oxygen species (ROS) content, and increase antioxidant enzyme activities and non-enzymatic antioxidant contents. For example, 80 mM $CaCl_2$ significantly increased the activities of antioxidant enzymes peroxidase (POD), superoxide dismutase (SOD), and catalase (CAT) in the 'Kyoho' grapevine [15]. With the 'Nanguo' pear, the activity and gene expression of polyphenol oxidase (PPO) were inhibited, and the activity and gene expression of SOD were promoted under 2% $CaCl_2$ treatment [18]. $CaCl_2$ treatments maintained higher levels of phenolics, flavonoids, anthocyanins, ascorbic acid, and total antioxidant capacity of cornelian cherry [43]. A total of 10 g L^{-1} $CaCl_2$ reduced the levels of superoxide anion and hydrogen peroxide, preserved the activities of SOD, CAT and POD, as well as the levels of ascorbic acid and glutathione of jujube [17]. However, the effect of $CaCl_2$ treatment on the oxidation resistance of litchi pericarp requires further research.

4.3. $CaCl_2$ Treatment Delayed the Decay of Litchi Fruit

Softening of the fruit is conducive to disease infestation. $CaCl_2$ treatment effectively delays fruit ripening, senescence, and softening, which is closely related to the role of calcium in maintaining the cell wall [11]. $CaCl_2$ treatment inhibited the softening of New Queen melon and papaya, slowed down the growth of microorganisms, and reduced the fruit decay rate [26,44]. $CaCl_2$ treatment can induce disease resistance. In Hami melon, it resisted penicillium infection by causing calcium-dependent protein kinase (CDPK) activity and upregulating HmCDPK2 gene expression [33]. Moreover, $CaCl_2$ can directly inhibit the growth of pathogenic microorganisms [45]. In our study, $CaCl_2$ treatment significantly inhibited cell wall degradation in litchi pericarp and pulp in order to resist disease infestation.

5. Conclusions

$CaCl_2$ treatment significantly strengthened the cell wall structure and delayed the cell wall degradation of litchi fruit. Therefore, pericarp browning, pulp softening, and disease incidence were inhibited, effectively maintaining the quality of litchi fruit. Since the effect of vacuum infiltration on the pulp is limited, further research on $CaCl_2$ treatment without vacuum infiltration can be considered for production applications.

Supplementary Materials: The following supporting information can be downloaded at: https://www.mdpi.com/article/10.3390/foods12132478/s1, Figure S1: Effects of $CaCl_2$ treatment on total soluble solids content (a), titratable acid content (b), ascorbic acid content (c), respiration rate (d), and weight loss (e).

Author Contributions: Conceptualization, X.G. and Z.W.; data curation, X.G. and Z.W.; formal analysis, X.G.; funding acquisition, Z.W. and T.L.; investigation, X.G. and Q.L.; methodology, X.G. and Q.L.; project administration, X.G. and Z.W.; resources, Z.W. and T.L.; supervision, X.G. and Z.W.; validation, X.G. and Q.L.; visualization, X.G.; writing—original draft, X.G. and Q.L.; writing—review and editing, X.G., Q.L., Z.W., T.L., D.H. and D.Z. All authors have read and agreed to the published version of the manuscript.

Funding: This research was funded by the Earmarked Fund for China Agricultural Research System-Litchi and Longan, grant number CARS-32; Guangdong Basic and Applied Basic Research Foundation, grant number 2022A1515012049; and the project "Basic substances as an environmentally friendly alternative to synthetic pesticides for plant protection (BasicS)", grant number 2020-C-353.

Data Availability Statement: Data is contained within the article.

Conflicts of Interest: The authors declare no conflict of interest.

References

1. Hu, G.B.; Feng, J.T.; Xiang, X.; Wang, J.B.; Salojarvi, J.; Liu, C.M.; Wu, Z.X.; Zhang, J.S.; Liang, X.M.; Jiang, Z.D.; et al. Two divergent haplotypes from a highly heterozygous lychee genome suggest independent domestication events for early and late-maturing cultivars. *Nat. Genet.* **2022**, *54*, 73–83. [CrossRef]
2. Sivakumar, D.; Terry, L.A.; Korsten, L. An overview on litchi fruit quality and alternative postharvest treatments to replace sulfur dioxide fumigation. *Food Rev. Int.* **2010**, *26*, 162–188. [CrossRef]
3. Guo, X.; Luo, T.; Han, D.; Zhu, D.; Jiang, Z.; Wu, Z. Integrated transcriptomics, proteomics, and metabolomics analysis reveals the mechanism of litchi pulp deterioration during long-term cold storage. *Postharvest Biol. Technol.* **2023**, *195*, 112140. [CrossRef]
4. Shen, W.; Li, W.; Shao, Y.; Zeng, J. Proanthocyanidin delays litchi peel browning by inhibiting ethylene biosynthesis, respiratory metabolism, and phenol oxidase activities. *Sci. Hortic.* **2023**, *309*, 111677. [CrossRef]
5. Xie, J.; Qin, Z.; Pan, J.; Li, J.; Li, X.; Khoo, H.E.; Dong, X. Melatonin treatment improves postharvest quality and regulates reactive oxygen species metabolism in "Feizixiao" litchi based on principal component analysis. *Front. Plant Sci.* **2022**, *13*, 965345. [CrossRef]
6. Xiang, Y.; Su, Z.; Li, T.; Xu, D.; Hu, M.; Sun, J.; Jiang, Y.; Zhang, Z. Inhibitory activity and action mode of apple polyphenols against *Peronophythora litchii* that causes litchi downy blight. *Postharvest Biol. Technol.* **2022**, *194*, 112095. [CrossRef]
7. He, M.; Xie, L.; Wu, Y.; Hong, M.; Jiang, Y.; Li, T. Fluopyram inhibits tricarboxylic acid cycle but induces γ-aminobutyric acid-shunt pathway to prolong postharvest shelf life of litchi fruit. *Postharvest Biol. Technol.* **2022**, *192*, 112036. [CrossRef]
8. Deshi, V.; Homa, F.; Ghatak, A.; Aftab, M.A.; Mir, H.; Ozturk, B.; Siddiqui, M.W. Exogenous methyl jasmonate modulates antioxidant activities and delays pericarp browning in litchi. *Physiol. Mol. Biol. Plants* **2022**, *28*, 1561–1569. [CrossRef]

9. Xu, D.; Xi, P.; Lin, Z.; Huang, J.; Lu, S.; Jiang, Z.; Qiao, F. Efficacy and potential mechanisms of benzothiadiazole inhibition on postharvest litchi downy blight. *Postharvest Biol. Technol.* **2021**, *181*, 111660. [CrossRef]
10. He, M.; Wu, Y.; Hong, M.; Yun, Z.; Li, T.; Jiang, Y. α-Lipoic acid treatment alleviates postharvest pericarp browning of litchi fruit by regulating antioxidant ability and energy metabolism. *Postharvest Biol. Technol.* **2021**, *180*, 111629. [CrossRef]
11. Gao, Q.Y.; Xiong, T.T.; Li, X.P.; Chen, W.X.; Zhu, X.Y. Calcium and calcium sensors in fruit development and ripening. *Sci. Hortic.* **2019**, *253*, 412–421. [CrossRef]
12. Shehata, S.A.; Abdelrahman, S.Z.; Megahed, M.M.A.; Abdeldaym, E.A.; El-Mogy, M.M.; Abdelgawad, K.F. Extending shelf life and maintaining quality of tomato fruit by calcium chloride, hydrogen peroxide, chitosan, and ozonated water. *Horticulturae* **2021**, *7*, 309. [CrossRef]
13. Atia, A.; Abdelkarim, D.; Younis, M.; Alhamdan, A. Effects of calcium chloride and salicylic acid postharvest treatments on the quality of Khalal Barhi dates at different ripening levels during cold storage. *J. Food Meas. Charact.* **2018**, *12*, 1156–1166. [CrossRef]
14. Gago, C.M.L.; Guerreiro, A.C.; Miguel, G.; Panagopoulos, T.; da Silva, M.M.; Antunes, M.D.C. Effect of Calcium chloride and 1-MCP (Smartfresh^{TM}) postharvest treatment on 'Golden Delicious' apple cold storage physiological disorders. *Sci. Hortic.* **2016**, *211*, 440–448. [CrossRef]
15. Fu, W.; Zhang, M.; Zhang, P.; Liu, Z.; Dong, T.; Zhang, S.; Ren, Y.; Jia, H.; Fang, J. Transcriptional and metabolite analysis reveal a shift in fruit quality in response to calcium chloride treatment on "Kyoho" grapevine. *J. Food Sci. Technol.-Mysore* **2021**, *58*, 2246–2257. [CrossRef] [PubMed]
16. Wei, D.; Zhao, X.H. Calcium maintained higher quality and enhanced resistance against chilling stress by regulating enzymes in reactive oxygen and biofilm metabolism of Chinese winter jujube fruit. *J. Food Biochem.* **2020**, *44*, e13161. [CrossRef]
17. Zhi, H.H.; Liu, Q.Q.; Xu, J.; Dong, Y.; Liu, M.P.; Zong, W. Ultrasound enhances calcium absorption of jujube fruit by regulating the cellular calcium distribution and metabolism of cell wall polysaccharides. *J. Sci. Food Agric.* **2017**, *97*, 5202–5210. [CrossRef]
18. Zhang, L.; Wang, J.W.; Zhou, B.; Li, G.D.; Liu, Y.F.; Xia, X.L.; Xiao, Z.G.; Lu, F.; Ji, S.J. Calcium inhibited peel browning by regulating enzymes in membrane metabolism of 'Nanguo' pears during post-ripeness after refrigerated storage. *Sci. Hortic.* **2019**, *244*, 15–21. [CrossRef]
19. Al-Qurashi, A.D.; Awad, M.A. Postharvest gibberellic acid, 6-benzylaminopurine and calcium chloride dipping affect quality, antioxidant compounds, radical scavenging capacity and enzymes activities of 'Grand Nain' bananas during shelf life. *Sci. Hortic.* **2019**, *253*, 187–194. [CrossRef]
20. Sabir, F.; Sabir, A.; Ozcelik, S.; Kucukbasmaci, A. Maintenance of postharvest quality of blackberry (*Rubus fructicosus* L.) fruits through salicylic acid and $CaCl_2$ immersions. *Acta Sci. Pol.-Hortorum Cultus* **2019**, *18*, 121–128. [CrossRef]
21. Lata, D.; Aftab, M.A.; Homa, F.; Ahmad, M.S.; Siddiqui, M.W. Effect of eco-safe compounds on postharvest quality preservation of papaya (*Carica papaya* L.). *Acta Physiol. Plant.* **2018**, *40*, 8. [CrossRef]
22. Yu, J.; Zhu, M.T.; Bai, M.; Xu, Y.S.; Fan, S.G.; Yang, G.S. Effect of calcium on relieving berry cracking in grape (*Vitis vinifera* L.) 'Xiangfei'. *PeerJ* **2020**, *8*, e9896. [CrossRef] [PubMed]
23. Mohebbi, S.; Babalar, M.; Zamani, Z.; Askari, M.A. Influence of early season boron spraying and postharvest calcium dip treatment on cell-wall degrading enzymes and fruit firmness in 'Starking Delicious' apple during storage. *Sci. Hortic.* **2020**, *259*, 108822. [CrossRef]
24. Lv, J.; Han, X.; Bai, L.; Xu, D.; Ding, S.; Ge, Y.; Li, C.; Li, J. Effects of calcium chloride treatment on softening in red raspberry fruit during low-temperature storage. *J. Food Biochem.* **2020**, *44*, e13419. [CrossRef]
25. Gao, Q.; Tan, Q.; Song, Z.; Chen, W.; Li, X.; Zhu, X. Calcium chloride postharvest treatment delays the ripening and softening of papaya fruit. *J. Food Process. Preserv.* **2020**, *44*, e14604. [CrossRef]
26. Zhang, Q.; Dai, W.; Jin, X.; Li, J. Calcium chloride and 1-methylcyclopropene treatments delay postharvest and reduce decay of New Queen melon. *Sci. Rep.* **2019**, *9*, 13563. [CrossRef] [PubMed]
27. Liu, H.; Chen, F.S.; Lai, S.J.; Tao, J.R.; Yang, H.S.; Jiao, Z.G. Effects of calcium treatment and low temperature storage on cell wall polysaccharide nanostructures and quality of postharvest apricot (*Prunus armeniaca*). *Food Chem.* **2017**, *225*, 87–97. [CrossRef]
28. Belge, B.; Goulao, L.F.; Comabella, E.; Graell, J.; Lara, I. Refrigerated storage and calcium dips of ripe 'Celeste' sweet cherry fruit: Combined effects on cell wall metabolism. *Sci. Hortic.* **2017**, *219*, 182–190. [CrossRef]
29. Antunes, M.D.; Guimaraes, A.C.; Gago, C.; Guerreiro, A.; Panagopoulos, J.; Boas, E.V.; Miguel, M.G. Membrane fatty acids and physiological disorders in cold-stored 'Golden Delicious' apples treated with 1-MCP and calcium chloride. *Horticulturae* **2022**, *8*, 162. [CrossRef]
30. Xie, B.; Ling, C.; Hu, S.; Hou, Y.; Zheng, Y.; Jin, P. CaM enhances chilling tolerance of peach fruit by regulating energy and GABA metabolism. *Postharvest Biol. Technol.* **2021**, *181*, 111691. [CrossRef]
31. Li, Z.; Wang, L.; Xie, B.; Hu, S.; Zheng, Y.; Jin, P. Effects of exogenous calcium and calcium chelant on cold tolerance of postharvest loquat fruit. *Sci. Hortic.* **2020**, *269*, 109391. [CrossRef]
32. Jiao, W.; Xi, Y.; Cao, J.; Fan, X.; Jiang, W. Regulatory effects of $CaCl_2$, sodium isoascorbate, and 1-methylcyclopropene on chilling injury of banana fruit at two ripening stages and the mechanisms involved. *J. Food Process. Preserv.* **2018**, *42*, e13442. [CrossRef]
33. Ning, M.; Tang, F.; Zhang, Q.; Zhao, X.; Yang, L.; Cai, W.; Shan, C. Effects of *Penicillium* infection on the expression and activity of CDPK2 in postharvest Hami melon treated with calcium chloride. *Physiol. Mol. Plant Pathol.* **2019**, *106*, 175–181. [CrossRef]
34. Guo, X.; Luo, T.; Han, D.; Zhu, D.; Li, Z.; Wu, Z.; Wu, Z. Multi-omics analysis revealed room temperature storage affected the quality of litchi by altering carbohydrate metabolism. *Sci. Hortic.* **2022**, *293*, 110663. [CrossRef]

35. Erdogan, S.; Erdemoglu, S.B.; Kaya, S. Optimisation of microwave digestion for determination of Fe, Zn, Mn and Cu in various legumes by flame atomic absorption spectrometry. *J. Sci. Food Agric.* **2006**, *86*, 226–232. [CrossRef]
36. Lin, Y.; Lin, Y.; Lin, Y.; Lin, M.; Chen, Y.; Wang, H.; Lin, H. A novel chitosan alleviates pulp breakdown of harvested longan fruit by suppressing disassembly of cell wall polysaccharides. *Carbohydr. Polym.* **2019**, *217*, 126–134. [CrossRef] [PubMed]
37. Chen, Y.; Hung, Y.C.; Chen, M.; Lin, H. Effects of acidic electrolyzed oxidizing water on retarding cell wall degradation and delaying softening of blueberries during postharvest storage. *LWT-Food Sci. Technol.* **2017**, *84*, 650–657. [CrossRef]
38. Bradford, M.M. A rapid and sensitive method for the quantitation of microgram quantities of protein utilizing the principle of protein-dye binding. *Anal. Biochem.* **1976**, *72*, 248–254. [CrossRef] [PubMed]
39. Brummell, D.A. Cell wall disassembly in ripening fruit. *Funct. Plant Biol.* **2006**, *33*, 103–119. [CrossRef] [PubMed]
40. Wang, J.; Mujumdar, A.S.; Deng, L.Z.; Gao, Z.J.; Xiao, H.W.; Raghavan, G.S.V. High-humidity hot air impingement blanching alters texture, cell-wall polysaccharides, water status and distribution of seedless grape. *Carbohydr. Polym.* **2018**, *194*, 9–17. [CrossRef]
41. Chen, F.; Liu, H.; Yang, H.; Lai, S.; Cheng, X.; Xin, Y.; Yang, B.; Hou, H.; Yao, Y.; Zhang, S.; et al. Quality attributes and cell wall properties of strawberries (*Fragaria annanassa* Duch.) under calcium chloride treatment. *Food Chem.* **2011**, *126*, 450–459. [CrossRef]
42. Jiang, Y.M.; Duan, X.W.; Joyce, D.; Zhang, Z.Q.; Li, J.R. Advances in understanding of enzymatic browning in harvested litchi fruit. *Food Chem.* **2004**, *88*, 443–446. [CrossRef]
43. Aghdam, M.S.; Dokhanieh, A.Y.; Hassanpour, H.; Fard, J.R. Enhancement of antioxidant capacity of cornelian cherry (*Cornus mas*) fruit by postharvest calcium treatment. *Sci. Hortic.* **2013**, *161*, 160–164. [CrossRef]
44. Mahmud, T.M.M.; Al Eryani-Raqeeb, A.; Syed Omar, S.R.; Mohamed Zaki, A.R.; Al Eryani, A.-R. Effects of different concentrations and applications of calcium on storage life and physicochemical characteristics of papaya (*Carica papaya* L.). *Am. J. Agri. Biol. Sci.* **2008**, *3*, 526–533. [CrossRef]
45. Gomes, A.M.A.; Silveira, E.B.; Mariano, R.L.R. Postharvest treatments with calcium and microorganisms to control soft rot in tomato. *Hortic. Bras.* **2005**, *23*, 108–111. [CrossRef]

foods MDPI

Article

Effect of Controlled Atmosphere Packaging on the Physiology and Quality of Fresh-Cut *Dictyophora rubrovolvata*

Ziqian Xia [1], Rui Wang [1,*], Chao Ma [1], Jiangkuo Li [2], Jiqing Lei [1], Ning Ji [1], Xianxing Pan [1] and Tongjie Chen [3]

1 College of Food and Pharmaceutical Engineering, Guiyang University, Guiyang 550000, China; xiaziqian0603@163.com (Z.X.); gyuchaoma@163.com (C.M.); jiqinglei@gmail.com (J.L.); jining552100@163.com (N.J.); pxx991025@163.com (X.P.)
2 Tianjin Key Laboratory of Postharvest Physiology and Storage of Agricultural Products, National Engineering and Technology Research Center for Preservation of Agricultural Produce, Tianjin 301699, China; lijkuo@sina.com
3 Gui Zhou Mei Wei Xian Dictyophora Industry Company Limited, Zhijin 552100, China; wo2427713193@163.com
* Correspondence: wangrui__gyu@163.com

Abstract: *Dictyophora rubrovolvata* is a typical edible fungus of Guizhou Province and is very popular due to its unique taste and texture. In this study, the effect of a controlled atmosphere (CA) on fresh-cut *D. rubrovolvata* shelf life was investigated. Firstly, this study addresses the influence of different O_2 concentrations (5%, 20%, 35%, 50%, 65%, 80%, or 95%) with N_2 balance on fresh-cut *D. rubrovolvata* quality while stored at 4 ± 1 °C for 7 d. Then, on the basis of the determined O_2 concentration (5%), CO_2 (0%, 5%, 10%, 15%, or 20%) was involved and stored for 8 d at 4 ± 1 °C. Evaluations of physiology parameters, texture, browning degree, nutritional, umami, volatile components, and total colony numbers were determined in fresh-cut *D. rubrovolvata.* From the results of water migration, the sample of 5% O_2/5% CO_2/90% N_2 was closer to 0 d than other groups at 8 days. Meanwhile, the polyphenol oxidase (2.26 ± 0.07 U/(g·min)), and catalase activity (4.66 ± 0.08 U/(g·min·FW)) were superior to the samples of other treatment groups on the eighth day (3.04 ± 0.06 to 3.84 ± 0.10 U/(g·min), 4.02 ± 0.07 to 4.07 ± 0.07 U/(g·min·FW)). Therefore, we found that a gas environment with 5% O_2/5% CO_2/90% N_2 could ensure the membrane integrity, oxidation, and prevent the browning of fresh-cut *D. rubrovolvata*, thus better maintaining the physiological parameters. Meanwhile, it also maintained the samples' texture, color, nutritional value, and umami taste. Furthermore, it inhibited the increase in total colony numbers. The volatile components were closer to the initial level compared with other groups. The results indicate that fresh-cut *D. rubrovolvata* could maintain its shelf life and quality when stored in 5% O_2/5% CO_2/90% N_2 at 4 ± 1 °C.

Keywords: controlled atmosphere; fresh-cut *D. rubrovolvata*; physiology; nutrition; umami

1. Introduction

Edible fungi have become an important part of the human diet due to having a high quality of protein, vitamins, minerals, low calorie content, and other components [1]. Vegetarians are also fond of them because of their essential amino acids from animal protein and natural sources of vitamin D_2 [2]. Consequently, the production and consumption have increased significantly as they are incorporated into consumers' diets. China is one of the world's largest producers and exporters of edible fungi [3]. In terms of value, the mushroom industry ranks after grain, vegetable, fruit, and edible oil plantations, and is higher than sugar, cotton, and tobacco.

D. rubrovolvata, which belongs to the *Phallaceae* family and roots in the forest or by broad-leaved trees on the ground, has been used for centuries as a food and medicine [4–6].

It is highly nutritious and rich in polysaccharides, polyphenols, flavonoids, and other bioactive substances. It improves immunity; has anticancer, antioxidant, and anti-inflammatory medicinal properties; and is deeply valued by consumers in China [7]. It was planted in 1983 [6]. As a result of its unique environmental conditions and climate conditions, Guizhou is an ideal place for *D. rubrovolvata* to grow. Meanwhile, because it provides high nutritional, medical, and economic value, *D. rubrovolvata* is currently one of the most popular edible fungi grown commercially in Guizhou Province, China, and is widely cultivated [5]. As a key edible fungus in Guizhou Province during the 13th Five-Year-Plan period, *D. rubrovolvata* played an important role in providing employment, stimulating entrepreneurship, and improving the economic level. In 2021, the annual output of *D. rubrovolvata* in Guizhou Province reached 8800 tons. At present, it is cultivated on a large scale in Zhijin County, Annong County, and other places in Guizhou Province. However, the fruiting body of *D. rubrovolvata* will soften and decay rapidly after harvesting. Because of the high respiration rate, significant water loss, and susceptibility to microorganisms during shelf life, there is a shorter shelf life for edible fungi than for fruits and vegetables [8]. Therefore, post-harvest treatment will become important. Furthermore, there are no reports related to the preservation of *D. rubrovolvata*.

In recent years, as consumers' awareness of health has increased, more and more people have enjoyed fresh edible fungi, resulting in increased production and consumption. Since fresh-cut edible fungi contain many nutritional ingredients such as vitamins, minerals, and bio-active compounds, they have become extremely popular among consumers [9]. Generally, the shelf life of an edible fungus under normal temperature is 1–3 days [10]. Following cutting, the shelf life decreases due to increased respiration rates, which may result in the consumption of nutrient substances. As a result of mechanical damage during processing, there are a number of physiological and biochemical changes that occur in edible fungi, such as browning, nutrient loss, water loss, and microbial pollution, which would shorten their shelf life and, therefore, affect sales and consumption. The deterioration rate of fresh-cut products is much higher than that of unprocessed ones. Therefore, an effective storage and preservation technology is essential for this emerging industry, but a huge challenge.

In order to prolong the shelf life of fresh-cut fruits, vegetables, and edible fungi, inhibit microorganisms, and maintain product quality, controlled atmosphere (CA) packaging has been studied and employed. Notably, a CA has been widely accepted by manufacturers and consumers because this technology extends the seasonable availability of produce, maintains the physicochemical and functional quality, and can reduce consumer costs [11]. Furthermore, the use of natural gas ingredients guarantees food safety. As a result, the processed products are free of toxic residues. In the beginning, the application of a CA generally consists of decreasing O_2 and increasing CO_2. Amodio et al. [12] confirmed that a CA of 3% O_2/20% CO_2/77% N_2 could effectively slow down browning and hardness decline and inhibit the growth of mold in fresh-cut *Pleurotus eryngii*. A study found that fresh-cut figs stored in a CA of 3% O_2/18% CO_2/79% N_2 had significantly lower respiration rates, weight loss, and microbial growth than those stored under normoxic conditions [13]. In Esmaeili et al.'s study [14], 5% O_2/10% CO_2/85% N_2 bound to the aloe vera concentrated gel could effectively maintain the anthocyanin and vitamin C content and inhibit the growth of microorganisms in strawberries. Recently, a CA with high O_2 has also been proven effective for the produce. According to Li et al. [15], a CA of 80% O_2/ 20% CO_2 increased the shelf life of *Agaricus bisporus* by retaining higher energy, suppressing ROS levels, and enhancing enzymatic activity. Another study found that CA treatment of 80% O_2/20% CO_2 inhibited polyphenol oxidase (PPO) and delayed browning of *Agaricus bisporus* pericarp and flesh. Additionally, it was able to maintain the integrity of cell membranes and provided good sensory properties [16]. A study found that high O_2 and CO_2 (80% O_2, 20% CO_2) treatment would improve wound healing and maintain quality by enhancing the activity of antioxidant enzymes, including SOD, CAT, APX, and GR, thus reducing water loss and browning for fresh-cut *Agaricus bisporus* [17,18]. It was in no doubt that CA packaging proved to be effective for preserving edible fungi, fruit, and vegetables for a long time.

The color and the texture were the main factors that influenced consumers' favorites for edible fungi, which would affect the product value. We found that the tissues of fresh-cut *D. rubrovolvata* were exposed to the environment due to cutting, leading to faster decay and browning. The resulting decay and browning severely limited the consumption of *D. rubrovolvata*. However, so far, there is no available literature on the storage of fresh-cut *D. rubrovolvata* under a CA. Therefore, the present study was performed according to a two-step screening. Firstly, the effect of O_2 concentration on the fresh-cut *D. rubrovolvata* was investigated. Then, on the basis of the O_2 concentration being determined, CO_2 was involved. A series of physiological and biochemical changes in fresh-cut *D. rubrovolvata* under CA conditions was investigated. Through the study of a series of physiological and biochemical color, texture, umami, flavor compounds, and other indicators, the best storage conditions of fresh-cut *D. rubrovolvata* were determined.

2. Materials and Methods

2.1. Materials

D. rubrovolvata was collected from a farm of Anlong Count (Guizhou Province, China, UTMX: 104°59′–105°41′ UTMY: 24°55′–25°33′). The fruiting bodies were harvested on 7 September and 17 September 2021, respectively. Afterward, samples were transported to the laboratory within 3 h after harvest. The polyamide package was purchased from Shijiazhuang Xilong Packaging Co., Ltd. (Shijiazhuang, China), the hydrogen peroxide, ethanol absolute, sodium bicarbonate, etc., from China National Medicines Co., Ltd. (Beijing, China), the amino acid standard from Sigma-Aldrich Trading Co., Ltd. (Shanghai, China), 5′-inosine monophosphate (5′-IMP) standard, 5′-guanosine monophosphate (5′-GMP) standard, 5′-xanthosine monophosphate (5′-XMP) standard and 5′-adenoshine monophosphate (5′-AMP) standard from Standford Chemicals Co. (Lake Forest, CA, USA).

2.2. Treatment and Storage

Fruit bodies similar in size (6–8 cm stalk length) and free of pests and mechanical damage were chosen for the experiment. After the fruit body was washed for 30 s in tap water and an electric fan was used to remove the *D. rubrovolvata* surface water, the stipe was cut perpendicularly into three sections and pre-cooled at 4 ± 1 °C for 12 h.

In a preliminary experiment, the pre-cooled fresh-cut *D. rubrovolvata* was packaged into polyamide packages (22 cm × 32 cm, 0.24 mm thickness, moisture permeability 3.12 g/(m^2 24 h), oxygen permeability 46 cm^3/(m^2 24 h 0.1 MPa)), with different gas compositions: 5% O_2, 20% O_2, 35% O_2, 50% O_2, 65% O_2, 80% O_2, and 95% O_2, respectively, and underwent N_2 balance and were stored at 4 ± 1 °C for seven days. Samples were taken out and analyzed at 0, 3, 5, and 7 d, with 6 bags taken for each treatment to assess the shelf life. Then, a total of 18 pieces of *D. rubrovolvata* were loaded into each bag with a weight of approximately 240 g. Samples were packaged in atmosphere with different gas compositions: 5% oxygen (O_2), 0% carbon dioxide (CO_2), 95% nitrogen (N_2) (5/0/95); 5% O_2, 5% CO_2, 90% N_2 (5/5/90); 5% O_2, 10% CO_2, 85% N_2 (5/10/85); 5% O_2, 15% CO_2, 80% N_2 (5/15/80); and 5% O_2, 20% CO_2, 75% N_2 (5/20/75), respectively, then heat-sealed with a gas packing machine. For simulation of sales scenarios, samples were stored at 4 ± 1 °C during the shelf life. Samples were taken out every two days (0, 2, 4, 6, and 8 days) during the shelf life, with 12 bags taken from each treatment and analyzed.

2.3. Decay Rate, Respiration Rate, Malondialdehyde (MDA) Content, Relative Conductivity, Ethanol Content, and Low-Field Nuclear Magnetic Resonance (LF-NMR) Transverse Relaxation Measurements

A decay rate analysis of fresh-cut *D. rubrovolvata* was performed in accordance with Rux et al., [19] with minor modifications. Quality evaluation based on visual observation of fresh-cut *D. rubrovolvata* for each of the CA packaging was performed at the end of storage by 10 panelists. The fruiting bodies with a surface browning and generating the off-flavor were identified as decay. The ratio of decayed samples to the total number of samples analyzed was stated as the decay rate (%).

The respiration rate of the fruiting body was determined by Zhang's method, with some modifications [20]. Before measurement, 120 g of fruiting bodies of *D. rubrovolvata* were placed in 1 L sealed plastic jars at 25 °C for 3 h. To analyze gas samples from the headspace of the plastic jars, gas syringes were used to draw 2 mL of gas from each and analyze them with infrared O_2 and CO_2 analyzers (Checkpoint II Portable residual oxygen meter, Dansensor, Rinsted, Denmark). In this study, the respiratory rate was expressed as mg/kg/h CO_2.

Huang's method was used to assay the MDA content, with slight modifications [21]. *D. rubrovolvata* (0.2 g, FW) was ground and homogenized in 12 mL of 10% (*w*/*v*) trichloroacetic acid (TCA) and filtered to obtain a sample extract. We added 5 mL filtrate and 5 mL 0.6% thiobarbituric acid (TBA) to the tube. The solution was vibrated and mixed before it was boiled for 15 min and cooled to room temperature. A final measurement was made of the supernatant's absorbance at 450 nm, 532 nm, and 600 nm. The MDA content of the fresh sample weight was measured as mmol/g.

Based on He et al.'s method [22], the relative conductivity was analyzed in the fruiting body, with a slight modification. A total of 40 mL of deionized water was added to the triangular flask its conductivity (P_0) was measured. The fruiting body of *D. rubrovolvata* was cut, mixed (2 × 2 mm) and transferred into triangular flasks. A conductivity meter was used to measure the conductivity of deionized water and fresh samples (P_1). It was oscillated for 10 min and stood still at 25 °C. Triangular flasks were cooled to room temperature (about 25 °C) after being placed in boiling water (100 °C) for 10 min and their conductivity (P_2) was measured. The relative conductivity was expressed as a percentage.

$$\text{Relative conductivity } (\%) = P_1 - P_0/(P_2 - P_0) \times 100\%$$

According to Ventura-Aguilar's description, headspace gas chromatography was used to determine *D. rubrovolvata*'s ethanol concentration [23]. Results were expressed as μg/g fresh weight.

An LF-NMR measurement was performed using a method described by Cheng et al. [24].

2.4. Texture, Browning Degree, Polyphenol Oxidase (PPO) Activity, Catalase (CAT) Activity, and Total Phenolic (TP) Content

According to Jahanbakhshi et al.'s report, the shear force of the fruiting body was determined [25].

Using Liu et al.'s method with minor modifications, the browning degree of fruiting bodies was measured [26]. A mixture of approximately 2 g of tissue was ground in an ice bath with 8 mL of 0.2 mol/L sodium phosphate buffer (pH 6.8) and stored at 4 °C for 15 min, and the mixture was centrifuged at 24,200× *g*. A UV-vis spectrophotometer was used to measure the 420 nm absorbance (Cary 60, Agilent, Santa Clara, CA, USA). The browning degree was 420 nm.

In order to determine the PPO activity, Liu et al.'s method was used [26]. The PPO activity was calculated as U/(g·min) fresh weight.

A modified method was used to determine the CAT activity [18]. After 2 g of powdered fruiting bodies was homogenized with 3.2 mL phosphate buffer (0.05 mol/L, pH 7.0), it was centrifuged at 24,200× *g* and 4 °C for 10 min. Next, 1.6 mL Tris-HCl solution and 2.72 mL deionized water were added to the test tube with 1 mL supernatant. After immersion in 37 °C water for 3 min, the mixture was transferred into a colorimetric dish, with 0.4 mL of hydrogen peroxide solution. Every 30 s for 3 min, spectrophotometric measurements at 240 nm were taken to determine the absorbance. CAT activity is expressed in units of U/(g·min·FW).

Using a Folin–Ciocalteu assay, the total phenolic (TP) content was determined [27]. The total phenolic content of the sample was calculated using a gallic acid standard curve and expressed in mg of gallic acid equivalent (GAE)/g of the sample.

2.5. Polysaccharides Content

Polysaccharides were determined using a method described by Wang et al. [28]. The polysaccharides content was expressed as g/(100 g).

2.6. Free Amino Acid

According to Wang et al.'s study [29], the free amino acid content was detected for fruiting bodies. Initially, 0.5 g of fresh samples were crushed and extracted with 5 mL of 0.01 mol/L hydrochloric acid. The sample was then heated in a boiling water bath for 30 min, followed by 10 min of centrifugation at 24,200× *g*. After separating the supernatant, we added 2 mL of 0.01 mol/L hydrochloric acid, followed by 5 min of sonication. We centrifuged the mixture, combined it with a supernatant, and added solvent until the volume was 10 mL. Determination was performed using a 0.22 μm filter membrane. Analysis was performed using an Agilent 1100 liquid chromatograph (with VWD detector) (1100, Agilent, Santa Clara, CA, USA) containing a ZORBAX Eclipse AAA column (4.6 × 150 mm, 3.5 μm). The HPLC conditions were as follows: mobile phase A: 40 mmol/L sodium dihydrogen phosphate (pH 7.8); mobile phase B: acetonitrile/methanol/water = 45/45/10; flow rate: 1.0 mL/min; UV detection wavelength, 338 nm (0 to 19 min), 266 nm (19.01 to 25 min); oven temperature: 40 °C; injection volume: 1 μL. Identification and quantification of each amino acid were conducted using a standard mixture of 17 amino acids (Sigma-Aldrich CN Inc., Shanghai, China).

2.7. 5′-Nucleotides

Wang et al.'s procedure was used to measure 5′-nucleotides, with minor modifications [29]. In 25 mL deionized water, 0.2 g of mushroom powder was boiled for 5 min, cooled, and centrifuged for 15 min at 24,200× *g*. A second extraction was carried out by adding 25 mL of deionized water after filtration. Prior to HPLC analysis using a RIGOL L-3000 high-performance liquid chromatography system (RIGOL L-3000, Puyuan, Beijing, China) equipped with a Diamonsil C18 column (250 × 4.6 mm, 5 μm), a 0.45 μm cellulose membrane was used to filter the extraction. The HPLC conditions were as follows: mobile phase, 0.01 mol/L KH_2PO_4 (pH 4.68); flow rate, 1.0 mL/min; UV detection wavelength, 254 nm; oven temperature, 25 °C; and injection volume, 20 μL. Standards for 5′-nucleotides were used to identify and quantify each nucleotide (Sigma-Aldrich CN Inc., Shanghai, China).

2.8. Equivalent Umami Concentration (EUC)

The equivalent umami concentration, or EUC (g MSG/kg), is defined as the MSG concentration corresponding to the umami intensity created by the synergy between umami amino acids and 5′-nucleotides, and calculated as follows:

$$Y = \sum a_i b_i + 12.18\left(\sum a_i b_i\right)\left(\sum a_j b_j\right), \quad (1)$$

where Y is the EUC value of the mixture in terms of g MSG/kg; a_i is the concentration (g/kg) of each umami amino acid (aspartic acid (Asp) or glutamic acid (Glu)); a_j is the concentration (g/kg) of each 5′-nucleotides (5′-IMP, 5′-GMP, 5′-XMP, or 5′-AMP); b_i is the relative umami concentration (RUC) for each umami amino acids with respect to MSG (Asp = 0.077, Glu = 1); b_j is the RUC for each umami 5′-nucleotides constant (5′-GMP = 2.3, 5′-IMP = 1, 5′-XMP = 0.61, 5′-AMP = 0.18); and 12.18 is a synergistic constant.

2.9. Electronic Nose

Following the measurement of Pei et al. [30], an electronic nose analysis of fresh-cut *D. rubrovolvata* was conducted.

2.10. Total Colony Numbers

As per Shen et al. [31], the total number of colonies was determined by aerobic plate counting. A colony-forming unit (CFU) represents the total number of colonies.

2.11. Statistical Analysis

The analyses were repeated three times, and the results were expressed as mean ± standard deviation. IBM SPSS 25 statistical software (Armonk, NY, USA) was used for the data processing and analysis. Origin 2017 was used for mapping and Duncan's method was used for multiple comparisons.

3. Results and Discussion

3.1. Effect of O_2 Concentration (%) on the Shelf Life and Quality of Fresh-Cut D. rubrovolvata

To evaluate the effect of O_2 concentrations (5%, 20%, 35%, 50%, 65%, 80%, 95%) on the shelf life of *D. rubrovolvata*, the shear force, relative conductivity, and decay rate were employed. According to Figure 1, the relative conductivity, shear force, and decay rate of all samples increased over its shelf life. In the first three days of storage, all three indicators of samples in a high O_2 (95%, 50%, 65%) environment were close to the initial value. However, with the storage time extended, the three indicators of samples in a low O_2 environment were significantly lower than those in a high O_2 environment. On the seventh day, the relative conductivity, shear force, and decay rate of samples under a 5% O_2 environment were 26.91%, 8.20 ± 0.19 N, and 17.67%, respectively, which were 19.77%, 17.98%, and 33.75% lower, respectively, than ones at 95% O_2/5% N_2 ($p < 0.05$). The above result suggested that a low O_2 concentration (5%) was more efficient at inhibiting the changes in the relative conductivity, shear force, and decay rate during the shelf life of fresh-cut *D. rubrovolvata*. The results of Li and Lyn's papers confirm the same findings [32,33].

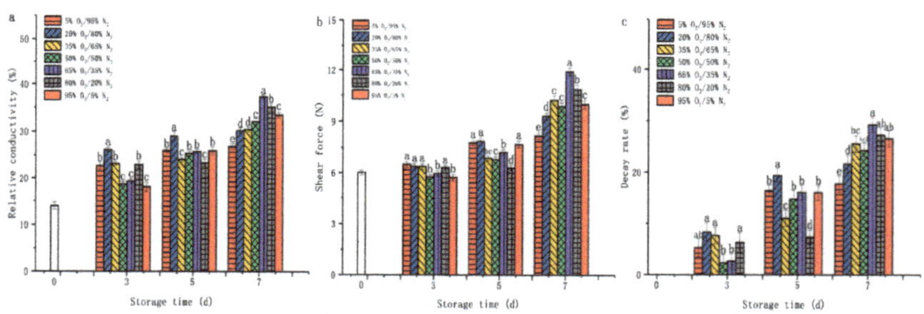

Figure 1. Effect of O_2 concentration (%) on relative conductivity (**a**), shear force (**b**), and decay rate (**c**) of fresh-cut *Dictyophora rubrovolvata*. Differences at $p < 0.05$ are indicated by different letters and data are expressed as the mean ± SD.

3.2. Effect of CO_2 Concentration (%) with 5% O_2 on the Shelf Life of Fresh-Cut D. rubrovolvata

3.2.1. Changes in Physiological Parameters

The appropriate concentration of CO_2 is also very important. An excessive CO_2 concentration (CO_2 concentration of more than 12%) could cause physiological damage to edible fungi [32]. At the optimal O_2 concentration (5%), the effects of CO_2 concentrations (0%, 5%, 10%, 15%, 20%) were introduced and investigated. Figure 2 illustrates the relationship between shelf life and physiological parameters in fresh-cut *D. rubrovolvata*. Clearly, the higher the concentrations of CO_2 (10%, 15%, or 20%), the higher the decay rate of the samples (Figure 2a). The highest decay rate in all treated fruiting bodies was observed at 8 d. A sample kept in a CA of 5% O_2/5% CO_2/90% N_2 provided a decay rate of 12.93%; samples under O_2 group alone treatments and other CO_2 groups were 1.45 times, 1.76 times, 2.43 times, and 2.03 times higher than that of 5% O_2/5% CO_2/90% N_2, respectively ($p < 0.05$).

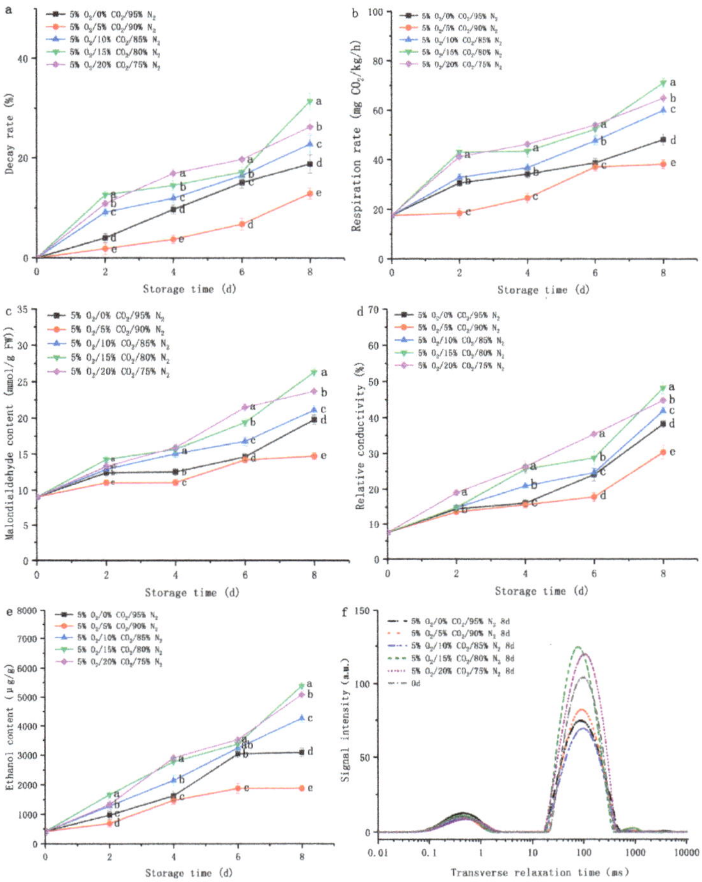

Figure 2. Effect of CO_2 concentration (%) with 5% O_2 on decay rate (**a**), respiration rate (**b**), malondialdehyde content (**c**), relative conductivity (**d**), ethanol content (**e**), and transverse relaxation time (**f**) of fresh-cut *Dictyophora rubrovolvata*. Differences at $p < 0.05$ are indicated by different letters and data are expressed as the mean ± SD.

The respiration of edible fungus after harvest largely determines the quality changes in the sample. Plant cells use energy reserves more efficiently when their respiration rate increases, resulting in reducing their quality and shelf life [20]. Capotorto et al. [34] showed that the combination of low O_2 and high CO_2 (10% O_2/10% CO_2/80% N_2) could effectively inhibit the increase in the respiratory rate of fresh-cut artichokes, thereby maintaining the nutrition and quality of the sample. In our study, the respiration rate of a sample under a 5% O_2/5% CO_2/90% N_2 gas environment was 38.26 ± 1.78 mg CO_2/kg/h ($p < 0.05$), while the respiration rate of samples under a CO_2 (0%, 10%, 15%, or 20%) storage environment was 48.14 ± 2.23, 60.16 ± 1.70, 71.32 ± 1.91, and 65.08 ± 1.76 mg CO_2/kg/h on the eighth day, respectively. The respiratory rate showed a similar trend to the decay rate (Figure 2b); the higher the respiratory rate, the higher the decay rate of the sample.

The MDA and relative conductivity are widely used to evaluate the membrane and tissue integrity of edible fungi, fruits, and vegetables, which are an important product of membrane lipid peroxidation. The fact that the content of the MDA changed is an important indicator of the degree of damage to the membrane system. The higher the content, the more serious the degree of damage to the membrane system [20]. The MDA content in a fruiting body at storage was 9.06 ± 0.38 mmol/g FW for the initial value and obviously increased during shelf life (Figure 2c). The MDA content in all samples was 62.82 to 190.71% higher than that on the initial day, and the minimum value was observed under 5% O_2/5% CO_2/90% N_2 (14.75 ± 0.48 mmol/g FW) conditions after 8 d ($p < 0.05$). Li et al. [32] found that 2% O_2/30% CO_2/68% N_2 could result in better inhibition of MDA accumulation in *Pleurotus eryngii* and it also alleviated lipid and membrane peroxidation in mushrooms to the maximum extent in comparison with other groups. This result could demonstrate that an application of low O_2 combined with CO_2 treatment would be a better way to preserve postharvest products. The relative conductivity increased in all groups over time. The increase in relative conductivity indicated a decrease in cell membrane integrity. It was found that fruiting bodies in higher CO_2 (10%, 15%, or 20%) presented a significantly higher level than 5% O_2/5% CO_2/90% N_2 after 4 d (Figure 2d). On the eighth day, the relative conductivity under CO_2-treated (10%, 15%, or 20%) fruiting bodies was 38.16 to 59.08% higher than a CA of 5% O_2/5% CO_2/90% N_2 ($p < 0.05$). This might be due to the fact that compared with other groups, 5% O_2/5% CO_2/90% N_2 could well inhibit the respiration rate (Figure 2e) of the sample, thus effectively protecting the integrity of the cell membrane.

As a volatile substance, ethanol can be accumulated during anaerobic conditions, which affects the normal metabolism of fruiting body cells [23]. In the present study, treatments stored under higher CO_2 were higher in ethanol than those stored under lower CO_2, and the level of ethanol increased obviously during the shelf life as well (Figure 2e). When stored until the eighth day, the ethanol content of the 5% O_2/5% CO_2/90% N_2 treatment group was the lowest (1882.22 ± 102.42 μg/g), and significantly lower than that of the other treatment groups (3087.84 ± 145.21 to 5394.45 ± 120.67 μg/g) ($p < 0.05$). Similar to our research results, Wang et al. [35] found that a low O_2 and high CO_2 (3% O_2/7% CO_2/90 N_2) storage environment could effectively inhibit the accumulation of ethanol and maintain good sensory quality of garlic scapes.

The moisture content and existing state of fresh fruiting bodies were important physical indexes of freshness. Water molecule dynamics and environment during edible fungi storage degradation were assessed by measuring the spin-spin relaxation time (T_2) of the sample. Figure 2f shows the distribution of transverse proton relaxation data and two proton components in fresh-cut *D. rubrovolvata* samples. Unlike the results of other researchers, where pools of water were attributed to the vacuole, cytoplasm, and cellular wall, only two peaks were observed for water relaxation times in our experiment [24]. It might be due to the destruction of cell tissue caused by cutting, which made the enzyme in different organelles contact the substrate to initiate the wound response induced by the signal. At the same time, due to the destruction of the subcellular structure, the solutes retained in the organelles were released, which were bound to water through hydrogen bonds [35]. The maximum amplitude T_2 value (95.48 ms) of the sample under the 5% O_2/5% CO_2/90% N_2 environment was found on the eighth day of storage, the same as on day 0. As CO_2 concentrations increased, the maximum amplitude T_2 value of the samples continued to change, of which the maximum amplitude T_2 values of the samples under a CO_2 (15%, 20%) environment were 77.53 ms and 109.70 ms. Compared with the initial level, they changed by 18.80% and 14.89%. Consistent with the results of MDA and relative conductivity (Figure 2c,d), a 5% CO_2 gas environment could reduce the oxidative stress reaction caused by mechanical injury and the oxidative damage of the membrane.

Overall, low O_2 and plausible CO_2 (5% O_2/5% CO_2/90% N_2) could delay the physiological and biochemical changes of fruiting bodies at 4 °C, inhibit their metabolism, delay the increased of cell membrane permeability, and delay aging, leading to improved storage quality of fresh-cut *D. rubrovolvata*.

3.2.2. Changes in Texture and Browning Degree

Subsequently, we evaluated the fresh-cut *D. rubrovolvata* samples' texture and browning degree due to the different CO_2 treatments (0%, 5%, 10%, 15%, or 20%) (Figure 3). The shear force and browning degree during shelf life significantly increased due to a higher

CO_2 concentration. Wang et al. [36] found that with the extension of storage time, under the condition of low O_2 and high CO_2, the hardness of the *Pleurotus eryngii* showed an increasing trend, which was consistent with our research. The reason for this phenomenon is that cutting and low-temperature storage increase the lignin content of edible mushrooms, leading to an increase in hardness [36]. The texture is the main attribute responsible for consumers' acceptance of fresh-cut *D. rubrovolvata*. During the first two days of storage, the shear force of fruiting bodies in each treatment showed no significant change ($p < 0.05$) (Figure 3a). On the eighth day of shelf life, the shear force of the fruiting body in the 5% O_2/5% CO_2/90% N_2 group increased from 5.72 ± 0.53 N at day 0 to 8.79 ± 0.55 N, significantly lower than other storage environments (12.68 ± 0.62 to 16.43 ± 0.75 N) ($p < 0.05$). From the results of the MDA and relative conductivity (Figure 2c,d), we concluded that an appropriate gas environment could inhibit the accumulation of the lignin, and thus the increase in shear force was delayed (Figure 3a).

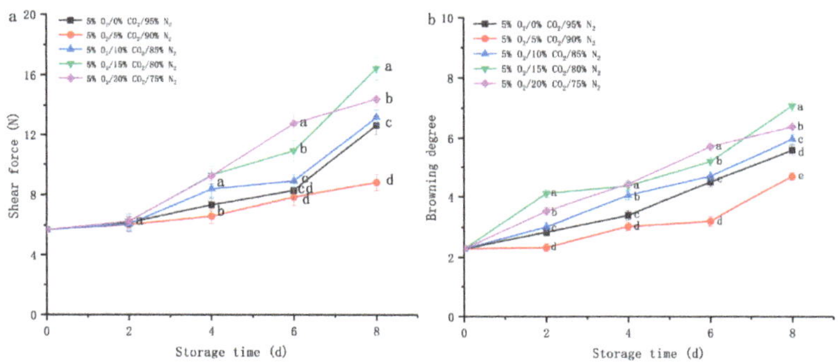

Figure 3. Effect of CO_2 concentration (%) with 5% O_2 on shear force (**a**), and browning degree (**b**) of fresh-cut *Dictyophora rubrovolvata*. Differences at $p < 0.05$ are indicated by different letters and data are expressed as the mean ± SD.

Among the quality attributes that affect consumers' purchasing behavior, color is the most intuitive influencing factor. A continual increase in the browning degree (Figure 3b) of the samples was observed during the shelf life. Interestingly, fresh-cut fruiting bodies did not turn brown after storage for 2 d at a CA of 5% O_2/5% CO_2/90% N_2, whereas slight browning was observed in the 5% O_2/15% CO_2/80% N_2 and 5% O_2/20% CO_2/75% N_2 treatment groups. The browning degree of the fruiting body under a CA of 5% O_2/5% CO_2/90% N_2 was lower (4.71 ± 0.13) than in other groups (5.60 ± 0.17 to 7.08 ± 0.18) after storage for eight days ($p < 0.05$). The reason for this phenomenon might be that a high concentration of CO_2 would cause CO_2 damage to the fruiting bodies or anaerobic respiration at the later stage of storage, aggravating the fruiting bodies' browning. Meanwhile, it also increased the relative conductivity and ethanol content (Figure 2d,e).

3.2.3. Changes in Enzyme Activity

As some researchers reported, the oxidative stress reaction of fresh-cut edible fungi, fruits, and vegetables generated from mechanical injury stress would cause membrane lipid peroxidation, damage the integrity of the cell membrane, and lead to a series of changes such as browning and aging [31]. PPO activity plays an important role in the browning reaction of fresh-cut products. Likewise, an antioxidant enzyme such as CAT can represent a plant's defense capacity. Figure 4 showed the changes in the enzyme activities of PPO and CAT during fresh-cut *D. rubrovolvata* storage at different CO_2 levels (0%, 5%, 10%, 15%, or 20%). The PPO activity of samples in different storage environments showed an upward trend (Figure 4a). Consistently, the PPO activity of samples stored under 5% O_2/5% CO_2/90% N_2 conditions differed significantly from that of other treatments during the shelf life ($p < 0.05$). After storage for eight days, the PPO activity of a sample in a 5% O_2/5% CO_2/90% N_2 storage environment only increased by 59.64% compared with that at the initial value and was lower than that in other gas environments (82.04 to 129.94%) ($p < 0.05$). Ye et al. [37] found that higher CO_2 accumulation induced a higher PPO activity in fresh *Lentinula edodes*.

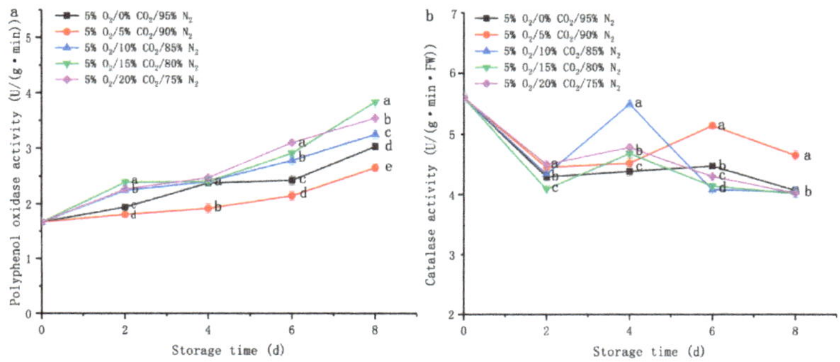

Figure 4. Effect of CO_2 concentration (%) with 5% O_2 on polyphenol oxidase activity (**a**), and catalase activity (**b**) of fresh-cut *Dictyophora rubrovolvata*. Differences at $p < 0.05$ are indicated by different letters and data are expressed as the mean ± SD.

Nevertheless, it is worth noting that the CAT activity of the samples first decreased, then increased, and finally decreased over the shelf life (Figure 4b). Over the course of the shelf life, the CAT activity of samples in a higher CO_2 storage environment (10%, 15%, or 20%) reached the maximum on the fourth day and then decreased. However, the CAT activity of samples under 5% O_2/0% CO_2/95% N_2 and 5% O_2/5% CO_2/90% N_2 storage environments reached the maximum on the sixth day, and the CAT activity of 5% O_2/5% CO_2/90% N_2 group (5.15 ± 0.05 U/(g·min·FW)) was higher than that of other groups (4.09 ± 0.07 to 4.48 ± 0.04 U/(g·min·FW)) ($p < 0.05$). After storage for eight days, the CAT activity of all groups decreased, but the CAT activity of the 5% O_2/5% CO_2/90% N_2 group was higher than that of other groups ($p < 0.05$).

The above results show that 5% O_2/5% CO_2/90% N_2 treatment is able to reduce the PPO activity, thus alleviating browning in fresh-cut *D. rubrovolvata* to the maximum extent in comparison with other treatments (Figure 4a). This result was consistent with the browning degree (Figure 3b). In our study, we found that the change in CAT activity was similar to that of the MDA content (Figure 2c); a 5% O_2/5% CO_2/90% N_2 treatment could well delay the time of the CAT peak (Figure 4b).

3.2.4. Changes in Nutrition

Total phenolic, polysaccharides, and free amino acids are important nutritional indicators in *D. rubrovolvata* [4–6]. The nutritional effect of different CO_2 treatments (0%, 5%, 10%, 15%, or 20%) on fresh-cut *D. rubrovolvata* was investigated (Figure 5). The total phenolic content of all samples decreased during the shelf life of eight days (Figure 5a). In spite of this, it is worth noting that total phenolic levels in fruiting bodies treated with 5% O_2/5% CO_2/90% N_2 were higher than in other gas environments ($p < 0.05$). On the eighth day of storage, the total phenolic content of the sample under the condition of 5% O_2/5% CO_2/90% N_2 was 59.19 ± 0.53 mg/g, while that of other treatment groups was 48.34 ± 0.60 to 56.80 ± 0.50 mg/g. Lower total phenolic levels were found in other treatments (8.01 to 18.33%) compared to that of 5% O_2/5% CO_2/90% N_2 samples on the eighth day ($p < 0.05$).

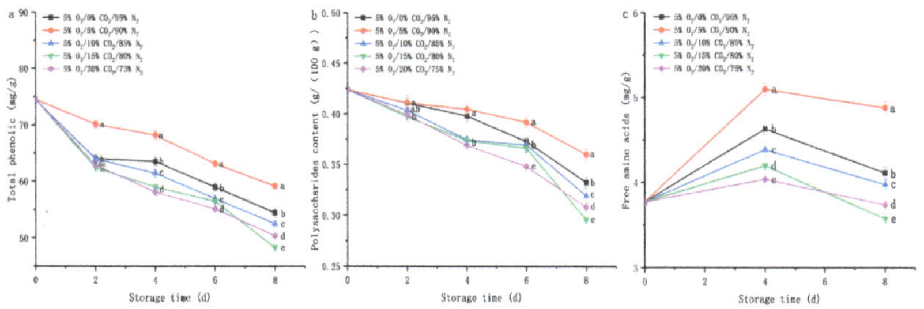

Figure 5. Effect of CO_2 concentration (%) with 5% O_2 on total phenolic (**a**), polysaccharides content (**b**), and free amino acids (**c**) of fresh-cut *Dictyophora rubrovolvata*. Differences at $p < 0.05$ are indicated by different letters and data are expressed as the mean ± SD.

A similar trend was observed for polysaccharides content in the samples (Figure 5b). During the shelf life, the lowest deterioration in polysaccharides was observed for 5% O_2/5% CO_2/90% N_2 in fresh-cut *D. rubrovolvata* compared with other groups. In the six days of storage, the polysaccharides content of the 5% O_2/15% CO_2/80% N_2 treatment group decreased slowly, then began to decline sharply after 6 d. The polysaccharides content of 5% O_2/5% CO_2/90% N_2 decreased slowly during the shelf life. On the eighth day of storage, the polysaccharides content of 5% O_2/5% CO_2/90% N_2 reached 0.37 ± 0.01 g/(100 g). This was significantly higher than in other treatments (0.30 ± 0.00 to 0.34 ± 0.01 g/(100 g)) ($p < 0.05$).

Unlike total phenolic and polysaccharides, free amino acids in the samples of each group first increased, then decreased (Figure 5c). On the eighth day, compared with other groups, the total content of free amino acids under 5% O_2/5% CO_2/90% N_2 was 4.88 ± 0.07 mg/g, 0.77, 0.9, 1.3, and 1.14 mg/g higher than in fresh-cut *D. rubrovolvata*, respectively ($p < 0.05$).

Li et al. [27]. found that the high respiratory rate indicated that the respiratory metabolism was vigorous, hence the nutrients stored in edible fungus would be rapidly consumed, leading to a rapid decline in storage quality. This result was coincident with our work; 5% O_2/5% CO_2/90% N_2 could significantly inhibit the increase of the respiratory rate of fruiting bodies (Figure 2b) and reduce the consumption of nutrients in samples.

3.2.5. Changes in Umami

5′-nucleotides, flavor amino acids (bitter, sweet, umami), and EUC can be used to assess the taste of edible fungus [29]. The effect of different CO_2 treatments (0%, 5%, 10%, 15%, or 20%) on the umami taste of fresh-cut *D. rubrovolvata* was investigated (Figure 6). Nucleotides are crucial to the umami taste of edible fungi. As shown in Figure 6a, all fresh-cut fruiting bodies contained four 5′-nucleotides (5′-GMP, 5′-AMP, 5′-IMP and 5′-XMP) in the first four days, which impacted the umami taste of mushrooms. However, only three nucleotides were detected in the 0%, 5%, and 10% CO_2 environments, and only two nucleotides were detected in the 15% and 20% CO_2 environments on the eighth day of storage. Compared to other groups, the fruiting bodies from 5% O_2/5% CO_2/90% N_2 environments had significantly higher contents of 5′-nucleotides ($p < 0.05$), with values of 0.58 ± 0.02 g/kg FW and 0.32 ± 0.01 g/kg FW after eight days as compared to 0.17 ± 0.00 to 0.42 ± 0.02 g/kg FW and 0.17 ± 0.00 to 0.29 ± 0.01 g/kg FW in other groups.

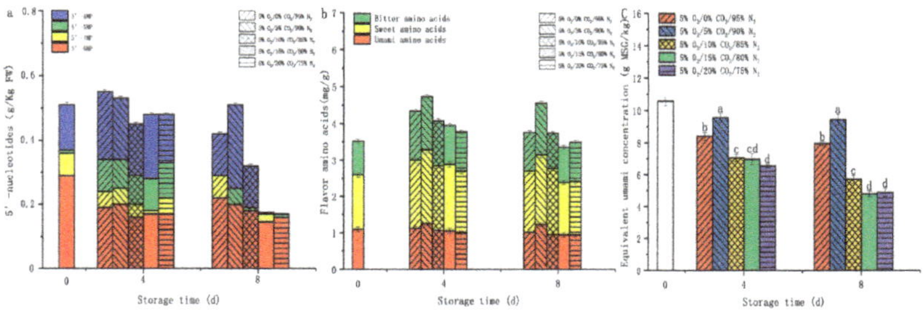

Figure 6. Effect of CO_2 concentration (%) with 5% O_2 on 5′-nucleotides (**a**), flavor amino acids (**b**), and equivalent umami concentration (**c**) of fresh-cut *Dictyophora rubrovolvata*. Differences at $p < 0.05$ are indicated by different letters and data are expressed as the mean ± SD.

In addition, flavor amino acids played a crucial role in the fruiting body taste. Among these, only umami amino acids such as aspartic (Asp) and glutamic (Glu) acids contribute to the characteristic umami taste [38]. The Asp and Glu contents in the 5% O_2/5% CO_2/90% N_2 treatment group were 0.10 ± 0.01 mg/g and 1.13 ± 0.03 mg/g, higher than in other gas environments by 111.11 to 142.86% and 100.89 to 129.89% on the eighth day, respectively ($p < 0.05$) (Figure 6b).

The combined action of 5′-nucleotides and umami amino acids (Asp and Glu) enhanced the umami flavor of the edible fungi [29]. In general, EUC represents their synergy. According to Figure 6c, the EUC values of a sample under a 5% O_2/5% CO_2/90% N_2 gas environment (9.43 ± 0.33 g MSG/kg) at eight days are significantly higher than those observed in other storage environments (4.79 ± 0.17 to 7.96 ± 0.11 g MSG/kg), and approximately 1.18 to 1.97 times higher than those of the fruiting bodies ($p < 0.05$).

3.2.6. E-Nose Analysis

The E-nose is sensitive to odor information obtained from the food samples, and slight changes in volatile compounds will cause different sensor responses. A radar chart was used for the statistical analysis to illustrate relationships and trends among variables. There was consistency in the signal values of the appearance change of 10 sensors during storage in these samples. However, significant differences were observed not only between the radar profiles of the samples of *D rubrovolvata* on days 0 and 8, but also between the edible fungi treated with different treatments, indicating that the volatile compounds in the samples changed significantly during the shelf life ($p < 0.05$). On the eighth day, the response value profile of the 5% O_2/5% CO_2/90% N_2 treatment group was the most similar to 0 d. This indicated that the volatile components were the closest to the initial level (Figure 7a). With the increase in CO_2 concentrations (10%, 15%, or 20%) in the storage environment, there was an obvious deviation between the response value profile of the samples and initial value. Figure 7b shows the results of the PCA analysis. On the eighth day, the distribution of *D. rubrovolvata* compared with the initial day could be observed by PCA under various conditions. The variance contribution rates of PC1 and PC2 were 68.6% and 18.9%, respectively. The first two PCs were 87.50% (more than 85%), which could determine most of the volatile substance information of the sample. At 8 d, the fruiting body under a 5% O_2/5% CO_2/90% N_2 environment compared with other samples was closest to initial level, and with an increase in the CO_2 concentration in the storage environment, the difference from initial level was increased. As shown in Figures 6c and 7, compared with other treatments, 5% O_2/5% CO_2/90% N_2 not only delayed the decline of sample EUC, but also preserved its odor.

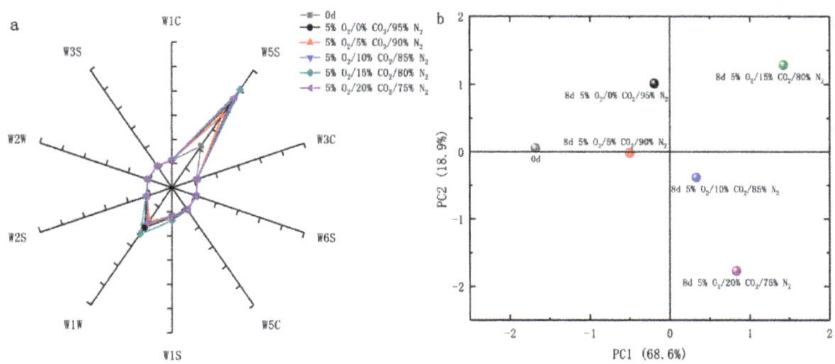

Figure 7. Effect of CO_2 concentration (%) with 5% O_2 on radar fingerprint charts (**a**), and principal component analysis (**b**) of fresh-cut *Dictyophora rubrovolvata*. Differences at $p < 0.05$ are indicated by different letters and data are expressed as the mean ± SD.

3.2.7. Changes in Total Colony Numbers

In general, CA with 2–5% O_2 or CO_2 was applied for fresh-cut products, which reduced the respiration rate and the growth rate of microorganisms [39]. Waghmare et al. [40] found that low O_2 combined with a suitable storage environment for CO_2 could effectively inhibit the growth of microorganisms in fresh-cut coriander. Consistent with the trend in the decay rate and respiration rate (Figure 2a,b), the total colony numbers of fresh-cut *D. rubrovolvata* gradually increased during the different gas environments for 8 d (Figure 8). Different treatments resulted in different rates of microbial growth. Interestingly, no microbial growth was observed in fresh-cut fruiting bodies after storage for 4 d at a CA of 5% O_2/5% CO_2/90% N_2, whereas microbial growth was observed in the other treatment groups ($p < 0.05$). The total colony numbers were significantly lower ($p < 0.05$) in the fruiting body at 5% O_2/5% CO_2/90% N_2 than the other groups, with values of 950 ± 141.42 CFU/g after eight days as compared to 2000 ± 141.42 to 5750 ± 212.13 CFU/g in other environments.

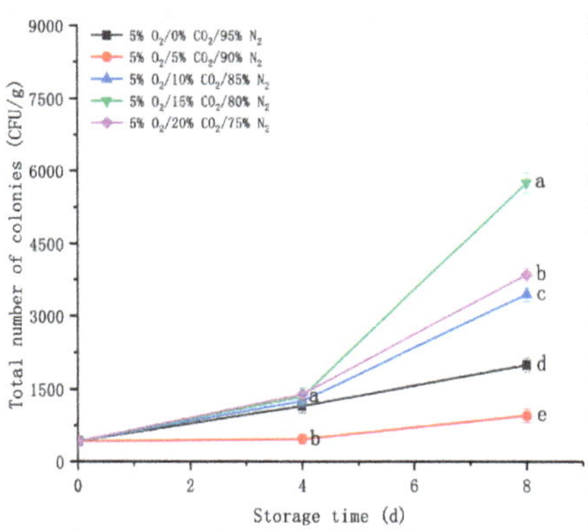

Figure 8. Effect of CO_2 concentration (%) with 5% O_2 on total colony numbers of fresh-cut *Dictyophora rubrovolvata*. Differences at $p < 0.05$ are indicated by different letters and data are expressed as the mean ± SD.

4. Conclusions

D. rubrovolvata is prone to problems such as the rapid softening and decay of fruiting bodies after harvest. According to the results of the experiment, we found that fresh-cut *D. rubrovolvata* was very sensitive to the concentration of O_2 and CO_2. Low O_2 (5%) combined with an appropriate concentration of CO_2 (5%) would be beneficial to the storage of fresh-cut *D. rubrovolvata*. Compared with other groups, 5% O_2/5% CO_2/90% N_2 could maintain water in the cells of fresh-cut *D. rubrovolvata* and inhibit the activity of PPO and CAT. Thus, it could effectively inhibit a rise in the decay rate, respiratory rate, relative conductivity, MDA content, ethanol content, shear force, browning degree, the total number of colonies, and retard the decline of total phenolic, polysaccharides, free amino acids, and EUC. The flavor of the sample could also be well maintained. In conclusion, the storage environment of 5% O_2/5% CO_2/90% N_2 could significantly delay the deterioration in the quality of *D. rubrovolvata* during the shelf life and have high food safety. Therefore, a CA with optimal $O_2/CO_2/N_2$ concentration could be a recommended technical method in other fresh-cut edible fungi.

Author Contributions: Conceptualization, Z.X. and R.W.; methodology, Z.X. and R.W.; investigation, Z.X., N.J., X.P., T.C., and R.W.; formal analysis, Z.X., J.L. (Jiangkuo Li) and J.L. (Jiqing Lei); data curation, Z.X. and C.M.; writing—original draft preparation, Z.X. and X.P.; writing—review and editing, Z.X. and R.W. All authors have read and agreed to the published version of the manuscript.

Funding: This research was funded by the Guizhou Province Key Technology Research and Development and Application of Innovation Base for Agricultural Products Primary Processing (Qian Ke Zhong Yin Di (2020) 4018) and the National Training Program of Innovation and Entrepreneurship for Undergraduates (No. 202110976016).

Data Availability Statement: Not applicable.

Conflicts of Interest: The authors declare no conflict of interest.

References

1. Castellanos-Reyes, K.; Villalobos-Carvajal, R.; Beldarrain-Iznaga, T. Fresh mushroom preservation techniques. *Foods* **2019**, *10*, 2126. [CrossRef] [PubMed]
2. Reis, F.S.; Martins, A.; Vasconcelos, M.H.; Morales, P.; Ferreira, I.C.F.R. Functional foods based on extracts or compounds derived from mushrooms. *Trends Food Sci. Technol.* **2017**, *66*, 48–62. [CrossRef]
3. Li, C.T.; Xu, S. Edible mushroom industry in China: Current state and perspectives. *Appl. Microbiol. Biotechnol.* **2022**, *106*, 3949–3955. [CrossRef]
4. Wang, Y.; Yin, J.; Huang, X.; Nie, S. Structural characteristics and rheological properties of high viscous glucan from fruit body of *Dictyophora rubrovolvata*. *Food Hydrocoll.* **2020**, *101*, 105514. [CrossRef]
5. Chen, X.; Zhou, X.; Zhao, J.; Tang, X.; Pasquali, M.; Migheli, Q.; Berg, G.; Cernava, T. Occurrence of green mold disease on *Dictyophora rubrovolvata* caused by *Trichoderma koningiopsis*. *J. Plant Pathol.* **2021**, *103*, 981–984. [CrossRef]
6. Sun, L.; Bao, C.; Chang, W.; Zhuang, Y. Preparation, characterisation, antioxidant and antiglycation activities of the novel polysaccharides from the pileus of *Dictyophora rubrovolvata*. *Int. J. Food Sci. Technol.* **2016**, *52*, 161–170. [CrossRef]
7. Wu, F.; Zhou, L.; Yang, Z.; Bau, T.; Li, T.; Dai, Y. Resource diversity of Chinese macrofungi: Edible, medicinal and poisonous species. *Fungal Divers.* **2019**, *98*, 1–76. [CrossRef]
8. Oms-Oliu, G.; Aguiló-Aguayo, I.; Martín-Belloso, O.; Soliva-Fortuny, R. Effects of pulsed light treatments on quality and antioxidant properties of fresh-cut mushrooms (*Agaricus bisporus*). *Postharvest Biol. Technol.* **2010**, *56*, 216–222. [CrossRef]
9. Kumar, K.; Mehra, R.; Guine, R.P.F.; Lima, M.J.; Lima, M.J.; Kumar, N.; Kaushik, R.; Ahmed, N.; Yadav, A.N.; Kumar, H. Edible mushrooms: A comprehensive review on bioactive compounds with health benefits and processing aspects. *Foods* **2021**, *10*, 2996. [CrossRef]
10. Sun, T.; Bian, J.; Wang, Y.; Hu, J.; Yun, X.; Chen, E.; Dong, T. One-step synthesis of poly(L-lactic acid)-based soft films with gas permselectivity for white mushrooms (*Agaricus bisporus*) preservation. *Foods* **2023**, *12*, 586. [CrossRef]
11. Falagan, N.; Terry, L. Recent advances in controlled and modified atmosphere of fresh produce postharvest technologies to reduce food waste and maintain fresh produce quality. *Johns. Matthey Technol. Rev.* **2018**, *62*, 107–117. [CrossRef]
12. Amodio, M.; Colelli, G. Controlled-atmosphere storage of fresh-cut 'cardoncello' mushrooms (*Pleurotus eryngii*). *Acta Hortic.* **2003**, *599*, 731–735. [CrossRef]
13. Hong, G.; Crisosto, C.; Cantwell, M. Quality and physiology of two cultivars of fresh-cut figs in relation to ripeness, storage temperature and controlled atmosphere. *Acta Hortic.* **2016**, *1141*, 213–219. [CrossRef]
14. Esmaeili, Y.; Zamindar, N.; Paidari, S.; Ibrahim, S.A.; Nafchi, A.M. The synergistic effects of aloe vera gel and modified atmosphere packaging on the quality of strawberry fruit. *J. Food Process. Preserv.* **2021**, *45*, e16003. [CrossRef]
15. Li, L.; Kitazawa, H.; Wang, X.; Sun, H. Regulation of respiratory pathway and electron transport chain in relation to senescence of postharvest white mushroom (*Agaricus bisporus*) under high O_2/CO_2 controlled atmospheres. *J. Agric. Food Chem.* **2017**, *65*, 3352–3360. [CrossRef]
16. Li, L.; Sun, H.; Kitazawa, H.; Wang, X. Effects of a high O_2 dynamic-controlled atmosphere technology on the browning of postharvest white mushroom (*Agaricus bisporus*) in relation to energy metabolism. *Food Sci. Technol. Int.* **2017**, *23*, 385–395. [CrossRef]

17. Wang, L.; Guo, Y.; Wang, X.; Zhang, X. Short-term O_2/CO_2 controlled atmosphere altered the water status and thus promoted phenolic biosynthesis during wound healing of fresh-cut white mushroom (*Agaricus bisporus*). *Postharvest Biol. Technol.* **2022**, *188*, 111879. [CrossRef]
18. Wang, L.; Han, M.; Cui, Y.; Wang, X.; Shang, X.; Wang, C. Pretreatment with high oxygen controlled atmosphere enhanced fresh-cut white mushroom (*Agaricus bisporus*) quality via activating wounding stress defenses. *J. Sci. Food Agric.* **2022**, *102*, 3359–3369. [CrossRef]
19. Rux, G.; Mahajan, P.V.; Geyer, M.; Linke, M.; Pant, A.; Saengerlaub, S.; Caleb, O.J. Application of humidity-regulating tray for packaging of mushrooms. *Postharvest Biol. Technol.* **2015**, *108*, 102–110. [CrossRef]
20. Zhang, Y.; Duo, Y.; Yan, R.; Xu, Y.; Xing, M.; Liao, S.; Wan, C.; Chen, C.; Zhu, L.; Kai, W.; et al. Amelioration of chilling injury by fucoidan in cold-stored cucumber via membrane lipid metabolism regulation. *Foods* **2023**, *12*, 301. [CrossRef]
21. Huang, Q.; Qian, X.; Jiang, T.; Zheng, X. Effect of chitosan and guar gum based composite edible coating on quality of mushroom (*Lentinus edodes*) during postharvest storage. *Sci. Hortic.* **2019**, *253*, 382–389. [CrossRef]
22. He, Y.; Fan, G.; Wu, C.; Kou, X.; Li, T.; Tian, F.; Gong, H. Influence of packaging materials on postharvest physiology and texture of garlic cloves during refrigeration storage. *Food Chem.* **2019**, *298*, 125019. [CrossRef]
23. Ventura-Aguilar, R.I.; Colinas-León, M.T.; Bautista-Baños, S. Combination of sodium erythorbate and citric acid with MAP, extended storage life of sliced oyster mushrooms. *LWT—Food Sci. Technol.* **2017**, *79*, 437–444. [CrossRef]
24. Cheng, S.; Ranran, L.; Yang, H.; Wang, S.; Lin, R.; Tian, M. Characterisation of moisture migration of shiitake mushroom (*Lentinula edodes*) during storage and its relationship to quality deterioration. *Int. J. Food Sci. Technol.* **2019**, *55*, 2132–2140. [CrossRef]
25. Jahanbakhshi, A.; Abbaspour-Gilandeh, Y.; Gundoshmian, T.M. Determination of physical and mechanical properties of carrot in order to reduce waste during harvesting and post-harvesting. *Food Sci. Nutr.* **2018**, *6*, 1898–1903. [CrossRef]
26. Liu, J.; Liu, S.; Zhang, X.; Kan, J.; Jin, C. Effect of gallic acid grafted chitosan film packaging on the postharvest quality of white button mushroom (*Agaricus bisporus*). *Postharvest Biol. Technol.* **2019**, *147*, 39–47. [CrossRef]
27. Plesoianu, A.M.; Nour, V. Effect of some polysaccharide-based edible coatings on fresh white button mushroom (*Agaricus bisporus*) quality during cold storage. *Agriculture* **2022**, *12*, 1491. [CrossRef]
28. Wang, Y.; Shi, X.; Yin, J.; Nie, S. Bioactive polysaccharide from edible *Dictyophora* spp.: Extraction, purification, structural features and bioactivities. *Bioact. Carbohydr. Diet. Fibre* **2018**, *14*, 25–32. [CrossRef]
29. Wang, J.; Li, W.; Li, Z.; Wu, W.; Tang, X. Analysis and evaluation of the characteristic taste components in portobello mushroom. *J. Food Sci.* **2018**, *83*, 1542–1551. [CrossRef]
30. Pei, F.; Yang, W.; Ma, N.; Fang, Y.; Zhao, L.; An, X.; Xin, Z.; Hu, Q. Effect of the two drying approaches on the volatile profiles of button mushroom (*Agaricus bisporus*) by headspace GC–MS and electronic nose. *LWT—Food Sci. Technol.* **2016**, *72*, 343–350. [CrossRef]
31. Shen, X.; Zhang, M.; Devahastin, S.; Guo, Z. Effects of pressurized argon and nitrogen treatments in combination with modified atmosphere on quality characteristics of fresh-cut potatoes. *Postharvest Biol. Technol.* **2019**, *149*, 159–165. [CrossRef]
32. Li, P.; Zhang, X.; Hu, H.; Sun, Y.; Wang, Y.; Zhao, Y. High carbon dioxide and low oxygen storage effects on reactive oxygen species metabolism in *Pleurotus eryngii*. *Postharvest Biol. Technol.* **2013**, *85*, 141–146. [CrossRef]
33. Lyn, F.H.; Adilah, Z.A.M.; Nor-Khaizura, M.A.R.; Jamilah, B.; Hanani, Z.A.N. Application of modified atmosphere and active packaging for oyster mushroom (*Pleurotus ostreatus*). *Food Packag. Shelf Life* **2020**, *23*, 100451. [CrossRef]
34. Capotorto, I.; Innamorato, V.; Cefola, M.; Cervellieri, S.; Lippolis, V.; Longobardi, F.; Logrieco, A.F.; Pace, B. High CO_2 short-term treatment to preserve quality and volatiles profile of fresh-cut artichokes during cold storage. *Postharvest Biol. Technol.* **2020**, *160*, 111056. [CrossRef]
35. Wang, H.; Xu, N.; Peng, Y.; Zhang, X. The determination of low oxygen threshold in garlic scapes during controlled atmosphere storage. *J. Sci. Food Agric.* **2022**, *102*, 4949–4954. [CrossRef]
36. Wang, Y.; Mo, Y.; Li, D.; Xiang, C.; Jiang, Z.; Wang, J. The main factors inducing postharvest lignification in king oyster mushrooms (*Pleurotus eryngii*): Wounding and ROS-mediated senescence. *Food Chem.* **2019**, *301*, 125224. [CrossRef] [PubMed]
37. Ye, J.; Li, J.; Han, X.; Zhang, L.; Jiang, T.; Xia, M. Effects of active modified atmosphere packaging on postharvest quality of shiitake mushrooms (*Lentinula edodes*) stored at cold storage. *J. Integr. Agric.* **2012**, *11*, 474–482. [CrossRef]
38. Harada-Padermo, S.; Dias-Faceto, L.; Selani, M.; Conti-Silva, A.; Vieira, T. Umami ingredient, a newly developed flavor enhancer from shiitake byproducts, in low-sodium products: A study case of application in corn extruded snacks. *LWT—Food Sci. Technol.* **2021**, *138*, 110806. [CrossRef]
39. Feng, L.; Zhang, M.; Adhikari, B.; Guo, Z. Effect of ultrasound combined with controlled atmosphere on postharvest storage quality of cucumbers (*Cucumis sativus* L.). *Food Bioprocess Technol.* **2018**, *11*, 1328–1338. [CrossRef]
40. Waghmare, R.B.; Annapure, U.S. Integrated effect of sodium hypochlorite and modified atmosphere packaging on quality and shelf life of fresh-cut cilantro. *Food Packag. Shelf Life* **2015**, *3*, 62–69. [CrossRef]

foods MDPI

Article

Melatonin Treatment of Strawberry Fruit during Storage Extends Its Post-Harvest Quality and Reduces Infection Caused by *Botrytis cinerea*

Surassawadee Promyou [1,*], Yenjit Raruang [2] and Zhi-Yuan Chen [2]

[1] Department of Agriculture and Resources, Faculty of Natural Resources and Agro-Industry, Chalermphrakiat Sakon Nakhon Province Campus, Kasetsart University, Sakon Nakhon 47000, Thailand
[2] Department of Plant Pathology and Crop Physiology, Louisiana State University Agricultural Center, Baton Rouge, LA 70803, USA; yraruang@agcenter.lsu.edu (Y.R.); zchen@agcenter.lsu.edu (Z.-Y.C.)
* Correspondence: csnsrwd@ku.ac.th

Abstract: Gray mold is a main disease of strawberry fruit (*Fragaria* × *xananassa* cv. Camino Real) caused by *Botrytis cinerea*, which leads to marketable value losses in the supply chain. The purpose of this study was to investigate the effects of exogenous melatonin (MT) on the physicochemical quality, antioxidant defense system, and disease resistance of strawberry fruit to *B. cinerea* infection. The results revealed that strawberry fruit immersed in 100 μM MT for 15 min effectively maintained its brightness and delayed the change in fruit color. MT also maintained the level of titratable acidity and slowed down the increase of total soluble solids in strawberry fruit. Moreover, strawberries immersed in MT maintained a fresh weight and fruit firmness, as well as reduced *B. cinerea* infection when compared to the untreated control fruit and fruit treated with 5% NaOCl. In addition, MT increased the accumulation of DPPH scavenging capacity and the activity of antioxidant enzymes (SOD, POD, and APX) with the exception of CAT. The same effect was also observed in strawberry fruit after immersion in MT and followed by *B. cinerea* inoculation. These findings demonstrated that exogenous MT could effectively maintain the postharvest quality of strawberries, even when the fruit was inoculated with *B. cinerea*.

Keywords: melatonin; disease resistance; strawberry; gray mold; antioxidant defense system

1. Introduction

Strawberries (*Fragaria* × *ananassa* cv. Camino Real) are a commercial non-climacteric fruit that belong to the Rosales family and are well known as a rich source of different types of nutrients. They contain many beneficial health compounds, such as high levels of antioxidants (polyphenolics, flavonoids, vitamin C, and anthocyanins), folate, potassium, and fiber [1]. Strawberries are produced in many countries. In 2020, the United States produced about 4510 thousand tons of strawberries for the market. Recently, strawberries have risen to become the third most important agricultural commodity, in terms of fruit, in the U.S. after grapes and oranges. Interestingly, strawberries are the fifth most consumed fresh fruit by weight, after bananas, apples, oranges, and grapes [2]. The most serious postharvest problem that results in fruit deterioration during storage and that affects strawberry fruit quality is the gray mold disease caused by *Botrytis cinerea* [3].The initial symptoms of the infected strawberry fruit are light brown lesions and a loss of fruit firmness, which cover the entire fruit surface within 48 h. Gray mold can result in losses ranging from 25–35%, or approximately USD 10–100 million per year. Therefore, to reduce fruit postharvest decay caused by pathogens, various chemical fumigants, such as methyl bromide, sulfur dioxide, phosphine, and ethyl formate have been used to protect the fruit from pathogen infections [4]. However, fumigation using chemicals has a negative impact on the environment and these chemicals are toxic to human health. Nowadays, the research into postharvest technology focuses on applying environmentally friendly technologies to replace the use of chemicals to control pathogens and maintain the quality of fruits and vegetables. Melatonin (*N*-acetyl-5-methoxytryptamine) is widespread in plants, animals, and humans [5,6]. The function of melatonin in plants has been widely reported [7]. For instance, melatonin, as a natural antioxidant, may play important regulatory roles in various physiological activities in plants, such as fruit development [8], fruit ripening [9], delaying fruit senescence [10], disease resistance, exerting antioxidant effects, regulating growth and development, and facilitating plant adaption to stress conditions [11]. Moreover, melatonin has been applied as pre-treatment for preventing fungal infections and reducing disease symptoms in fruits [12–14]. Such an application increases the resistance of apples to Marssonina blot, by promoting the expression of chitinase genes, as well as regulating hydrogen peroxide and pathogenesis-related proteins [14]. Similar effects of melatonin on inhibiting pathogen infections and prolonging the shelf-life of post-harvest fruits were also observed in apples [15], strawberries [13], and citrus fruits [12]. As a safe and effective treatment, melatonin acts not only as a signaling molecule for enhancing the resistance of plants to biotic and abiotic stresses, but also as a strong free radical scavenger in many plants [5]. Recent studies have demonstrated that the function of melatonin treatment on the enhancement of fruit disease resistance could be associated with the induction of various defense-related genes, such as those encoding peroxidase (POD), lipid-transfer proteins, chitinases, β-1,3-glucanase, and pathogenesis-related (PR) proteins [16,17]. Cao et al. [15] and Hu et al. [18] reported that exogenous melatonin treatment enhanced disease resistance in apples, peaches, and kiwis during storage by continuously increasing the activity of POD, superoxide dismutase (SOD), catalase (CAT), and ascorbate peroxidase (APX). In addition, a decrease in H_2O_2 content has also been reported in strawberries [13], citrus fruits [12], and peaches [19]. Although the roles and functions of melatonin in the antioxidant defense mechanism have been widely studied under different stress conditions in fruits from different plant species, the detailed role of exogenous melatonin application against biotic stress from *B. cinerea* infection in strawberries during storage is still limited. The objective of the current study was to investigate the novelty effect of exogenous melatonin treatment on the antioxidant defense system and on grey mold disease in strawberries caused by *B. cinerea*. We focused our attention on strawberries because it is one of the most economically important agricultural commodities, and gray mold disease in strawberries causes severe fruit decay both before and after harvest. Moreover, this study can provide new insights into the improvement of pathogen resistance by melatonin in non-climacteric fruits during postharvest storage. In this study, we summarize the novel beneficial effects of melatonin treatment as a useful technique for improving postharvest preservation of fruit, even in non-climacteric fruits in the future.

2. Materials and Methods

2.1. Fruit Material

Strawberries (*Fragaria* × *ananassa* cv. Camino Real) used in the experiments were harvested from an orchard located in Ponchatoula, Louisiana, USA, at the commercial harvesting stage (approximately 80% ripened) during the spring season (March through May), and they were delivered to the laboratory at the Department of Plant Pathology and Crop Physiology, at the Louisiana State University Agricultural Center, Baton Rouge, USA, within 1 h of harvesting. The fruit were selected for uniformity and were free of any physical damage and apparent diseases.

2.2. Exogenous Melatonin (MT) Treatment and Botrytis cinerea Artificial Inoculation

The selected strawberry fruits were cleaned with tap water before treatment. The strawberries that were surface disinfected with 5% sodium hypochlorite (NaOCl) for 15 min were labelled as the positive control, and the negative controls were strawberries immersed in distilled water for 15 min. The strawberries immersed in 100 μM of MT for 15 min were labelled as MT-treated.

This study included the following six treatments: (1) Control (no treatment and non-inoculated); (2) Inoculated control (no treatment + *B. cinerea*); (3) Positive control (treated with 5%NaOCl); (4) Inoculated positive control (5% NaOCl + *B. cinerea*); (5) MT-treated (100 μM MT); and (6) MT-treated and inoculated (100 μM MT + *B. cinerea*).

Artificial inoculations were conducted 24 h after MT or 5% NaOCl treatment. The *Botrytis cinerea* isolate was collected from infected strawberries that showed typical symptoms of gray mold. The *Botrytis cinerea* isolate was used for inoculum production and was cultivated on potato dextrose agar (PDA) for 5 days at 25 °C under 12 h cycles of dark and blue-red lights (440–460 nm and 620–630 nm). Spores were suspended from the petri dishes with 50 mL of sterile distilled water supplemented with 0.1% (*v*/*v*) Tween 80 using a scalpel. The spore concentration of the inoculum suspension was adjusted at 2×10^5 spores mL^{-1} after counting the spores using a hemocytometer. A wound-injection method by Testempasis et al. [20] with a slight modification was used to infect the strawberry fruits with *B. cinerea* to ensure consistent inoculum and to avoid the unintended destruction of the fruit. Each strawberry fruit was injected with 50 μL of spore suspension using a sterile syringe needle at a depth of about 5 mm. Following artificial inoculation, the fruits were placed in plastic trays and stored at room temperature (25 ± 2 °C) with 80–85% relative humidity for 6 days. The non-inoculated controls were injected with 50 μL of sterile distilled water supplemented with 0.1% (*v*/*v*) Tween 80, and were stored under the same conditions as the inoculated ones. On each sampling date, the treatment was replicated in triplicate, and each replication consisted of ten strawberries (30 randomly sampled fruits were taken from each treatment).

The physicochemical quality, including visual appearance, disease assessment, fruit color, weight loss, fruit firmness, total soluble solids content (TSS), and titratable acidity (TA), 2,2-diphenyl-1-picrylhydrazyl (DPPH) scavenging capacity, and antioxidant enzyme activities were determined every two days during storage. The whole fruits from the same treatment in each parameter study were pooled, frozen in liquid nitrogen, and stored at −80 °C for later analysis. Fruits were also regularly collected for chemical analysis from the replicated treatments. This study was repeated twice.

2.3. Disease Assessment

The severity of gray mold disease was evaluated by observing the decay area on the fruit surface. Each fruit was rated from 1 to 4 based on the following disease rating scores: 1 = no occurrence of fruit surface infection; 2 = moderate (1–35% of the fruit surface is infected); 3 = severe (36–70% of the fruit surface is infected); and 4 = very severe (71–100% of the fruit surface is infected), as shown in Figure 1. The gray mold disease index (DI) was calculated as $\sum$ (disease scores × number of fruit at that score)/total number of fruits in each group (30 fruits in the present experiments).

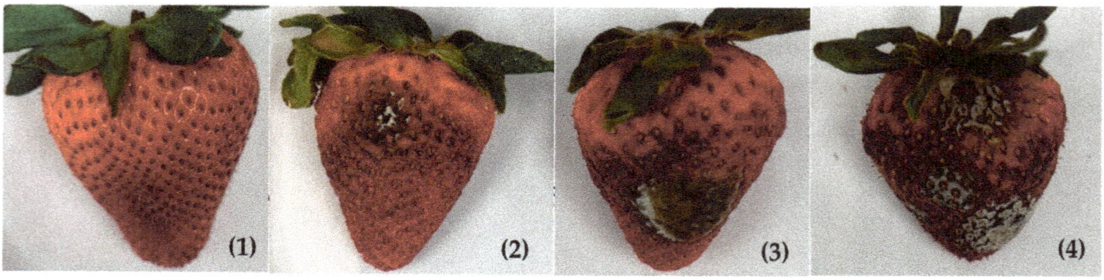

Figure 1. Gray mold disease rating scores (**1–4**) of strawberry fruit.

2.4. qPCR Quantification of B. cinerea Biomass in Strawberry Fruit

DNA was extracted from fresh fruit using a modified CTAB method, as described by Doyle and Doyle [21]. The yield and purity of the extracted DNA were determined using a NanoDrop ND-1000 spectrophotometer (Thermo Scientific, Wilmington, DE, United States). The extracted DNAs were adjusted to the same concentration (100 ng μL^{-1}) and used as templates for the PCR. The sequences of specific *Bc3* primers targeting the pathogen intergenic spacer region between 28S rDNA and 18S rDNA are as follows: 5′-GCTGTAATTTCAATGTGCAGAATCC-3′ (forward) and 5′- GGAGCAACAATTAATCGC ATTTC-3′ (reverse). The reaction was prepared at 1X final concentration in a 20 μL volume containing 1 μL of each primer (10 μM) and 5 μL of the diluted genomic DNA as the template. The primer pair (St-EF-forward, 5′-TGCTGTTGGAGTCATCAAGAATG-3′ and St-EF-reverse, 5′-TTGGCTGCAGACTTGGTCAC-3′ as reverse)] that amplified the elongation factor gene of the strawberry was used as an internal control to normalize the level of *B. cinerea* biomass.

2.5. Physicochemical Quality Measurements

The skin color of the fruit in the center of the fruit surface was measured using a chromometer (CTI color Analyzer Digital Precise Colorimeter & Color Difference Meter, China). Color measurements contained the three-parameter color system formulated by the International Commission on Illumination (CIE): L*, a* and b*. L* value represents lightness to darkness (100–0), the a * value represents redness or greenness (−greenness to +redness), and the b* value represents blueness or yellowness (−blueness to +yellowness).

Total soluble solids (TSS) content measurement was performed with fruit juice using a digital refractometer (Milwaukee's Instruments MA871, Milwaukee Instrument, Inc., Rocky Mount, NC, USA) and the TSS data were presented as °Brix.

Titratable acidity (TA) was determined using the titrimetric method [22]. The result was expressed as g of citric acid per 100 g of fruit fresh weight.

Fresh weight loss of the strawberry fruit was measured before treatment and at different sampling days. The percentage of weight loss during storage was calculated by comparing the initial weight.

Fruit firmness was measured at two opposite sides of each individual fruit using an Effegi firmness penetrometer (model NEWTRY GY-2 Fruit Firmness Penetrometer, Milwaukee Instrument, Inc. NC, USA) with a 0.5-cm cylindrical plunger. The plunger was inserted at a depth of 0.5 cm, and the data were expressed as Newton (N). Each fruit was punctured twice in the middle part of the fruit.

2.6. Antioxidant Capacity Measurement

The 2,2-diphenyl-1-picrylhydrazyl (DPPH) scavenging capacity was determined according to Brand-Williams et al. [23]. For the DPPH scavenging capacity, frozen fruit powder (2 g) was homogenized with 3 mL of 85% methanol, and then the filtrated fruit extract was centrifuged at 6000× *g* for 10 min at 4 °C. A seventy μL supernatant of the strawberry fruit extract was added to a test tube and mixed with 2930 μL of 200 μM DPPH solution in methanol. The reaction mixture was incubated in the dark at room temperature for 30 min. The absorbance of the solution was read at 517 nm, and calculated using the following formula: DPPH scavenging capacity (%) = $[1 - (A0/A1)] \times 100$. Where A0 is the absorption of the sample, and A1 is the absorption of the blank DPPH solution.

2.7. Antioxidant Enzyme Activity Measurement

Two grams of strawberry fruit were homogenized in 20 mL of 50 mM sodium phosphate buffer (pH 7.8) containing 1.0 mM EDTA, 0.3% (*v*/*v*) Triton X-100, and 1% (*w*/*v*) polyvinyl polypyrrolidone (PVPP) for 3 min. Following the filtration of the homogenate through cheesecloth, the filtrate was centrifuged at 15,000× *g* for 20 min at 4 °C. The supernatant liquid was used as the crude extract of antioxidant enzymes. The activities of the antioxidant enzymes were assayed using a modified method described by

Suiubon et al. [24]. In order to calculate the specific enzyme activity, the protein content was measured using the method described by Bradford [25] using bovine serum albumin as a standard.

The SOD (EC 1.15.1.1) activity was assayed according to the ability to inhibit the photochemical reduction of nitro blue tetrazolium (NBT). The final assay volume was 2 mL, that consisted of 0.8 mL of distilled water, 0.5 mL of 50 mM sodium phosphate buffer (pH 5.0), 0.1 mL of 20 μM NBT dissolved in 70% ethanol, 0.2 mL of 22 μM methionine, 0.2 mL of 0.1%(v/v) Triton-X, 0.1 mL of 0.6 μM riboflavin as an enzyme substrate, and 0.1 mL of the above crude extract. The absorbance of the mixture was measured at 560 nm. One unit of SOD activity was defined as the amount of enzyme that inhibited 50% NBT at 560 nm. Total enzyme activity was expressed as units mg^{-1} protein.

The CAT (EC 1.11.1.6) activity was assayed in a 3 mL reaction volume that consisted of 0.1 mL of crude enzyme extract and 2.9 mL of 25 mM H_2O_2. The absorbance was measured at 240 nm every 30 s for 3 min with distilled water as the bank. One unit of CAT activity was defined as a decrease in absorbance at 240 nm of 0.01 per minute and the data were expressed as units mg^{-1} protein.

The POD (EC 1.11.1.7) activity was determined based on the ability to convert guaiacol to tetraguaiacol. The reaction mixture consisted of 1 mL of sodium phosphate buffer (50 mM, pH 5.5), 1 mL of 25 mM guaiacol, 0.5 mL of 2% (v/v) H_2O_2, and 0.5 mL of crude enzyme extract. The absorbance was measured at 470 nm every 30 s for 3 min using distilled water as the bank. One unit of POD activity was defined as the amount of the enzyme that caused an absorbance change of 0.01 per minute and the data were expressed as units mg^{-1} protein.

The APX (EC 1.11.1.11) activity was measured based on the decrease in ascorbic acid. The assay reaction solution (2.0 mL final volume) contained 0.5 mL of 100 mM potassium phosphate buffer (pH 7.0), 0.5 mL of 1 mM ascorbic acid, 0.5 mL of 0.4 mM EDTA, 0.02 mL of 10 mM H_2O_2, 0.38 mL of distilled water, and 0.1 mL of crude enzyme extract. The absorbance of the mixture was measured at 290 nm. One unit of enzyme activity was defined as the increase of absorbance per minute and the data were expressed as units mg^{-1} protein.

2.8. Statistical Analysis

The experiment data are the average of three replications ± SD. An analysis of variance (ANOVA) was calculated, and the means separations were determined using Duncan's multiple range test (DMRT) at $p < 0.05$. All experiments were repeated twice at later dates.

3. Results

3.1. Visual Appearance and Disease Development

The visual appearances of the strawberries immersed in MT solution, NaOCl, and the control with or without being inoculated with *B. cinerea* before they were stored at room temperature for 6 days, are shown in Figure 2. The incidence of gray mold disease appeared on day 2 of storage and continued to increase thereafter in all treated fruits except in the MT-treated fruit with or without artificial inoculation with *B. cinerea*. Interestingly, the symptoms of gray mold disease did not appear on the positive control (NaOCl), MT, and MT + *B. cinerea* treated fruits during the 6 days of storage at room temperature, whereas the inoculated positive control (NaOCl + *B. cinerea*) fruits were found to be infected, indicating that MT was more effective for controlling gray mold than NaOCl.

Figure 2. Effect of exogenous melatonin (MT) on the visual appearance of strawberry fruits inoculated with and without *Botrytis cinerea*, and stored at room temperature (25 ± 2 °C, 80–85% RH) for 6 days.

The result of the calculated gray mold disease index shown in Figure 3A reveals that the highest disease index score was found first in the inoculated control and then in the control and inoculated positive control (NaOCl + *B. cinerea*), respectively, whilst no difference was found in the disease index on the strawberry fruits immersed in the positive control (NaOCl), MT solutions, and MT + *B. cinerea* stored for 4 and 6 days. At the end of storage, the control fruits and inoculated control (+*B. cinerea*) showed visible fruit decay from *B. cinerea* infection with a disease index score of 2.0–3.0.

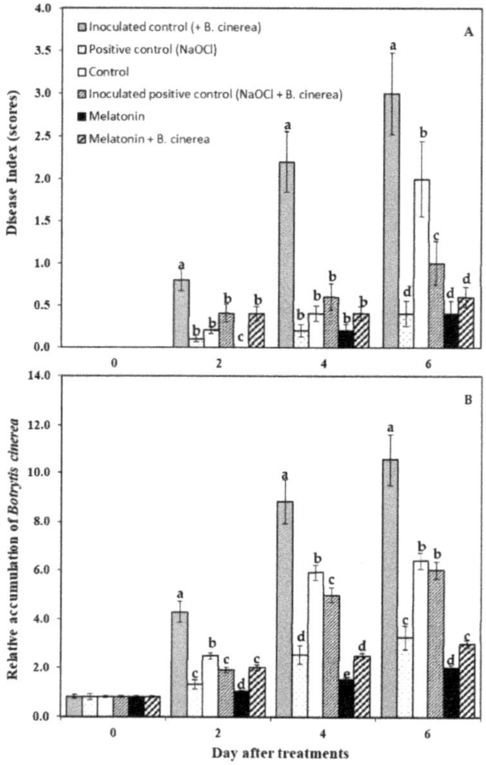

Figure 3. Disease index (**A**) and biomass of *Botrytis cinerea* were measured using real time PCR (**B**) in strawberry fruits treated with exogenous melatonin and inoculated with and without *Botrytis cinerea*, and stored at room temperature (25 ± 2 °C, 80–85% RH) for 6 days. Values are the average of three replicates ± SD. Different letters indicate significant differences according to Duncan's multiple range test ($p < 0.05$).

To validate these results, the biomass of *B. cinerea* was quantified based on the relative level of the *B. cinerea* genomic DNA compared to the strawberry DNA, as shown in Figure 3B. *B. cinerea* was detected in strawberry fruit one day after inoculation with no visual appearance of gray mold infection. The inoculated control (+*B. cinerea*) strawberry fruit had more *B. cinerea* biomass than the other treated fruits during 6 days of storage. Whereas MT-treated fruit had significantly less *B. cinerea* biomass than other treatments during storage for 4 and 6 days. On day 6 after treatment, the fungal biomass in strawberry fruit immersed in MT then inoculated with or without *B. cinerea* were reduced by 5.28- and 3.52-fold, respectively, compared to the inoculated control (+*B. cinerea*) fruit. The fungal biomass in the fruit immersed in NaOCl then inoculated with or without *B. cinerea* were reduced by 1.76- and 3.24-fold, respectively, when compared to the inoculated control (+*B. cinerea*) fruit. Moreover, the results found that there were no significant differences in

B. cinerea biomass between the control and inoculated positive control (NaOCl + *B. cinerea*) treated fruit, whereas the *B. cinerea* biomass in MT+ *B. cinerea* treated fruit was 3.19-fold less than that in the control fruit (Figure 3B).

3.2. Physicochemical Qualities: Fruit Colour, TSS, TA, Fresh Weight Loss, and Fruit Firmness

The color parameters (L*, a* and b* values) of the strawberry fruit stored for 6 days were also determined (Table 1). The L* value of the strawberry fruit refers to the lightness and is normally used for evaluating the darkness of the postharvest fruit. The initial average value of L* of the strawberry fruits used in this study was 30.48, which continuously decreased in all fruit samples, indicating darkening of the fruit with time. Following 2 days of storage, MT and MT + *B. cinerea* treated fruits had the highest L* value when compared to the other treatments. However, the strawberry fruit from the control and inoculated control (+*B. cinerea*) groups had the lowest L* value and that significantly decreased after 4 and 6 days of storage ($p < 0.05$). The L* value of the positive control (NaOCl) fruit was significantly lower than that of the MT and MT+ *B. cinerea* treated fruit, and it was similar to that of the control fruit on the last day of storage. This indicates that MT was effective in maintaining the brightness of the strawberry fruit during storage, even though the fruits were infected with *B. cinerea*.

Table 1. Effects of melatonin on fruit color and other physicochemical qualities (total soluble solids, TSS; total titratable acidity, TA) of strawberry fruits treated with exogenous melatonin, and inoculated with or without *Botrytis cinerea* during a six-day storage at room temperature (25 ± 2 °C, 80–85% RH).

Day after Treatment	Treatments	Color §			TSS (°Brix)	TA (% Citric Acid)
		L*	a*	b*		
Day 0		30.48 a	22.68 c	23.67 b	6.14 c	1.02 a
Day 2	Inoculated control (+*B. cinerea*)	24.69 c	26.32 a	23.99 b	6.02 c	0.72 b
	Positive control (NaOCl)	25.89 c	26.96 a	24.60 b	6.32 c	0.70 b
	Control	25.53 c	26.22 a	25.35 a	6.56 c	0.65 b
	Inoculated positive control (NaOCl + *B. cinerea*)	24.49 c	26.48 a	25.29 a	6.82 c	0.63 b
	Melatonin	27.16 b	23.53 b	24.30 b	6.54 c	0.78 b
	Melatonin + *B. cinerea*	26.19 b	23.24 b	24.00 b	6.22 c	0.71 b
Day 4	Inoculated control (+*B. cinerea*)	22.30 d	26.91 a	23.95 b	8.00 a	0.68 b
	Positive control (NaOCl)	24.70 c	26.75 a	24.49 b	7.42 b	0.66 b
	Control	23.55 d	27.21 a	25.53 a	8.10 a	0.60 b
	Inoculated positive control (NaOCl + *B. cinerea*)	24.45 c	26.44 a	25.37 a	7.86 b	0.59 b
	Melatonin	26.55 b	24.56 b	24.70 b	7.64 b	0.64 b
	Melatonin + *B. cinerea*	26.96 b	24.79 b	24.50 b	7.38 b	0.68 b
Day 6	Inoculated control (+*B. cinerea*)	19.74 e	28.54 a	25.15 a	8.30 a	0.50 c
	Positive control (NaOCl)	23.19 d	26.65 a	25.64 a	7.46 b	0.51 c
	Control	22.09 d	28.67 a	26.57 a	8.22 a	0.55 c
	Inoculated positive control (NaOCl + *B. cinerea*)	22.24 d	28.54 a	25.77 a	7.76 b	0.53 c
	Melatonin	25.55 c	25.36 b	25.35 a	7.32 b	0.61 b
	Melatonin + *B. cinerea*	25.65 c	25.74 b	26.59 a	7.98 b	0.68 b

Data are the means of three biological replications, each consisting of ten fruits. § Within each column, a value followed with the same letter does not differ significantly at $p < 0.05$.

The initial average value of a* and b* of the strawberries used in this study were 22.68 and 23.67, respectively. The a* and b* values of the strawberries significantly increased during the 6 days of storage ($p < 0.05$). On day 6, the a* values of the MT and MT+ *B. cinerea* treated fruits were significantly lower than those of all other treated strawberries, indicating a delayed change in the fruit color. The b* values of the fruits from all treatments were similar.

It is widely accepted that TSS and TA of the fruits are related to the ripening process. TSS content of the strawberries increased during the 6-day storage (Table 1). A rapid increase in TSS was found in the control and inoculated control (+*B. cinerea*) fruits, and it was significantly higher than in other treatments ($p < 0.05$), indicating that MT or NaOCl treatment delays fruit ripening. However, the TSS of the strawberries immersed in NaOCl and MT solutions with or without *B. cinerea* inoculation, increased slightly and did not differ significantly among the treatments on days 4 and 6 of storage. In this study, the initial TA of the strawberries was 1.02% and it decreased slightly during storage. No significant difference in TA was observed among the treatments within the first 4 days of storage. At the end of storage, TA of MT and MT+ *B. cinerea* treated fruits remained constant and was higher than in the other treatments.

Generally, the accelerated fruit weight losses are the result of an increased rate of transpiration and respiration. The results in Figure 4A showed that the weight loss of the strawberry fruits increased continuously during the 6 days of storage at room temperature in all treatments. The weight loss of the control and inoculated control (+*B. cinerea*) fruits increased dramatically, while that of NaOCl and MT with and without *B. cinerea* inoculation increased only slightly. At the end of storage, the highest weight loss was found in the inoculated control (+*B. cinerea*) fruit (25.25%), followed by the control, MT+ *B. cinerea*, positive control (NaOCl), inoculated positive control (NaOCl + *B. cinerea*), and MT fruits (22.60, 15.86, 13.29, 12.16, and 12.11%, respectively).

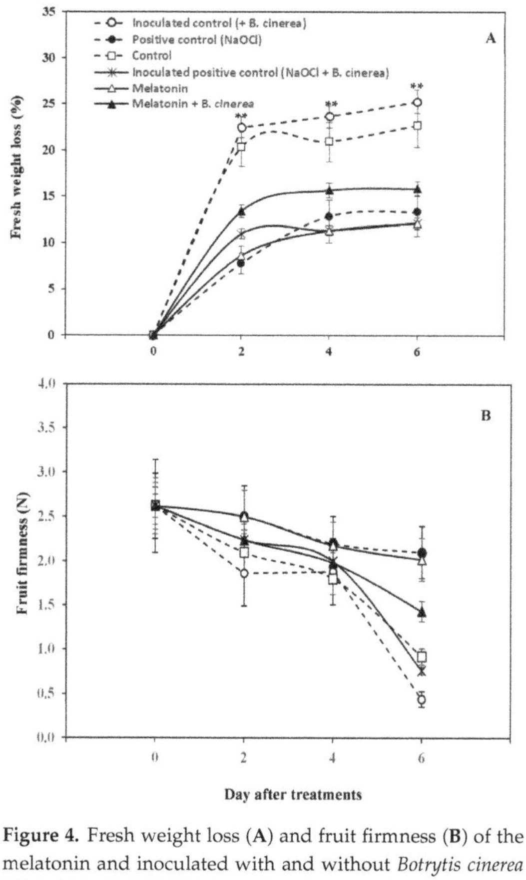

Figure 4. Fresh weight loss (**A**) and fruit firmness (**B**) of the strawberries treated with exogenous melatonin and inoculated with and without *Botrytis cinerea* during storage at room temperature (25 ± 2 °C, 80–85% RH) for 6 days. Data are the means ± S.D. of three replications (three sets of ten fruits). Significant differences between treatments are indicated with asterisks (** ($p < 0.01$)).

Furthermore, the fruit firmness decreased significantly for most of the treatments except that of the positive control (NaOCl) and MT solution-treated fruits, which remained fairly stable during the 6 days of storage (Figure 4B). There was no significant difference in the firmness between the control and inoculated control (+*B. cinerea*) until day 6 when a rapid decrease in fruit firmness in the inoculated control (+*B. cinerea*) fruit was observed.

3.3. Antioxidant Capacity and Antioxidant Enzymes

The non-enzyme antioxidant or antioxidant capacity of the strawberries during storage was expressed as the DPPH scavenging capacity. The antioxidant capacity of exogenous MT treatment was similar to the MT+ *B. cinerea* and increased significantly ($p < 0.05$) at days 4 and 6 during storage compared to other treatments (Figure 5). The antioxidant capacity of the control, inoculated control (+*B. cinerea*), positive control (NaOCl), and inoculated positive control (NaOCl+ *B. cinerea*) fruits remained constant throughout the storage period and were not found to be significantly different from one another (Figure 5).

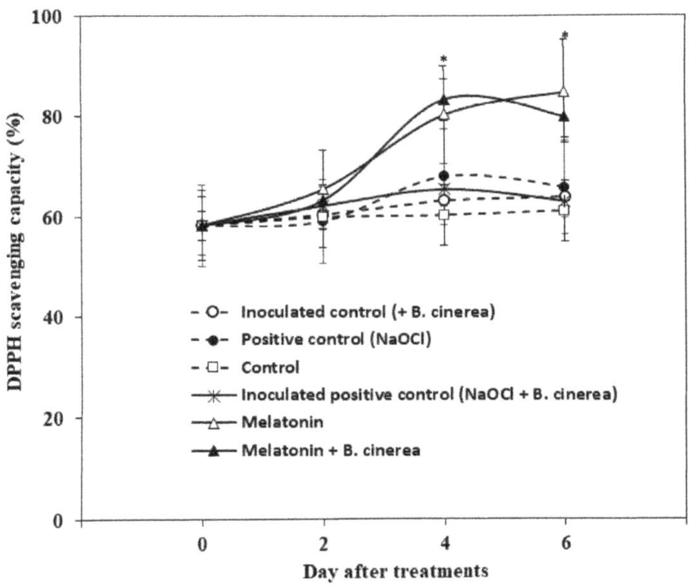

Figure 5. DPPH scavenging capacity of the strawberries treated with exogenous melatonin and inoculated with and without *Botrytis cinerea* during storage at room temperature (25 ± 2 °C, 80–85% RH) for 6 days. Data are the means ± S.D. of three replications (three sets of ten fruits). Significant differences between treatments are indicated with asterisks (* ($p < 0.05$)).

The activities of SOD, CAT, POD, and APX were determined in strawberry fruits during the storage period and are shown in Figure 6A–D. The results reveal that SOD activities of the strawberry fruits in the control and inoculated control (+*B. cinerea*) remained constant throughout storage, while that in the positive control (NaOCl) and inoculated positive control (NaOCl + *B. cinerea*) fruits increased slightly, but less than that of the MT and MT+ *B. cinerea*. The highest level of SOD activity was detected in the MT-treated and MT-treated and inoculated fruits, which was significantly higher than that of the other treatments ($p < 0.05$) from day 2 to day 6 (Figure 6A).

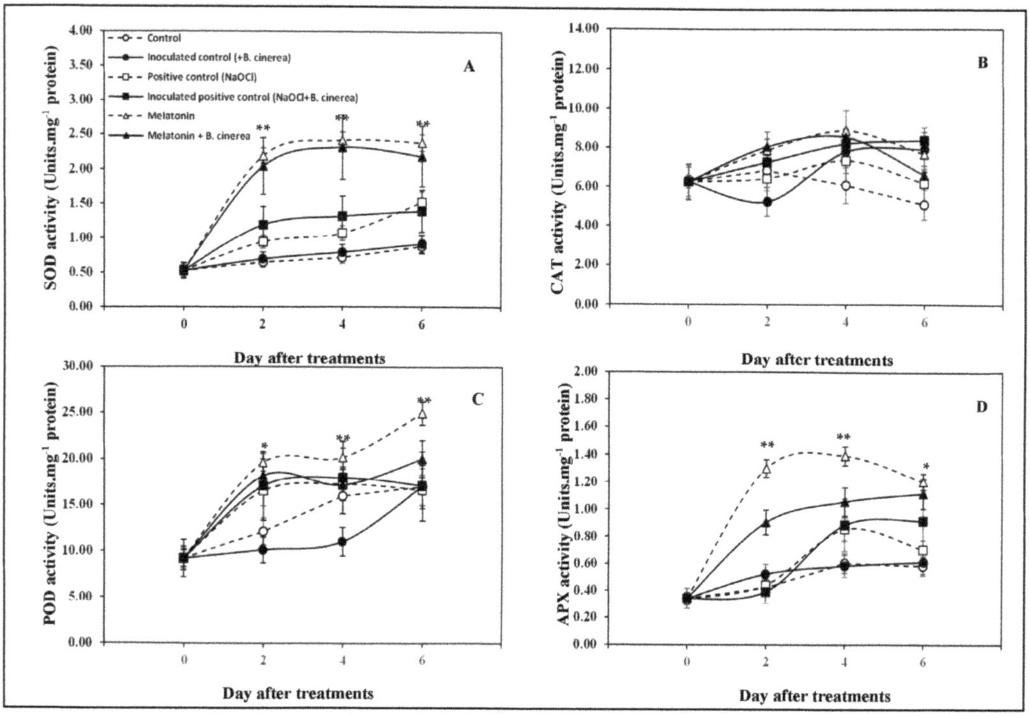

Figure 6. SOD (**A**), CAT (**B**), POD (**C**), and APX (**D**) activity of the strawberry fruits treated with exogenous melatonin and inoculated with and without *Botrytis cinerea* during storage at room temperature (25 ± 2 °C, 80–85% RH) for 6 days. Data are the means ± S.D. of three replications (three sets of ten fruits). Significant differences between treatments are indicated with asterisks (** ($p < 0.01$); * ($p < 0.05$)).

The CAT activity of the strawberry fruits in all treatments fluctuated during storage. At days 4 and 6 of storage, the control fruits were found to have the lowest CAT activities while the MT treatment appeared to significantly induce the CAT activities compared to the control from days 2 to 4 of storage ($p < 0.05$) (Figure 6B). As for the POD, its activity increased in all treatments during storage (Figure 6C). Compared to the control, a significant increase in POD activity was observed at day 2 of storage in either NaOCl or MT-treated fruits with or without *B. cinerea* inoculation. This significance was maintained at day 4 and day 6 for the MT-treated group, and at day 6 for MT and MT + *B. cinerea* treated fruits. In addition, *B. cinerea* inoculation appeared to significantly reduce the POD activity ($p < 0.05$) compared to the control during the initial four days of storage (Figure 6C).

The MT treatment also significantly induced the APX activity of strawberry fruits during storage with or without *B. cinerea* inoculation when compared to the control (Figure 6D). The APX activity of strawberries treated with exogenous MT increased sharply and reached its maximum level between days 2 and 4 of storage and this was significantly higher than the other treatments (Figure 6D). Later, APX activity of MT-treated fruit slightly decreased, and no differences were found when compared with the MT +*B. cinerea* fruit at day 6 of storage ($p < 0.05$). APX activities of the control and inoculated control (+*B. cinerea*) fruits remained constant throughout in the duration of storage (Figure 6D). The NaOCl treatment also seems to increase the APX activity, however, at a slower pace and for a shorter duration.

4. Discussion

Exogenously applied melatonin markedly reduced disease incidence in both naturally infected and artificially *B. cinerea* inoculated strawberry fruits. It is more effective than the commonly used NaOCl method, which can only reduce natural infection in strawberry fruits during storage (Figures 2 and 3). Recently, exogenous MT has been evaluated as an effective postharvest treatment to improve fruit quality and disease resistance in various fruits during storage, such as peaches [19], apples [15], strawberries [13,26], kiwis [18], and plums [27]. However, exogenously applied MT treatment had no significant effect on the postharvest green mold in citrus fruits caused by *Penicillium digitatum* [12], and anthracnose disease in banana fruits caused by *Collectotrichum musae* [17]. Moreover, Lee et al. [28] reported that Arabidopsis serotonin N-acetyltransferase (*SNAT*) knockout mutant plants exhibit a decreased MT synthesis resulting in the increased susceptibility to the pathogen, suggesting a direct involvement of MT in modulating plant disease resistance.

In the present study, strawberry fruits treated with MT exhibited a delay in fruit color development, and this has been reported in several kinds of fruits, such as bananas [10], kiwis [18], and mangoes [29]. Moreover, MT has been found to slow down the increase of TSS in bananas [10], cherries [30], and mangoes [29], which delays the ripening of climacteric fruit. These results revealed that MT could not only inhibit the ripening process of climacteric fruits, but also delay the physicochemical changes of non-climacteric fruits during storage, as shown in the strawberries in the present study. Onik et al. [31] reported that the application of MT was effective in reducing the respiration rate of apples, which resulted in slowed fruit weight loss. Therefore, it appears that MT has a high efficiency in maintaining weight loss and fruit firmness of strawberries during storage. These agree with the findings of previous studies, which showed that exogenous MT maintained the fruit firmness in bananas [10], pears [32], and mangoes [29]. Reducing weight loss and maintaining fruit firmness by exogenous MT application might be due to maintaining and improving membrane function, and reducing the activity of cell wall hydrolases resulted in delaying degradation of the cell wall [33].

Reactive oxygen species (ROS) are the early signaling molecules in plant infections by pathogens and play a role in fruit deterioration during the postharvest storage. For scavenging ROS and alleviate ROS toxicity, fruits have developed broad and efficient enzyme and non-enzyme antioxidant defense systems to induce disease resistance and protect against oxidative damage [34]. The present data show that the exogenous application of MT enhanced the antioxidant capacity in strawberry fruits infected by *B. cinerea* during storage at room temperature. In similar studies, the enhanced antioxidant capacity by exogenously applied MT was reported in strawberry fruits during storage at 4 °C [13,26], in pomegranates [35], and mangoes [36].

It is widely recognized that the enzymatic antioxidant system is an important regulator of ROS in plant cells. SOD, CAT, POD, and APX are the key antioxidant enzymes for scavenging ROS. Plants produce a series of ROS under stress. It is widely recognized that SOD removes O^{2-} whilst CAT, POD, and APX remove H_2O_2 to prevent oxidative damage to plant cells [37]. In our experiments, MT consistently induced the SOD, POD, and APX activities in strawberries during storage and also against the damage induced by *B. cinerea* infection. However, MT did not seem to affect the CAT activity of strawberries during storage, which did not vary notably among different treatments. This is unlike the results with peaches and kiwis, where MT treatment induced the activities of CAT, POD, and APX during storage at the postharvest senescence stage [18,19].

These results demonstrate that exogenous MT could be used to improve plant disease resistance to reduce *B. cinerea* infection by upregulating antioxidant defense systems. This is in agreement with the study by Cao et al. [15] who found that the MT treatment reduced pathogen infection in apples by continuously increasing the activities of antioxidant enzymes. MT upregulated the activity of antioxidant enzymes in peaches and reduced the levels of O^{2-} and H_2O_2 in different peach varieties, therefore maintaining the metabolism balance of ROS and delaying fruit deterioration [19]. Our data also indicated that MT is able to mobilize the defense-related antioxidant enzymes to scavenge the produced ROS through the increase the enzymatic activities of SOD, POD, and APX enzymes to at least partially convert H_2O_2 to H_2O, resulting in reduced toxicity to plant cells.

5. Conclusions

The current study demonstrated that exogenously applied MT is an effective alternative to NaOCl treatment in reducing the symptoms of gray mold disease caused by *B. cinerea* in strawberries during storage at room temperature. This is likely due to the higher induction of activities of the enzymatic antioxidants, such as SOD, POD, and APX, to detoxify the reactive oxygen compounds produced during storage. In addition, in the case of strawberries (non-climacteric fruit), MT treatment maintained the fruit quality based on visual appearance, fruit color, TSS, TA, fresh weight, and fruit firmness. This study provides a better alternative approach for the post-harvest handling of strawberries to extend their shelf life during storage at room temperature.

Author Contributions: Conceptualization, S.P. and Z.-Y.C.; Data curation, S.P., Y.R., and Z.-Y.C.; Formal analysis, S.P. and Y.R.; Funding acquisition, S.P. and Z.-Y.C.; Investigation, S.P. and Y.R.; Methodology, S.P. and Y.R.; Project administration, Z.-Y.C.; Resources, Z.-Y.C.; Supervision, Z.-Y.C.; Writing—original draft, S.P.; Writing—review & editing, Y.R. and Z.-Y.C. All authors have read and agreed to the published version of the manuscript.

Funding: This work was supported by: The Office of the Ministry of Higher Education, Science, Research and Innovation; and the Thailand Science Research and Innovation through the Kasetsart University Reinventing University Program 2021. Grant No. [RUP1/ Con (1.1)-CASAF 014].

Institutional Review Board Statement: Not applicable for studies not involving humans.

Informed Consent Statement: Not applicable.

Data Availability Statement: Data is contained within the article.

Acknowledgments: This work was financially supported by the Office of the Ministry of Higher Education, Science, Research and Innovation; and the Thailand Science Research and Innovation through the Kasetsart University Reinventing University Program 2021. The authors would like to acknowledge the Department of Plant Pathology and Crop Physiology, Louisiana State University Agricultural Center, Baton Rouge, LA, USA for certain facility support.

Conflicts of Interest: The authors declare no conflict of interest.

References

1. Tulipani, S.; Marzban, G.; Herndl, A.; Laimer, M.; Mezzetti, B.; Battino, M. Influence of environmental and genetic factors on health-related compounds in strawberry. *Food Chem.* **2011**, *124*, 906–913. [CrossRef]
2. National Agricultural Statistics Service (NASS). Economics, Statistics and Market Information System. 2021. Available online: https://usda.library.cornell.edu/concern/publications/zs25x846c?locale=en (accessed on 6 June 2022).
3. Williamson, B.; Tudzynski, B.; Tudzynski, P.; VanKan, J.A.L. *Botrytis cinerea*: The causeof grey mold disease. *Mol. Plant Pathol.* **2007**, *8*, 561–580. [CrossRef] [PubMed]
4. Jamieson, L.E.; Page-Weir, N.E.M.; Chagan, A.; Brash, D.W.; Klementz, D.; Bycroft, B.L.; Connolly, P.G.; Waddell, B.C.; Gilbertson, R.; Bollen, F.; et al. Phosphine fumigation to disinfest kiwifruit. *NZ Plant Prot.* **2012**, *65*, 35–43. [CrossRef]
5. Arnao, M.B.; Hernández-Ruiz, J. Functions of melatonin in plants: A review. *J. Pineal Res.* **2015**, *59*, 133–150. [CrossRef] [PubMed]
6. Reiter, R.J.; Tan, D.X.; Zhou, Z.; Cruz, M.H.; Fuentes-Broto, L.; Galano, A. Phytomelatonin: Assisting plants to survive and thrive. *Molecules* **2015**, *20*, 7396–7437. [CrossRef]
7. Sun, C.; Liu, L.; Wang, L.; Li, B.; Jin, C.; Lin, X. Melatonin: A master regulator of plant development and stress responses. *J. Exp. Bot.* **2020**, *63*, 126–145. [CrossRef]
8. Wang, C.; Yin, L.Y.; Shi, X.Y.; Xiao, H.; Kang, K.; Liu, X.Y.; Zhan, J.C.; Huang, W.D. Effect of cultivar, temperature, and environment conditions on thedynamic change of melatonin in mulberry fruit development and wine fermentation. *J. Food Sci.* **2016**, *81*, 958–967. [CrossRef]
9. Sun, Q.; Zhang, N.; Wang, J.; Zhang, H.; Li, D.; Shi, J.; Li, R.; Weeda, S.; Zhao, B.; Ron, S.; et al. Melatonin promotes ripening and improves quality of tomato fruit during postharvest life. *J. Exp. Bot.* **2015**, *66*, 657–668. [CrossRef]
10. Hu, W.; Yang, H.; Tie, W.W.; Yan, Y.; Ding, Z.H.; Liu, Y. Natural variationin banana varieties highlights the role of melatonin in post-harvest ripening and quality. *J. Agric. Food Chem.* **2017**, *65*, 9987–9994. [CrossRef]

11. Xu, Y.; Charles, M.T.; Luo, Z.; Mimee, B.; Tong, Z.; Veronneau, P.Y.; Roussel, D.; Rolland, D. Ultraviolet-C primming of strawberry leaves against subsequent *Mycosphaerella fragariac* infection involves the action of reactive oxygen species, plant hormones and terpenes. *Plant Cell Environ.* **2019**, *42*, 815–831. [CrossRef]
12. Lin, Y.; Fan, L.; Xia, X.; Wang, Z.; Yin, Y.; Cheng, Y.; Li, Z. Melatonin decreases resistance to postharvest green mold on citrus fruit by scavenging defense-relatedreactive oxygen species. *Postharvest Biol. Technol.* **2019**, *153*, 21–30. [CrossRef]
13. Liu, C.; Zheng, H.; Sheng, K.; Liu, W.; Zheng, L. Effects of melatonin treatment on thepostharvest quality of strawberry fruit. *Postharvest Biol. Technol.* **2018**, *139*, 47–55. [CrossRef]
14. Yin, L.; Wang, P.; Li, M.; Ke, X.; Li, C.; Liang, D.; Xu, S.; Ma, X.; Li, C.; Zou, Y.; et al. Exogenous melatonin improves Malus resistance to Marssonina apple blotch. *J. Pineal Res.* **2013**, *54*, 426–434. [CrossRef] [PubMed]
15. Cao, J.; Yu, Z.; Zhang, Y.; Li, B.; Liang, W.; Wang, C. Control efficiency ofexogenous melatonin against postharvest apple grey mold and its influence on theactivity of defensive enzymes. *Plant Physiol.* **2017**, *53*, 1753–1760.
16. Gao, S.; Ma, W.; Lyu, X.; Cao, X.; Yao, Y. Melatonin may increase diseaseresistance and flavonoid biosynthesis through effects on DNA methylation and geneexpression in grape berries. *BMC Plant Biol.* **2020**, *20*, 231. [CrossRef]
17. Li, T.; Wu, Q.; Zhu, H.; Zhou, Y.; Jiang, Y.; Gao, H.; Yun, Z. Comparative transcriptomicand metabolic analysis reveals the effect of melatonin on delaying anthracnoseincidence upon postharvest banana fruit peel. *BMC Plant Biol.* **2019**, *19*, 289. [CrossRef]
18. Hu, M.; Li, J.; Rao, J. Effect of melatonin on ripening and senescence of postharvest kiwifruits. *Food Sci.* **2018**, *39*, 226–232.
19. Gao, H.; Zhang, Z.K.; Chai, H.K.; Cheng, N.; Yang, Y.; Wang, D.N. Melatonin treatment delays postharvest senescence and regulates reactive oxygenspecies metabolism in peach fruit. *Postharvest Biol. Technol.* **2016**, *118*, 103–110. [CrossRef]
20. Testempasis, S.; Tanou, G.; Minas, I.; Samiotaki, M.; Molassiotis, A.; Karaoglanidis, G. Unraveling interactions of the necrotrophic fungal species Botrytis cinerea with 1methylcyclopropene or ozone-Treated apple fruit using proteomic analysis. *Front. Plant Sci.* **2021**, *12*, 644255.21. [CrossRef]
21. Doyle, J.J.; Doyle, J.L. A rapid DNA isolation procedure for small quantities of fresh leaf tissue. *Phytochem. Bullet.* **1987**, *19*, 11–15.
22. AOAC. *Official Methods of Analysis*, 16th ed.; Association of OfficialAnalytical Chemists International: Washington, DC, USA, 1995.
23. Brand-Williams, W.; Cuvelier, M.E.; Berset, C. Use of a free radical methodto evaluate antioxidant activity. *LWT-Food Sci. Technol.* **1995**, *28*, 25–30. [CrossRef]
24. Suiubon, S.; Supapvanich, S.; Promyou, S. Postharvest quality maintenance oflongan fruit by ultraviolet-C incorporated with salicylic acid application. *Emir. J. Food Agric.* **2017**, *29*, 179–187. [CrossRef]
25. Bradford, M.M. A rapid and sensitive method for the quantitation ofmicrogram quantities of protein utilizing the principle of protein-dye binding. *Anal. Biochem.* **1976**, *72*, 248–254. [CrossRef] [PubMed]
26. Aghdam, M.S.; Fard, J.R. Melatonin treatment attenuates postharvest decayand maintains nutrition quality of strawberry fruit (*Fragaria x anannasa* cv. Selva) by enhancing GABA shunt activity. *Food Chem.* **2017**, *221*, 1650–1657. [CrossRef] [PubMed]
27. Bal, E. Physicochemical changes in 'Santa Rosa' plum fruit treated withmelatonin during cold storage. *J. Food Meas. Charact.* **2019**, *13*, 1713–1720. [CrossRef]
28. Lee, H.Y.; Byeon, Y.; Tan, D.X.; Reiter, R.J.; Back, K. *Arabidopsis serotonin*N-acetyltransferase knockout mutant plants exhibit decreased melatonin andsalicylic acid levels resulting in susceptibility to an avirulent pathogen. *J. Pineal Res.* **2015**, *58*, 291–299. [CrossRef]
29. Liu, S.; Huang, H.; Huber, D.J.; Pan, Y.; Shi, X.; Zhang, Z. Delay of ripening and softeningin 'Guifei' mango fruit by postharvest application of melatonin. *Postharvest Biol.Technol.* **2020**, *163*, 111136. [CrossRef]
30. Tijero, V.; Munoz, P.; Munne-Bosch, S. Melatonin as an inhibitor of sweet cherriesripening in orchard trees. *Plant Physiol. Biochem.* **2019**, *140*, 88–95. [CrossRef]
31. Onik, J.C.; Wai, S.C.; Li, A.; Lin, Q.; Sun, Q.; Wang, Z.; Duan, Y. Melatonin treatmentreduces ethylene production and maintains fruit quality in apple during postharveststorage. *Food Chem.* **2020**, *337*, 127753. [CrossRef]
32. Liu, J.; Yang, J.; Zhang, H.; Cong, L.; Zhai, R.; Yang, C.; Wang, Z.; Ma, Z.; Xu, L. Melatonininhibits ethylene synthesis via nitric oxide regulation to delay postharvest senescencein pears. *J. Agric. Food Chem.* **2019**, *67*, 2279–2288. [CrossRef]
33. Wang, F.; Zhang, X.; Yang, Q.; Zhao, Q. Exogenous melatonin delays postharvest fruitsenescence and maintains the quality of sweet cherries. *Food Chem.* **2019**, *301*, 125–311. [CrossRef]
34. Jiang, Y.; Duan, X.; Joyce, D.; Zhang, Z.; Li, J. Advances in understanding of enzymatic browning in harvested litchi fruit. *Food Chem.* **2004**, *88*, 443–446. [CrossRef]
35. Jannatizadeh, A. Exogenous melatonin applying confers chilling tolerance inpomegranate fruit during storage. *Sci. Hortic.* **2019**, *246*, 544–549. [CrossRef]
36. Rastegar, S.; Khankandani, H.H.; Rahimzadeh, M. Effects of melatonin treatment onthe biochemical changes and antioxidant enzyme activity of mango fruit duringstorage. *Sci. Hortic.* **2020**, *259*, 108835. [CrossRef]
37. Apel, K.; Hirt, H. Reactive Oxygen Species: Metabolism, Oxidative Stress, and Signal Transduction. *Annu. Rev. Plant Biol.* **2004**, *55*, 373–379. [CrossRef] [PubMed]

foods

MDPI

Article

Integrating Metabolomics and Proteomics Technologies Provides Insights into the Flavor Precursor Changes at Different Maturity Stages of Arabica Coffee Cherries

Zelin Li [1], Bin Zhou [1], Tingting Zheng [1], Chunyan Zhao [1], Xiaojing Shen [1], Xuefeng Wang [1], Minghua Qiu [2] and Jiangping Fan [1,*]

[1] College of Food Science and Technology, Yunnan Agricultural University, Kunming 650201, China
[2] State Key Laboratory of Phytochemistry and Plant Resources in West China, Kunming Institute of Botany, Chinese Academy of Sciences, Kunming 650201, China
* Correspondence: 1993033@ynau.edu.cn; Tel.: +86-871-65227843

Abstract: The metabolic modulation of major flavor precursors during coffee cherry ripening is critical for the characteristic coffee flavor formation. However, the formation mechanism of flavor precursors during coffee cherry ripening remains unknown. In the present study, a colorimeter was employed to distinguish different maturity stages of coffee cherry based on the coffee cherry skin colors, and proteomics and metabolomics profiles were integrated to comprehensively investigate the flavor precursor dynamics involved in Arabica coffee cherry ripening. The data obtained in the present study provide an integral view of the critical pathways involved in flavor precursor changes during coffee cherry ripening. Moreover, the contributions of critical events in regulating the development of flavor precursors during the four ripening stages of coffee cherries, including the biosynthesis and metabolism pathways of organic acids, amino acids, flavonoids, and sugars, are discussed. Overall, a total of 456 difference express metabolites were selected, and they were identified as being concentrated in the four maturity stages of coffee cherries; furthermore, 76 crucial enzymes from the biosynthesis and metabolism of sugars, organic acids, amino acids, and flavonoids contributed to flavor precursor formation. Among these enzymes, 45 difference express proteins that could regulate 40 primary amino acids and organic acids flavor precursors were confirmed. This confirmation indicates that the metabolic pathways of amino acids and organic acids played a significant role in the flavor formation of Arabica coffee cherries during ripening. These results provide new insights into the protease modulation of flavor precursor changes in Arabica coffee cherry ripening.

Keywords: Arabica coffee; ripening; flavor precursors; metabolomics; proteomics

1. Introduction

Coffea arabica L., a genus of the family *Rubiaceae*, accounts for 60% of global coffee production, owing to the unique taste and pleasant aroma of the beverage [1]. The complex and distinct flavor of coffee beverages is mainly dependent not only on the roasting process conditions of coffee beans but also on the chemical compounds present in the fresh coffee cherry, which act as aroma precursors [2]. Primary coffee cherry components include carbohydrates, proteins, amino acids, trigonelline, caffeine, lipids, organic acids, flavonoid, and glycosides [3]. Among the pool of green coffee constituents, however, carbohydrates (sugars), proteins, acids fraction, flavonoids, and lipids have been suggested to be the principal flavor precursors [4]. Carbohydrates represent 40–65% of dry green coffee beans, and Arabica coffee contains more sugars than Robusta coffee [5]. During the roasting process, 12–24% of the sugars are degraded during light roasting, while 35–40% of sugars are degraded upon dark roasting [6], contributing to the caramel aroma, black color, and characteristic taste via the Maillard reaction and caramelization. The acid fractions, such as chlorogenic acids, could contribute to bitter and tart flavors, and amino acids serve as precursors for the Maillard reaction, which generates sweet, flower-like, and fruity flavors [7]. Wang et al. [8] found that 2-acetylthiazole, which contributes to fried chestnut, toasted cereal, roast meat, and nut flavors, could be formed by the reaction of D-glucose with L-cysteine. However, previous research mainly focused on the content of a single flavor precursor or the heat reaction changes in precursor compounds during coffee processing. To date, the integrated analysis of flavor precursors which might contribute to Arabica coffee cherry has not been reported.

In recent years, omics technology has been considered an effective way to reveal the relationship between metabolites, proteins, and genes [9]. The high-quality coffee genome-wide sequence has made it possible to apply transcriptomic and proteomic analysis techniques to detect the gene and protein compositions in coffee cherries [10,11], while applying transcriptomics alone could not accurately predict the protein abundance and protease activity [11]. Combined metabolomics and proteomics could be used for qualitative and quantitative analyses of low-molecular-weight metabolites and high-molecular-weight proteins, and the dynamic state of the cell, respectively [12]. To date, most proteomics and metabolomics profiling on fruits have mainly concentrated on exploring their developmental ripening process, carbohydrate metabolism, and stress response [12–15]. There is little research on applying metabolomics and proteomics to extend the flavor precursors changing coffee cherry ripening. Moreover, whether the flavor precursors change in metabolism pathways and during protease metabolism between coffee cherries and ripening behaviors remains to be studied. Thus, proteomics and metabolomics profiling were employed to understand the dynamic changes in metabolites and proteins that regulate the content of flavor precursors during the ripening of Arabica coffee cherries. The present study aims to present a dynamic map of Arabica coffee cherry ripening, concentrating on the processes of sugar, organic acids, and amino acids and the biosynthesis and metabolism of flavonoids to provide insight into the modulation of flavor precursor formation. These data would provide valuable information for further studies on the functions of these flavor precursors and influence coffee bean quality. The results would be interesting in understanding the flavor precursor formation in Arabica coffee cherry and in providing novel insights into the modulation of Arabica coffee cherry ripening.

2. Materials and Methods

2.1. Materials

Fresh *Coffea Arabica* L. cherries were used in this study and picked at an altitude of 1040 m in December 2020 from Baoshan City, Yunnan Province, China (25°1′13″ N, 98°49′30″ E). The coffee cherries sampled did not contain pests or carry diseases. The voucher specimens Coffea Arabica L. cherries (No. ACCS1-4) were obtained from the College of Food Science and Technology, Yunnan Agricultural University. The coffee cherries were divided into four groups (A, B, C, and D) based on their maturity level using color measurements ($n = 20$ per group). The samples from 20 cherries per group in one biological replicate were immediately frozen in liquid nitrogen for untargeted metabolomics ($n = 6$ per group) and proteomic analysis ($n = 3$ per group).

2.2. Reagents

Methyl alcohol, bovine serum albumin (BSA), TMT10plex™ (Thermo Fisher Scientific Co., Ltd., Waltham, MA, USA) isobaric label reagent set, and acetonitrile (≥99.0%) were purchased from Thermo Fisher Scientific Co., Ltd. (Waltham, MA, USA). 2-chlorophenylalanine, trichloroacetic acid, β-mercaptoethanol, and acetone were obtained from Aladdin Co., Ltd. (Los Angeles, CA, USA). Formic acid (LC-MS level) was obtained from Tokyo Chemical Industry (TCI) Co., Ltd. (Tokyo, Japan). Ammonium formate, ammonium bicarbonate, urophilia, dithiothreitol (DTT), iodoacetamide (IAM), and TEAB were purchased from Sigma-Aldrich Co., Ltd. (St. Louis, MO, USA). ZipTip and dd H_2O were purchased from Millipore Co., Ltd. (Burlington, MA, USA). A protein quantitative dye solution was obtained from Huaxingbio Co., Ltd. (Beijing, China). Tryptase was obtained from Promega Co., Ltd. (Madison, WI, USA).

2.3. Color Measurement

The exterior color of the coffee cherry samples (n = 20 per group) was determined using a colorimeter (NR10QC+, Shenzhen 3nh Technology Co., Ltd., Shenzhen, China). The lightness (L), redness (a), yellowness (b), chroma (c), hue angle (h°), and total chromatism (ΔE) were measured at two surfaces of each coffee cherry.

2.4. Untargeted Metabolomics Analysis

The untargeted metabolomics analysis was carried out following the protocol described by Tian et al. [16]. Briefly, six whole coffee cherries from each group were selected and ground at low temperatures with liquid nitrogen. A total of 200 ± 2 mg of each sample was homogenized in advance by a KZ-II tissue grinder (Servicebio Co., Ltd., Wuhan, China) and poured into a 2 mL Eppendorf (EP) tube. Then, 0.6 mL of 2-chlorophenylalanine (4 parts per million, PPM) in methanol was accurately added at −20 °C and vortexed for 30 s. Following this, 100 mg glass beads were added to an EP tube, and the samples were ground at 50 Hz for 90 s. Furthermore, the samples were placed in an ultrasonic bath at room temperature for 15 min and centrifuged at 12,000× g for 10 min. The supernatant (300 μL) was filtered through a 0.22 μm membrane and transferred to a vial for on-machine detection by triple quadrupole LC-MS/MS. For quality assurance purposes, 20 μL of each group of samples was mixed with a quality control sample (QC). A total of 5 QC samples were obtained using the same method for further analysis.

The metabolomics analysis of the coffee cherries was carried out using a Q Exactive Plus HPLC-MS/MS system (Thermo Fisher Scientific, Waltham, MA, USA). An Acquity UPLC HSS T3 column (150 mm × 2.1 mm, 1.8 μm, Waters, Manchester, UK) was employed for analyte separation. Gradient elution was carried out with 0.1% (v/v) formic acid in water (C) and 0.1% (v/v) formic acid in acetonitrile (D) in the positive model and 5 mmol/L ammonium formate in water (A) and acetonitrile (B) at a flow rate of 0.25 mL/min in the negative model. Following equilibration of the instrument, the coffee cherry sample extracts were injected (2 μL). Gradient elution was set as follows for the mobile phase: 0–1 min, 2% B/D; 1–9 min, 2–50% B/D; 9–12 min, 50–98% B/D; 12–13.5 min, 98% B/D; 13.5–14 min, 98–2% B/D; and 14–20 min, 2% D-positive model (14–17 min, 2% B-negative model). Full-scan data were acquired between the mass range of m/z 81–1000. The positive ion spray voltage was 4.20 kV; the negative ion spray voltage was 3.50 kV; and the sheath and auxiliary gas flows were 30 arb and 10 arb, respectively. The capillary temperature and collision voltages were 325 °C and 30 eV. Moreover, dynamic exclusion was used to remove unnecessary MS/MS information. A base peak chromatogram of a typical sample is presented in Supplementary Figure S1.

The raw data were pre-treated using Proteowizard software (v3.0.8789; State of California, USA, Cedars-Sinai Medical Center) and R (v3.3.2; Auckland, New Zealand, Ross Ihaka and Robert Gentleman). Furthermore, three-dimensional data sets were obtained, including m/z, retention time (RT), and intensities of the peaks. The RT–m/z pairs were used as the identifier for each ion [17]. For a better comparison, the peak area was normalized, and any peaks with missing values (ion intensity = 0) and pseudo-positive peaks were removed. Peaks with poor repeatability in the QC samples were deleted. The proportion of potential characteristic peaks with a relative standard deviation (RSD) was less than 30% in the QC samples, and more than 70% of the peaks demonstrated high quality (Supplementary Figure S2). The metabolite pathways were subsequently mapped by the online Kyoto encyclopedia of genes and genomes (KEGG) database [17].

2.5. Tandem Mass Tag (TMT) Labelling Proteomics Analysis

In this section, TMT labeling proteomics analysis was also carried out according to the procedure described by Tian et al. [16]. Three whole coffee cherries from each group were randomly selected and ground with liquid nitrogen. The samples were transferred into a 6 mL extraction solution (lysis solution/protease inhibitor = 50:1 v/v) containing 5 g of trichloroacetic acid and 35 μL of β-mercaptoethanol in four volumes of cold acetone for protein extraction. After incubating overnight at −20 °C, the protein precipitate was washed with 1 mL of acetone and collected by centrifugation at 14,000× g for 20 min at 4 °C. Furthermore, the protein concentration was determined with the Bradford method and exhibited specific protein profiles on an SDS-PAGE (Supplementary Figure S3). As previously reported, approximately 1 mg of each protein sample was digested overnight with 20 μg of trypsin at 37 °C [18]. After digestion, the peptides were desalted by a Strata X C18 SPE column, reconstituted in 0.5 mol/L TEAB, and processed according to the manufacturer's protocol for the TMT kit.

The peptides were dissolved in the mobile phase A of LC (0.1% formic acid) and separated using a RIGOL L-3000 HPLC system. Mobile phase A contained 0.1% formic acid, and mobile phase B was an aqueous solution containing 0.1% formic acid and 80% acetonitrile. The LC elution settings were 0–49 min, 7–40% B; 49–60 min, 100% B with 0.7 mL/min flow. The resulting peptides were subjected to tandem MS/MS after ionization of the Nanospray FlexTM (NSI) ion source on a Q Exactive HF-X mass spectrometer. The spray voltage was set to 2.4 kV, and the ion transfer tube temperature was set to 275 °C. The scanning range of the primary mass spectrum was set to 407–1500 m/z. The top 40 parent ions were selected for MS/MS with an automatic gain control (AGC) target of 3×10^6. The secondary MS resolution was set at 45,000 (200 m/z) with an AGC target of 5×10^4, and the maximum injection time was set at 86 ms. The normalized collision energy of 32% was used for the peptides chosen for MS/MS. The protein was quantified by using the razor and unique peptides.

The protein identification in this study was carried out using the protein database of *Coffea Arabica* L. in https://www.uniprot.org/ (accessed on 11 January 2021). The PDQuest 2.4 software (Bio-Rad Co., Ltd., Hercules, CA, USA) was used to search the database. Thermo Proteome Discoverer software version 1.4 (Thermo Fisher Scientific, Bremen, Germany) was used with the default settings to process the raw data. According to those reported in the UniProt database, the gene ontology (GO) annotations were performed for the identified proteins. A KEGG enrichment-pathway analysis was further performed for the identified proteins.

2.6. Statistics

Data analysis was carried out using SPSS Version 29.0 (IBM Corp., New York, NY, USA), R (v3.3.2) and GraphPad Prism 9 (GraphPad Software Inc., San Diego, CA, USA). One-way analysis of variance (ANOVA) followed by Duncan's post hoc test was carried out to perform multi-group comparisons. $p < 0.05$ was considered statistically significant. "**" represents the significant differences ($p < 0.01$). Multivariate statistical analysis was carried out using MetaboAnalyst Version 5.0 (Xia-lab, Montreal, QC, Canada).

3. Results and Discussion

3.1. Changes in Coffee Cherry Skin Color

Fruit color indicates maturity level and quality stage and is an essential parameter in fruit classification [19]. As the coffee cherry grows in different stages of maturity, changes in skin color would be observed, and flavor precursors will accumulate. Coffee maturity was divided into A–D ripening stages (Figure 1), and the color of the coffee cherries was presented as green–brown–red–deep red, respectively. Moreover, the values of a, b, c, h°, and (E) demonstrated changes at different ripening stages. Specifically, an "a" value (redness) of −7.37 ± 2.54 was below 0 in the A group, indicating that these samples had a primarily green color. However, the "a" value of group C was 33.43 ± 3.60, suggesting that the coffee cherries ripened from having green skin to red skin. Coffee cherry maturity was attributed to pigments accumulating outside the skin and the changing components in the inner tissue [20].

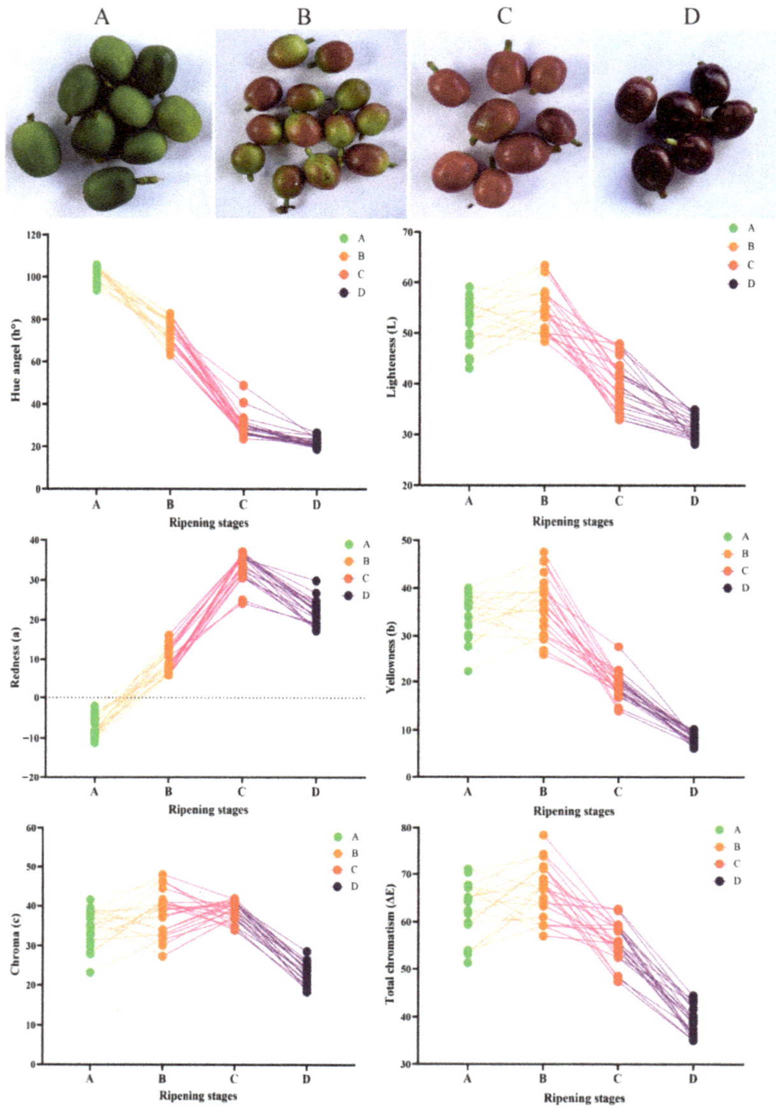

Figure 1. Changes in skin color during coffee cherry ripening.

3.2. Overview of Untargeted Metabolomics

There were 10,807 and 8376 detectable peaks in the positive and negative modes, respectively (Supplementary Data S1, sheets 1–2, and Supplementary Figure S1). In the positive and negative models, the QC samples gathered together, indicating that the data had few errors and higher quality. Furthermore, a total of 70.5% and 88.8% of the peaks with RSD less than 30% were in the positive and negative ion modes, respectively, which also well proved the high data quality (Supplementary Figure S2). After raw data pre-treatment, the positive (7618) and negative (7435) data were combined into a data set (Supplementary Data S1, sheet 3–4) and were imported into the R package (v3.2.2). Furthermore, a principal component analysis (PCA) and a partial least squares discriminant analysis (PLS-DA) were performed to visualize the metabolic differences among the samples from the four maturity stages. The differential clustering among the four stages of coffee cherries in the PCA (Figure 2a,b) and PLS-DA (Figure 2c,e) obtained similar results. The samples were separated into groups, and aggregation of the same group indicated that there were differences in the metabolite profiles at each maturity stage but similar metabolites at the same maturity stage. A permutation test was conducted to verify whether the PLS-DA model of the data had been over-fitted. The results showed that all blue Q2 points were lower than the original from left to right (Figure 2d,f). These results suggested that different maturity stages of coffee cherries had distinctly different metabolite characteristics.

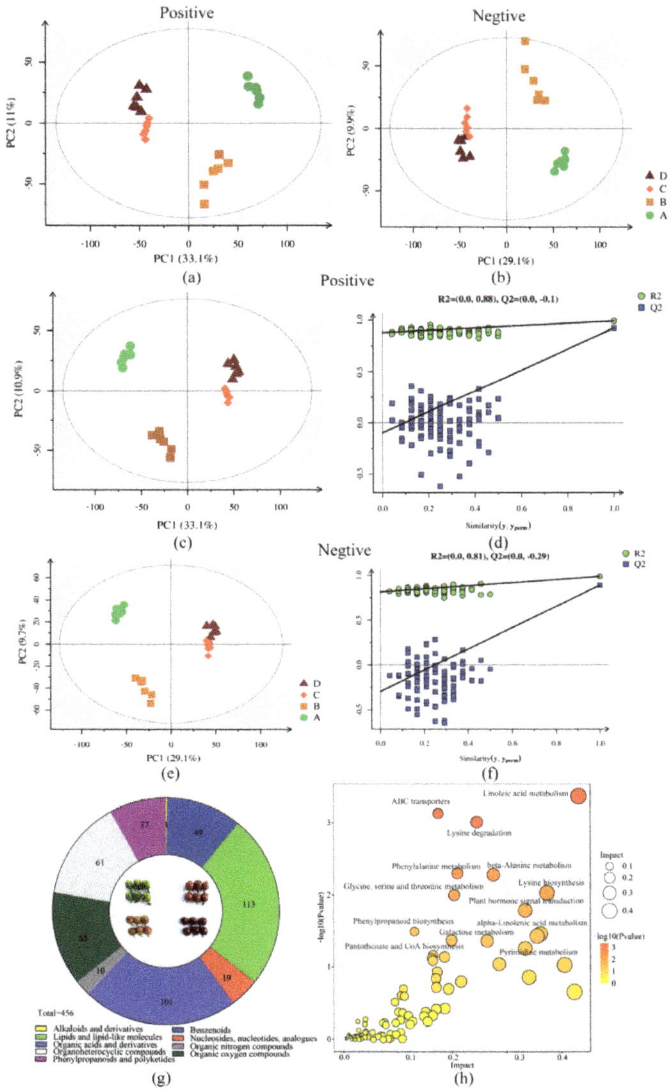

Figure 2. Principal component analysis (PCA) of peak areas detected in four groups of coffee samples in positive and negative ion models (**a**,**b**) (four maturity stages: A (•), B (■), C (♦), D (▲)). Partial least squares discriminant analysis (PLS-DA) and permutation test of peak areas in positive and negative ion models in four group samples (**c**–**f**). Classification of DEMs in four group samples (**g**). KEGG pathways of DEMs involved (**h**).

Based on MS/MS spectra, retention time, and metabolomics database, 2178 nonvolatile compounds were identified (Supplementary Data S1, sheet 5). Then, a variable importance in projection (VIP) analysis ranked the overall contribution of each variable to the PLS-DA model, and variables with VIP > 1.0 and $p < 0.05$ were classified as difference express metabolites (DEMs). Finally, a total of 456 DEMs were selected, and they were identified as being concentrated in the four maturity stages of coffee cherries, such as glucose, sucrose, tyrosine, cystathionine, palmitic acid, and quercetin. These DEMs were divided into super-classes of alkaloids and derivatives (trigonelline), benzenoids (49 metabolites), lipids and lipid-like molecules (113 metabolites), nucleosides, nucleotides and analogs (19 metabolites), organic acids and derivatives (101 metabolites), organic nitrogen compounds (10 metabolites), organic oxygen compounds (61 metabolites), organoheterocyclic compounds (65 metabolites), and phenylpropanoids and polyketides (37 metabolites) (Supplementary Data S1, sheet 6, and Figure 2g). DEMs were further grouped into different classes, including flavonoids, sugars, amino acids, and fatty acids. These DEMs were significantly enriched into 87 KEGG pathways such as linoleic acid metabolism, lysine biosynthesis, galactose metabolism, and flavonoid biosynthesis (Supplementary Data S1, sheet 7, and Figure 2h). Some DEMs were closely related to the characteristic flavor formation as important precursors during coffee cherry ripening. For example, mature coffee cherries could produce more concentrations of carbohydrate and aroma compounds of carbohydrate degradation with higher sensory quality scores than immature green cherries [21].

3.3. Overview Analysis of TMT Labelling Proteomics

Previous research observed that the metabolites expressed generally fall behind alterations in corresponding proteins, indicating that protein expression changes are vital factors in triggering metabolic changes [22]. A TMT labeling proteomics analysis was employed to reveal the underlying molecular mechanism for the changes in nonvolatile compounds during the different maturity stages of coffee cherries. A total of 11,309 peptides were obtained (Supplementary Data S2, sheet 1), and the length of the peptides ranged mainly from 14 to 26 (Supplementary Figure S4a). The range of the peptide-spectrum match (PSM) captured per peptide segment with a mass spectrum ranged between 1 and 5 (Supplementary Figure S4b), indicating that the lengths of the peptides were reasonable, with good accuracy and reliability for identifying peptides. The molecular weights of most identified proteins ranged from 10 to 100 kDa (Supplementary Figure S4c), suggesting that the proteins have a reasonable molecular weight. Furthermore, these peptides were matched with 2748 unique proteins in the samples from the four maturity stages. The proteins included beta-galactosidase (A0A6P6WN40), cysteine protease (A0A6P6W856), and aminomethyltransferase (A0A6P6U842) (Supplementary Data S2, sheet 2). The results of the PCA showed that the samples from the four maturity stages were clustered together in different clusters (Figure 3a). Furthermore, the condition of FC $\geq$ 1.5 and $p < 0.05$ were employed to select the expressed proteins (DEPs). A total of 223 (B vs. A), 81 (C vs. B), and 98 (D vs. C) DEPs were identified, including carbohydrate synthetic protease, amino acid synthetic protease, lipase, and enzymes related to the synthesis of flavonoid polyphenols (Supplementary Data S2, sheet 3–5). There were 125 (B vs. A), 34 (C vs. B), and 71 (D vs. C) proteins up-regulated and 98 (B vs. A), 47 (C vs. B), and 27 (D vs. C) proteins down-regulated (Figure 3b).

The highly represented GO classes of DEPs include the molecular function (molecular level activities such as catalytic or binding), the biological process (cellular and metabolic processes), and the cellular component (cell part and its extracellular environment) (Supplementary Figure S5). The 102 terms of the KEGG pathways were enriched in the four stages of coffee cherries, with metabolic pathways and biosynthesis being the most enriched terms. These enrichments of the KEGG pathways included amino sugar and nucleotide sugar metabolism, glutathione metabolism, photosynthesis, flavonoid biosynthesis, fructose and mannose metabolism, glycolysis/gluconeogenesis, galactose metabolism, inositol phosphate metabolism, and fatty acid degradation (Supplementary Data S2, sheet 6, and Figure 3c).

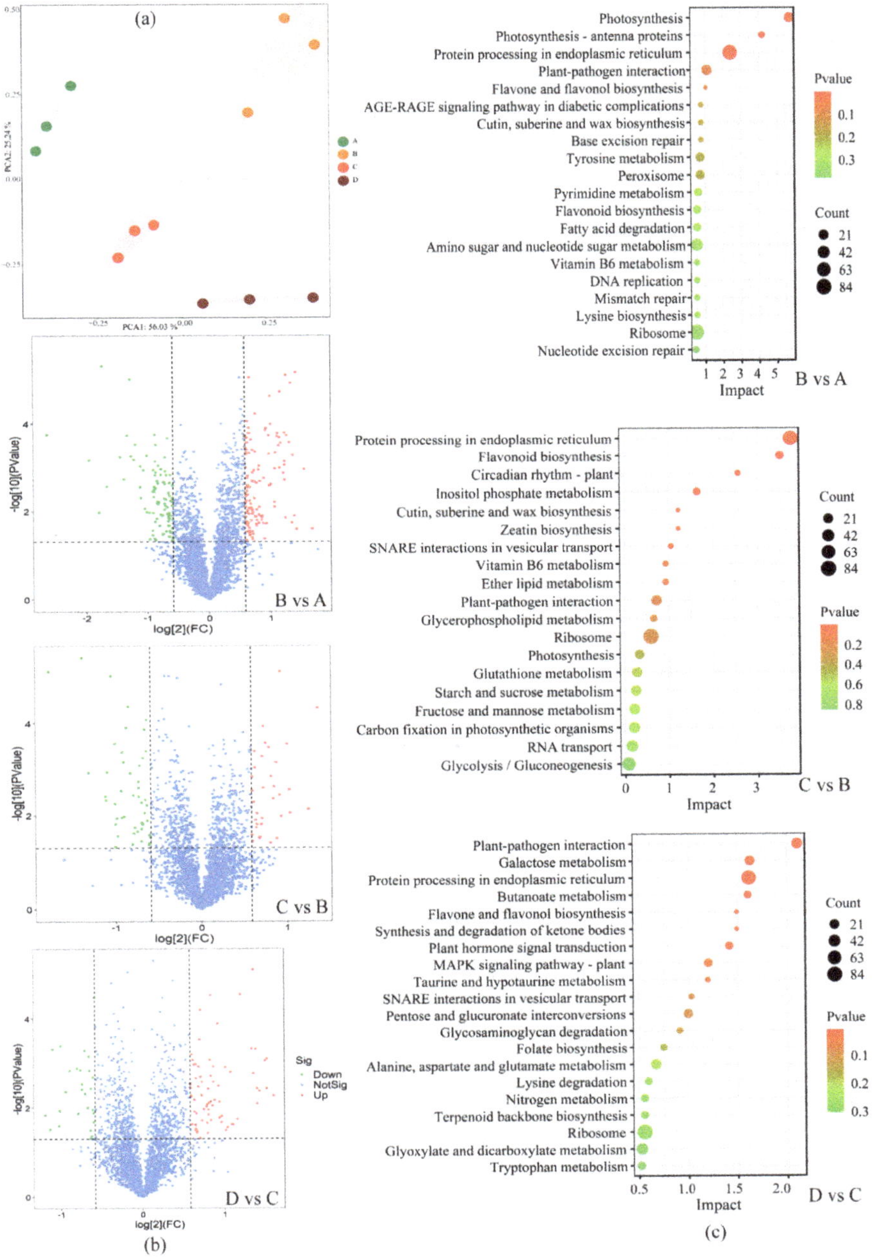

Figure 3. Results of proteomics overview analysis. PCA score of four coffee sample groups (**a**). Volcanic map analysis of DEPs. Red: up-regulation; green: down-regulation; blue: not significant. A total of 125, 34, and 71 proteins were up-regulated, while 98, 47, and 27 proteins were down-regulated (**b**). The enrichment analysis of the KEGG pathway in coffee at each maturation level and the top 20 pathways in each stage (**c**).

Because of the formation of the metabolites by enzyme regulation, most metabolites have been proven to contribute directly to flavor or to serve as flavor precursors [1,23]. For example, flavonoids could provide a bitter and astringent taste; free amino acids present an umami taste; sugar contributes to a sweet taste; and organic acids contribute to a sour taste [18]. Thus, the following analyses mainly focused on the biosynthesis and metabolism pathways of organic acids, amino acids, flavonoids, and sugars and the effect of relevant protease in the four stages of coffee cherry ripening.

3.4. Sugar-Related Flavor Precursors

The Maillard reaction was the main pathway for producing flavor during the roasting process, and amino acids and sugars are the most important precursors to the Maillard reaction [24]. However, different amino acids and sugars as precursors could generate different coffee flavors. A high sugar content not only increases the sweetness of coffee but also protects the beans from oxidation by forming a thin sugar film on the surface and facilitating the Maillard reaction [25]. Moreover, the higher sugar content (mainly sucrose, glucose, and fructose) could also increase the levels of pyrazines, furans, ketones, and heterocyclic nitrogen-containing compounds after roasting. These compounds mainly contribute to the flower, fruit, and caramel roast aroma of coffee [25].

The sugar contents of coffee cherries gradually increase as cherries mature, especially the content of fructose, sucrose, and galactose, reaching 15%, 3.2%, and 2.4%, respectively [26]. In this study, the contents of sucrose, fructose, glucose, mannose, trehalose, and galactose markedly increased from ripening stages A to D. During the early maturity stages, the concentrations of cellobiose, xylose, and ribulose significantly increased. However, the level of these sugars decreased at a later maturity stage (Figure 4). In addition, 21 important DEPs related to the biosynthesis and degradation of sugars were identified in this study. A previous study showed that sucrose is simultaneously regulated by synthetic and degradative enzymes during fruit ripening and potentially degraded to fructose through β-fructofuranosidase regulation [27]. Our proteomic results showed that the level of β-fructofuranosidase (A0A6P6VLM6, FFS) and α-glucosidase (A0A6P6UBP3, GSL) showed a positive relationship with sucrose and fructose (Figure 4), indicating that these enzymes lead to the accumulation of the two reducing sugars during the ripening process. However, the level of glucan 1,3-α-glucosidase (A0A6P6VAQ9, PG13G) increased in the first stage. Moreover, glucan 1,3-α-glucosidase (A0A6P6VAQ9, PG13G) levels decreased with time, suggesting a negative relationship with the accumulation of sucrose and fructose. Furthermore, fructose and mannose could achieve mutual transformation by regulating with mannose-6-phosphate isomerase (A0A6P6TMS7, M6PI) [16]. Glucose was reported as a product derived from both trehalose and α-lactose and correlated with the levels of relative enzymes [16]. In this study, the expression of α,α-trehalose phosphate synthase (A0A6P6VRB4, TPS) and glucose-6-phosphate isomerase (A0A6P6UUG7, G6PI) gradually decreased ($p < 0.05$) along with glucose formation. This incidence suggests that the total glucose pool could be enlarged by utilizing the two phosphate enzymes. The activities of endoglucanase (A0A6P6U1D9, EGL) and glucan endo-1,3-β-D-glucosidase (A0A6P6UPB4, GE13G) increased at all stages, further indicating cellobiose accumulation during coffee cherry ripening.

Studies proposed that sucrose-phosphate synthase participated in the cycle of sucrose synthesis and degradation during fruit ripening [16]. In the current study, the level of sucrose-phosphate synthase (A0A6P6WVX8, SPS) was the highest at stage C (Figure 4), indicating that this stage of sucrose synthesis and degradation was very active. Based on the flavor contribution, sucrose could also become one of the critical items that could indicate the optimal harvest time for coffee cherries.

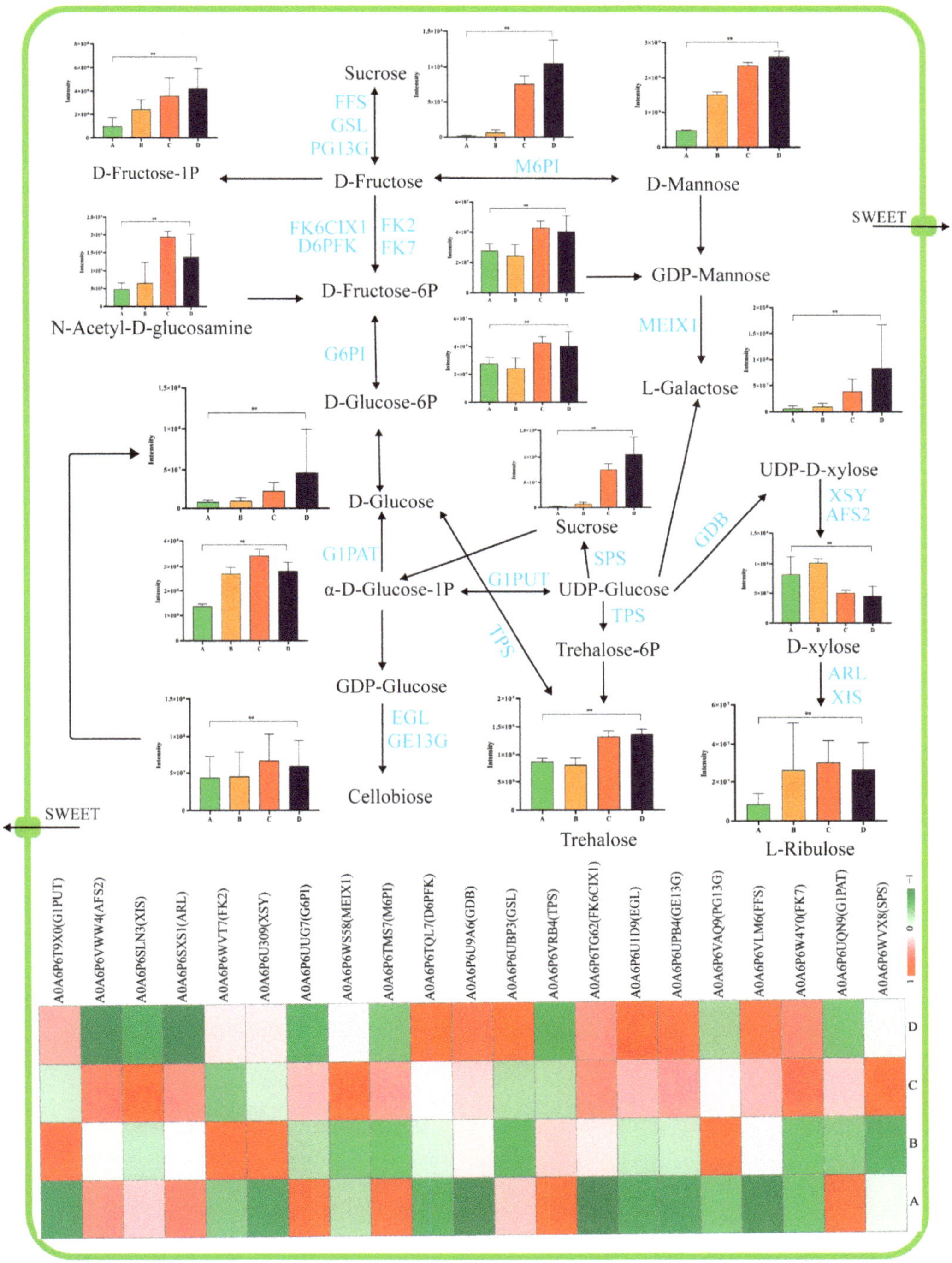

Figure 4. Pathway viewer of main sugar metabolism in coffee cherries. The ripening progression of protein expression from A to D is presented in four box strings. "**" represents the significant differences ($p < 0.01$). Putative protein families were from the *Coffea Arabica* L. genome database. G1PUT,

UTP-glucose-1-phosphate uridylyltransferase; AFS2, α-L-arabinofuranosidase 2; XIS, xylose isomerase; ARL, aldose reductase-like; FK2, fructokinase-2; XSY, UDP-D-xylose synthase 2; G6PI, glucose-6-phosphate isomerase; MEIX1, GDP-mannose 3,5-epimerase isoform X1; M6PI, mannose-6-phosphate isomerase; D6PFK, ATP-dependent 6-phosphofructokinase; GDB, UDP-glucuronate decarboxylase; GSL, α-glucosidase-like; TPS, α,α-trehalose-phosphate synthase; FK6CIX1, fructokinase-6, chloroplastic isoform X1; EGL, endoglucanase; GE13G, glucan endo-1,3-beta-D-glucosidase; FFS, β-fructofuranosidase; FK7, fructokinase-7; G1PAT, glucose-1-phosphate adenylyltransferase; SPS, sucrose-phosphate synthase.

3.5. Amino-Acid- and Organic-Acid-Related Flavor Precursors

Amino acids are among the main parameters for evaluating fruit quality. During the fruit ripening process, their levels and compositions could change substantially [28]. In addition to being precursors, amino acids could directly contribute to sweet, umami, or bitter tastes. For instance, alanine contributes to the sweetness of fruits. Aminotransferase plays a vital role in the synthesis or degradation of aspartate and pyruvic acid and could also catalyze the conversion of amino acids during fruit ripening [29]. In this study, 40 DEMs and 45 DEPs related to amino acids significantly contributed to flavor formation during coffee cherry ripening. Among them, the contents of aspartate and pyruvic acid increased but that of alanine decreased. The level of alanine-glyoxylate transaminase (A0A6P6TM10) decreased ($p < 0.05$) from stages A to D, suggesting that alanine could convert to other amino acids via the action of alanine-glyoxylate transaminase (A0A6P6TM10) (Figure 5). Cysteine, glutathione, and methionine are sulfur (S)-containing amino acids in green coffee beans and have been linked explicitly to the cooked meat flavor. This flavor has been attributed to alkyl pyrazines, diketones, 2-furfurylthiol, and other S-containing compounds decomposed by S-containing amino acids [30]. Cysteine could be synthesized by serine in the regulation of homoserine kinase and cysteine synthase, and methionine was negatively confirmed with the ethylene biosynthesis during fruit ripening [31,32]. In the present study, serine, cysteine, and cystathionine increased, while glutathione, methionine, and γ-glutamylcysteine decreased during coffee cherry ripening. The level of relevant proteases, including homoserine kinase (A0A6P6VCB8), cysteine synthase (A0A6P6WQV5), L-cysteine desulfhydrase (A0A6P6VZM9), farnesyl cysteine lyase (A0A6P6T9Z7), 5-methyl-tetrahydro pteroyl triglutamate homocysteine S-methyltransferase (A0A6P6SM43), S-adenosylmethionine synthase (A0A6P6SUD6), and glutathione synthetase (A0A6P6W069), were also upregulated (Figure 5). This result suggested that the above proteases regulated S-containing amino acid biosynthesis in coffee. The decrease in the content of methionine implied an ethylene biosynthesis increase, which was also connected to ripening and consistent with the findings of a previous study [32]. In addition, the amount of β-alanine, glutamine, phenylalanine, and the relevant enzymes increased during coffee cherry ripening. These constituents, which acted as precursors, participated in organic acid biosynthesis.

The organic acid fraction in green coffee is composed of volatile and nonvolatile aliphatic (citric, malic, and quinic acid) and phenolic acids (about 8%) [33]. These acids were highly relevant for the acidic taste in the cherries. The nonvolatile organic acids could serve as a precursor to form volatile phenols such as 2-methoxy phenol (guaiacol), 4-ethyl-2-methoxy phenol (4-ethyl guaiacol), and 4-vinyl-2-methoxy phenol (4-vinyl guaiacol). These compounds potentially provide the typical smoky and woody aroma in dark roasted coffee beans [34,35]. The tricarboxylic acid cycle (TCA) is a vital biological process involved in the biosynthesis and derivatization of organic acids [36]. Pyruvic acid is a critical precursor of the TCA cycle and a junction among sugars, organic acids, and amino acids, and its degradation is accompanied by a series of conversions between organic acids to meet the needs for intermediates and energy during fruit ripening [37,38]. Among metabolites in the TCA cycle, fumaric acid, malic acid, and succinic acid increased at all stages. On the other hand, pyruvic acid, citric acid, and isocitric acid declined (Figure 5), indicating that these organic acids dynamically changed during the ripening process.

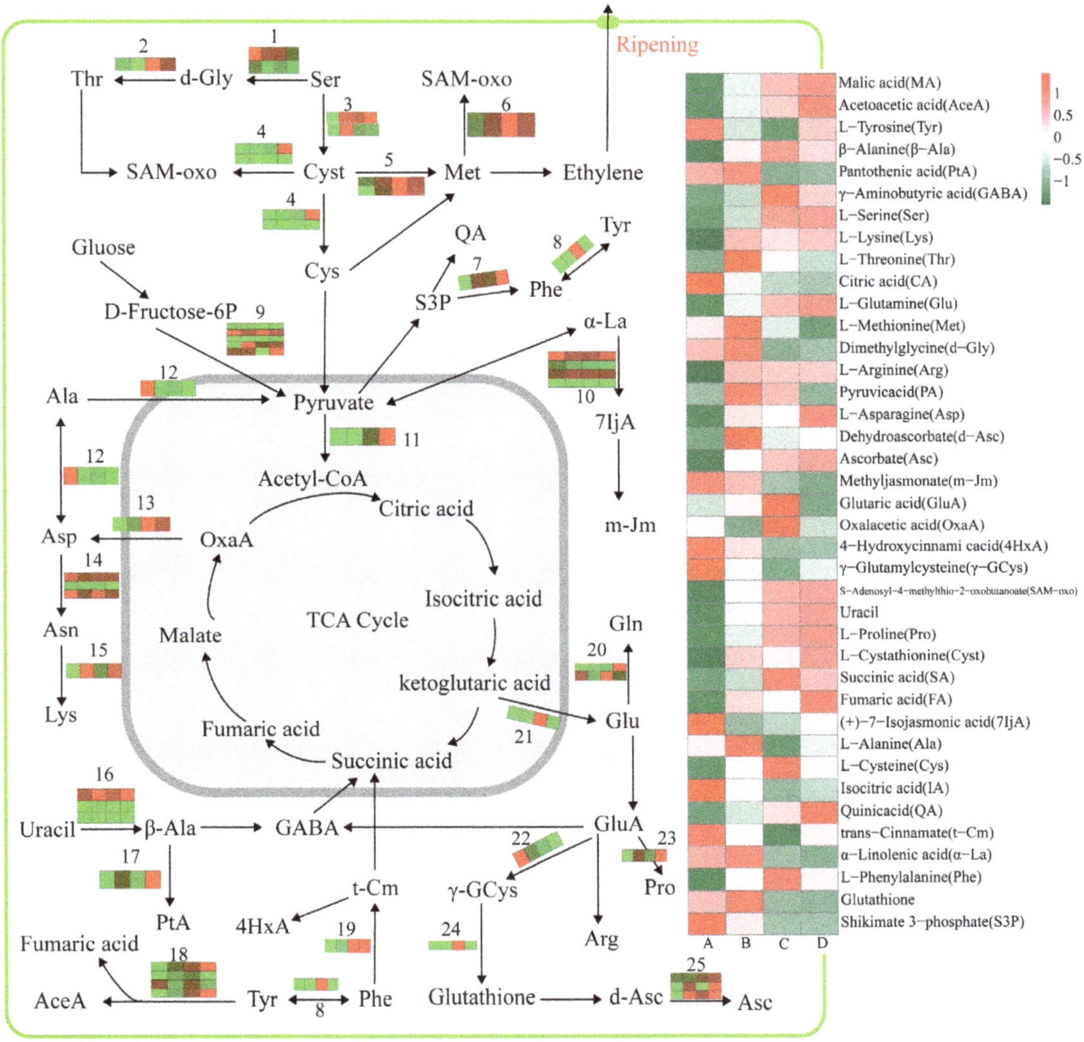

Figure 5. Pathway displays a subset of organic acid and amino acid metabolism proteases required in coffee cherries. The metabolites changing as ripening progresses from A to D are presented in four box strings. Heat maps, for each numbered reaction, were drawn using the TMT value. The numbered reactions were catalyzed by the following enzymes: 1, serine hydroxymethyltransferase (A0A6P6TLR0) and glutamate-glyoxylate aminotransferase 2 (A0A6P6VAJ1); 2, L-threonine aldolase 1 (A0A6P6VRS3); 3, homoserine kinase (A0A6P6VCB8) and cysteine synthase (A0A6P6WQV5); 4, L-cysteine desulfhydrase (A0A6P6VZM9) and farnesylcysteine lyase (A0A6P6T9Z7); 5, 5-methyltetrahydropteroyl triglutamate homocysteine S-methyltransferase (A0A6P6SM43); 6, S-adenosylmethionine synthase (A0A6P6SUD6); 7, shikimate dehydrogenase (A0A6P6X2E2); 8, arogenate dehydratase (A0A6P6W2X6); 9, NADP-dependent glyceraldehyde-3-phosphate dehydrogenase (A0A6P6SZ76), pyruvate phosphate dikinase (A0A6P6WRX8), phosphoglycerate mutase (A0A6P6SCM7), phosphopyruvate hydratase (A0A6P6T3Q6), and glyceraldehyde-3-phosphate dehydrogenase (A0A6P6VC84); 10, allene oxide synthase (A0A6P6UTE2), allene oxide synthase 1 (A0A6P6UPD2), allene-oxide cyclase (A0A6P6TAJ2), enoyl-CoA hydratase 2 (A0A6P6TYF9); 11, acetyl-CoA acetyltransferase (A0A6P6SED1); 12, alanine-glyoxylate transaminase (A0A6P6TM10);

13, alanine aminotransferase 2 (A0A6P6X499); 14, asparagine synthetase (A0A6P6TA86), L-asparaginase 1 (A0A6P6X331), probable L-asparaginase 3 (A0A6P6VNC7); 15, diaminopimelate decarboxylase 1 (A0A6P6VRS2); 16, dihydrothymine dehydrogenase (A0A6P6XK57), dihydropyrimidinase (A0A6P6SXN9), β-ureidopropionase (A0A6P6WDT2); 17, alanine-tRNA ligase (A0A6P6V5B4); 18, aspartate-prephenate aminotransferase (A0A6P6TCB7), tyrosine aminotransferase (A0A6P6S801), histidinol-phosphate aminotransferase (A0A6P6VSK7), fumarylacetoacetase (A0A6P6UAI6); 19, phenylalanine ammonia-lyase (A0A6P6VDX4); 20, glutamine synthetase (A0A6P6T536), glutamate synthase (A0A6P6W7J9), 21, glutamate dehydrogenase (A0A6P6SVM6); 22, γ-glutamylcyclotransferase family protein (A0A6P6UG48); 23, acetylglutamate kinase (A0A6P6TDN5); 24, glutathione synthetase (A0A6P6W069); 25, L-ascorbate peroxidase 2 (A0A6P6W0Y9), L-ascorbate peroxidase 3 (A0A6P6VST3), L-ascorbate peroxidase 6 (A0A6P6UR23).

3.6. Flavonoid-Related Flavor Precursors

Flavonoids, especially flavonols, contribute tremendously to the astringent taste, color, and aroma formation of fruits. The phenylpropanoid and flavonoid biosynthesis pathways were principally for the biosynthesis of flavonoids [18,39]. During fruit ripening, there was a dramatic change in flavonoid composition and content, which was associated with the transformation of fruit pigmentation and flavor [40]. In this study, 10 DEPs were annotated as crucial enzymes involved in the biosynthesis of phenylpropanoids and flavonoids. These levels of enzymes were upregulated besides trans-cinnamate 4-monooxygenase (A0A6P6ULM4, C4M), flavonol synthase-like (A0A6P6UIY9, FSL), and flavanol 3-O-glucosyltransferase (A0A6P6X835, UGT) during Arabica coffee cherry ripening (Figure 6). As the first rate-limiting enzyme in the flavanol biosynthesis pathway, phenylalanine ammonia-lyase directly catalyzes the degradation of phenylalanine to trans-cinnamic acid [18]. Phenylalanine ammonia-lyase (A0A6P6VPJ8, PAL) and phenylalanine were upregulated, and trans-cinnamic acid was downregulated. This result indicated that trans-cinnamic acid was the substrate that could synthesize chalcone and naringenin by combining with trans-cinnamate 4-monooxygenase (A0A6P6ULM4, C4M), chalcone synthase (A0A6P6WUD7, CHS), and chalcone-flavanone isomerase family protein (A0A6P6WAJ7, CFI) in the phenylpropanoids pathway. In addition, the content of naringenin and chalconaringenin continuously increased from stages A to D (Figures 5 and 6). These results further confirmed the above inference. The naringenin further decomposed and was involved in the synthesis of other flavonoids in the flavonoids biosynthesis pathway. Considering the nutrition and the astringent taste of flavonoids, maturity stage C could be considered a suitable harvest stage when flavor precursors were prepared. In addition, glycosidic compounds were also vital in green coffee beans because the aglycones could be released during the roasting process to enrich the aroma profile [41]. Cyanidin glycosides (n = 5), delphinidin glucosides (n = 2), kaempferol glycosides (n = 5), and quercetin glycosides (n = 2) were detected in this study. These glycosides exhibited various expression trends during ripening, mainly quercetin 3-O-glucoside, which possessed the highest ionic peak intensity at all four stages (Figure 6 and Supplementary Data S2, sheet 2).

In summary, sugar, amino acids, and organic acids metabolism in fresh coffee cherries could contribute to the flower, fruit, and caramel roast aroma and sweet taste as a flavor candidate. The flavonoid fraction could contribute tremendously to the astringent taste and skin color. In addition, the environment, such as altitude, soil, daylight hours, and precipitation, is also a potential factor that could affect the ripening of fresh coffee cherries. For instance, coffee cherry has a higher sensory quality when it ripens a month later at higher altitudes than at lower altitudes, [42,43]. The coffee cherry flavor is directly linked to the chemical composition of the raw bean and ultimately affects the flavor property of the coffee product. These results can complement knowledge of the formation of coffee flavor precursors during the ripening process and can increase the understanding of coffee flavor. However, this study design was based on coffee cherry's four maturity stages, which were divided based on coffee cherry skin color, and then, proteomics and metabolomics profiles were integrated to comprehensively investigate the flavor precursor involved in Arabica coffee cherry ripening. If the maturity stages of coffee cherry are evaluated by other sensory analysis indices, such as hardness, the difference express metabolites and crucial enzymes that contribute to flavor precursors formation might be slightly different. A future study will be attempted to evaluate coffee cherry flavor precursor changes during Arabica coffee cherry ripening based on the total sensory quality properties of cherries.

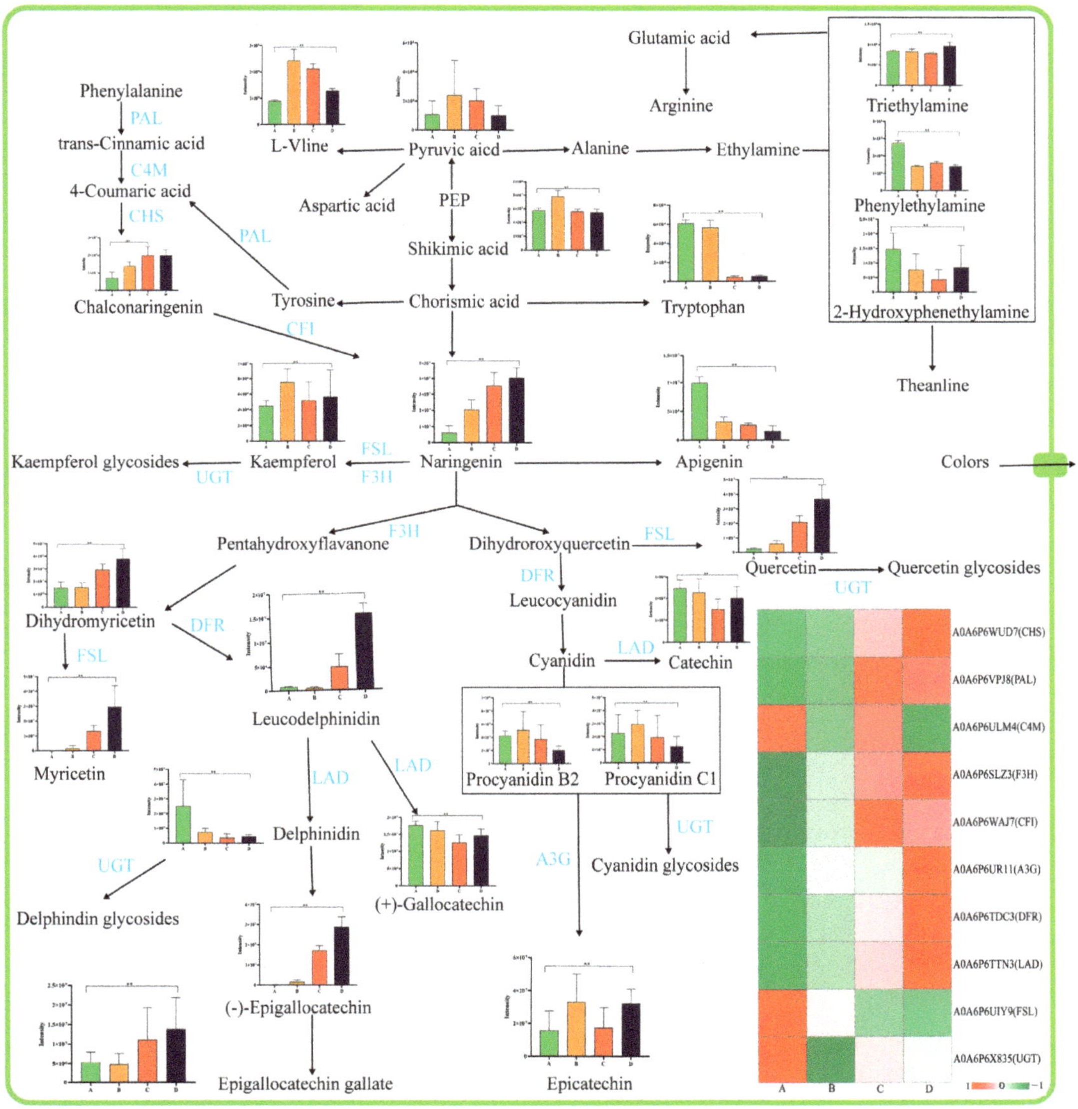

Figure 6. Diagram of phenylpropanoids and flavonoids biosynthesis pathways in the coffee cherries at different maturity stages. The ripening progression of protein expression from A to D is indicated in four box strings. "**" represents the significant differences ($p < 0.01$). CHS, chalcone synthase; PAL, phenylalanine ammonia-lyase; C4M, trans cinnamate 4-monooxygenase; F3H, flavanone 3-hydroxylase; CFI, chalcone flavanone isomerase; A3G, anthocyanidin 3-O-glucosyltransferase; DFR, dihydroflavonol 4-reductase; LAD, leucoanthocyanidin dioxygenase; FSL, flavonol synthase-like.

4. Conclusions

In the present study, the coffee cherry skin colors mainly presented green–brown–red–deep red changes from the A to D ripening stages, as detected by the NR10QC+ colorimeter. A comprehensive analysis of *coffea arabica* L. cherries integrating metabolomics and proteomics was performed on a large scale in whole cherries to understand the dynamic changes in metabolites and proteins that regulate the content of the flavor precursor during the maturity stage of the coffee. The metabolomics and proteomic results identified 456 common DEMs and 223 (B vs. A), 81 (C vs. B), and 98 (D vs. C) DEPs, respectively. Furthermore, the critical precursor metabolites and the relevant enzymes (especially amino acids and organic acids metabolism, flavonoid biosynthesis, and sugar metabolism) with their associated pathways were confirmed. Our results provide an integrated view on the significant pathways about the functions of metabolites and proteins required for coffee cherry flavor precursors and new insights into the processes underlying the fruit ripening traits.

Supplementary Materials: The following supporting information can be downloaded at https://www.mdpi.com/article/10.3390/foods12071432/s1, Supplementary Data S1: Original data of metabolomics; Supplementary Data S2: Original data of proteomics; Figure S1: Base peak chromatogram of typical sample in metabolomics; Figure S2: QC and RSD of coffee cherries metabolomics; Figure S3: SDS-page of coffee cherry protein in proteomics; Figure S4: Length distribution, special PSM numbers, and Mw of peptide of coffee cherries in proteomics; Figure S5: GO terms of DEPs in proteomics.

Author Contributions: Conceptualization, T.Z. and X.W.; methodology, X.S.; software, Z.L.; validation, Z.L. and B.Z.; formal analysis, X.S.; investigation, Z.L.; resources, Z.L. and C.Z.; data curation, Z.L.; writing—original draft preparation, Z.L. and C.Z.; writing—review and editing, J.F.; visualization, Z.L. and C.Z.; supervision, M.Q. and J.F.; project administration, J.F.; funding acquisition, J.F. All authors have read and agreed to the published version of the manuscript.

Funding: This research was supported by National Natural Science Foundation of China-Yunnan Joint Fund (CN), grant number U1902206.

Institutional Review Board Statement: Not applicable.

Informed Consent Statement: Not applicable.

Data Availability Statement: The data are contained within the article and the Supplemental Data, and the data presented in this study are available from the corresponding author upon request.

Conflicts of Interest: The authors confirm that they have no conflict of interest with respect to the work described in this manuscript. The funders had no role in the design of the study; in the collection, analyses, or interpretation of data; in the writing of the manuscript; or in the decision to publish the results.

References

1. Wang, Y.; Wang, X.; Hu, G.; Hong, D.; Bai, X.; Guo, T.; Zhou, H.; Li, J.; Qiu, M. Chemical ingredients characterization basing on 1H NMR and SHS-GC/MS in twelve cultivars of *coffea arabica* roasted beans. *Food Res. Int.* **2021**, *147*, 110544. [CrossRef] [PubMed]
2. De Rosso, M.; Lonzarich, V.; Navarini, L.; Flamini, R. Identification of new glycosidic terpenols and norisoprenoids (aroma precursors) in *C. arabica* L. green coffee by using a high-resolution mass spectrometry database developed in grape metabolomics. *Curr. Res. Food Sci.* **2022**, *5*, 336–344. [CrossRef] [PubMed]
3. Trinh, N.T.N.; Tuan, N.N.; Thang, T.D.; Kuo, P.; Thanh, N.B.; Tam, L.N.; Tuoi, L.H.; Nguyen, T.H.D.; Vu, D.C.; Ho, T.L.; et al. Chemical composition analysis and antioxidant activity of *coffea robusta* monofloral honeys from Vietnam. *Foods* **2022**, *11*, 388. [CrossRef] [PubMed]
4. Poisson, L.; Schmalzried, F.; Davidek, T.; Blank, I.; Kerler, J. Study on the role of precursors in coffee flavor formation using in-bean experiments. *J. Agric. Food Chem.* **2009**, *57*, 9923–9931. [CrossRef] [PubMed]
5. Vezzulli, F.; Rocchetti, G.; Lambri, M.; Lucini, L. Metabolomics combined with sensory analysis reveals the impact of different extraction methods on coffee beverages from *coffea arabica* and *coffea canephora* var. robusta. *Foods* **2022**, *11*, 807. [CrossRef] [PubMed]
6. Bradbury, A.G.W.; Halliday, D.J. Chemical structures of green coffee bean polysaccharides. *J. Agric. Food Chem.* **1990**, *38*, 389–392. [CrossRef]

7. Shakoor, A.; Zhang, C.; Xie, J.; Yang, X. Maillard reaction chemistry in formation of critical intermediates and flavour compounds and their antioxidant properties. *Food Chem.* **2022**, *393*, 133416. [CrossRef]
8. Wang, X.; Ma, Y.; Guo, Y.; Luo, X.; Du, M.; Dong, L.; Yu, P.; Xu, X. Reinvestigation of 2-acetylthiazole formation pathways in the Maillard reaction. *Food Chem.* **2021**, *345*, 128761. [CrossRef]
9. Tiwari, M.; Singh, B.; Min, D.; Jagadish, S.V.K. Omics path to increasing productivity in less-studied crops under changing climate-lentil a case study. *Front. Plant Sci.* **2022**, *13*, 813985. [CrossRef]
10. Alkimim, E.R.; Caixeta, E.T.; Sousa, T.V.; Resende, M.D.V.; Da Silva, F.L.; Sakiyama, N.S.; Zambolim, L. Selective efficiency of genome-wide selection in *Coffea canephora* breeding. *Tree Genet. Genomes* **2020**, *16*, 41. [CrossRef]
11. Huded, A.K.C.; Jingade, P.; Bychappa, M.; Mishra, M.K. Genetic diversity and population structure analysis of coffee (*coffea canephora*) germplasm collections in Indian gene bank employing SRAP and SCoT markers. *Int. J. Fruit Sci.* **2020**, *20*, S757–S784. [CrossRef]
12. Berni, R.; Charton, S.; Planchon, S.; Legay, S.; Romi, M.; Cantini, C.; Cai, G.; Hausman, J.; Renaut, J.; Guerriero, G. Molecular investigation of Tuscan sweet cherries sampled over three years: Gene expression analysis coupled to metabolomics and proteomics. *Hortic. Res.* **2021**, *8*, 12. [CrossRef]
13. Der Agopian, R.G.; Fabi, J.P.; Cordenunsi-Lysenko, B.R. Metabolome and proteome of ethylene-treated papayas reveal different pathways to volatile compounds biosynthesis. *Food Res. Int.* **2020**, *131*, 108975. [CrossRef] [PubMed]
14. Tang, H.; Zhang, X.; Gong, B.; Yan, Y.; Shi, Q. Proteomics and metabolomics analysis of tomato fruit at different maturity stages and under salt treatment. *Food Chem.* **2020**, *311*, 126009. [CrossRef] [PubMed]
15. Monribot-Villanueva, J.L.; Altúzar-Molina, A.; Aluja, M.; Zamora-Briseño, J.A.; Elizalde-Contreras, J.M.; Bautista-Valle, M.V.; Arellano De Los Santos, J.; Sánchez-Martínez, D.E.; Rivera-Reséndiz, F.J.; Vázquez-Rosas-Landa, M.; et al. Integrating proteomics and metabolomics approaches to elucidate the ripening process in white *Psidium guajava*. *Food Chem.* **2022**, *367*, 130656. [CrossRef] [PubMed]
16. Tian, X.; Zhu, L.; Yang, N.; Song, J.; Zhao, H.; Zhang, J.; Ma, F.; Li, M. Proteomics and metabolomics reveal the regulatory pathways of ripening and quality in post-harvest kiwifruits. *J. Agric. Food Chem.* **2021**, *69*, 824–835. [CrossRef]
17. Ma, Y.; Ling, T.; Su, X.; Jiang, B.; Nian, B.; Chen, L.; Liu, M.; Zhang, Z.; Wang, D.; Mu, Y.; et al. Integrated proteomics and metabolomics analysis of tea leaves fermented by *Aspergillus niger*, *Aspergillus tamarii* and *Aspergillus fumigatus*. *Food Chem.* **2021**, *334*, 127560. [CrossRef]
18. Chen, D.; Sun, Z.; Gao, J.; Peng, J.; Wang, Z.; Zhao, Y.; Lin, Z.; Dai, W. Metabolomics combined with proteomics provides a novel interpretation of the compound differences among Chinese tea cultivars (*Camellia sinensis* var. *sinensis*) with different manufacturing suitabilities. *Food Chem.* **2022**, *377*, 131976. [CrossRef]
19. Kapoor, L.; Simkin, A.J.; George Priya Doss, C.; Siva, R. Fruit ripening: Dynamics and integrated analysis of carotenoids and anthocyanins. *BMC Plant Biol.* **2022**, *22*, 27. [CrossRef]
20. Gonzali, S.; Perata, P. Fruit colour and novel mechanisms of genetic regulation of pigment production in tomato fruits. *Horticulturae* **2021**, *7*, 259. [CrossRef]
21. Velásquez, S.; Peña, N.; Bohórquez, J.C.; Gutierrez, N.; Sacks, G.L. Volatile and sensory characterization of roast coffees—Effects of cherry maturity. *Food Chem.* **2019**, *274*, 137–145. [CrossRef] [PubMed]
22. Chen, Q.; Shi, J.; Mu, B.; Chen, Z.; Dai, W.; Lin, Z. Metabolomics combined with proteomics provides a novel interpretation of the changes in nonvolatile compounds during white tea processing. *Food Chem.* **2020**, *332*, 127412. [CrossRef] [PubMed]
23. Lin, N.; Liu, X.; Zhu, W.; Cheng, X.; Wang, X.; Wan, X.; Liu, L. Ambient ultraviolet B signal modulates tea flavor characteristics via shifting a metabolic flux in flavonoid biosynthesis. *J. Agric. Food Chem.* **2021**, *69*, 3401–3414. [CrossRef] [PubMed]
24. Wu, H.; Lu, P.; Liu, Z.; Sharifi Rad, J.; Suleria, H.A.R. Impact of roasting on the phenolic and volatile compounds in coffee beans. *Food Sci. Nutr.* **2022**, *10*, 2408–2425. [CrossRef] [PubMed]
25. Liu, C.; Yang, N.; Yang, Q.; Ayed, C.; Linforth, R.; Fisk, I.D. Enhancing robusta coffee aroma by modifying flavour precursors in the green coffee bean. *Food Chem.* **2019**, *281*, 8–17. [CrossRef]
26. Hu, G.; Peng, X.; Wang, X.; Li, X.; Li, X.; Qiu, M. Excavation of coffee maturity markers and further research on their changes in coffee cherries of different maturity. *Food Res. Int.* **2020**, *132*, 109121. [CrossRef] [PubMed]
27. Wang, J.; Lv, M.; He, H.; Jiang, Y.; Yang, J.; Ji, S. Glycine betaine alleviated peel browning in cold-stored "Nanguo" pears during shelf life by regulating phenylpropanoid and soluble sugar metabolisms. *Sci. Hortic.* **2020**, *262*, 109100. [CrossRef]
28. Osorio, S.; Scossa, F.; Fernie, A.R. Molecular regulation of fruit ripening. *Front. Plant Sci.* **2013**, *4*, 198. [CrossRef]
29. Liu, J.; Chang, M.; Meng, J.; Feng, C.; Zhao, H.; Zhang, M. A comparative proteome approach reveals metabolic changes associated with *flammulina velutipes mycelia* in response to cold and light stress. *J. Agric. Food Chem.* **2018**, *66*, 3716–3725. [CrossRef]
30. Hall, R.D.; Trevisan, F.; de Vos, R.C.H. Coffee berry and green bean chemistry-opportunities for improving cup quality and crop circularity. *Food Res. Int.* **2022**, *151*, 110825. [CrossRef]
31. Veliz, C.G.; Criado, M.V.; Galotta, M.F.; Roberts, I.N.; Caputo, C. Regulation of senescence-associated protease genes by sulphur availability according to barley (*Hordeum vulgare* L.) phenological stage. *Ann. Bot.* **2020**, *126*, 435–444. [CrossRef] [PubMed]
32. Sohail, M.; Wills, R.B.H.; Bowyer, M.C.; Pristijono, P. Beneficial impact of exogenous arginine, cysteine and methionine on postharvest senescence of broccoli. *Food Chem.* **2021**, *338*, 128055. [CrossRef] [PubMed]
33. Wei, F.; Furihata, K.; Koda, M.; Hu, F.; Miyakawa, T.; Tanokura, M. Roasting process of coffee beans as studied by nuclear magnetic resonance: Time course of changes in composition. *J. Agric. Food Chem.* **2012**, *60*, 1005–1012. [CrossRef] [PubMed]

34. Schieber, A.; Wüst, M. Volatile phenols-important contributors to the aroma of plant-derived foods. *Molecules* **2020**, *25*, 4529. [CrossRef] [PubMed]
35. Bellumori, M.; Angeloni, G.; Guerrini, L.; Masella, P.; Calamai, L.; Mulinacci, N.; Parenti, A.; Innocenti, M. Effects of different stabilization techniques on the shelf life of cold brew coffee: Chemical composition, flavor profile and microbiological analysis. *LWT Food Sci. Technol.* **2021**, *142*, 111043. [CrossRef]
36. Medeiros, D.B.; Aarabi, F.; Martinez Rivas, F.J.; Fernie, A.R. The knowns and unknowns of intracellular partitioning of carbon and nitrogen, with focus on the organic acid-mediated interplay between mitochondrion and chloroplast. *J. Plant Physiol.* **2021**, *266*, 153521. [CrossRef]
37. Pratelli, R.; Pilot, G. Regulation of amino acid metabolic enzymes and transporters in plants. *J. Exp. Bot.* **2014**, *65*, 5535–5556. [CrossRef]
38. Li, N.; Wang, J.; Wang, B.; Huang, S.; Hu, J.; Yang, T.; Asmutola, P.; Lan, H.; Qinghui, Y. Identification of the carbohydrate and organic acid metabolism genes responsible for brix in tomato fruit by transcriptome and metabolome analysis. *Front. Genet.* **2021**, *12*, 714942. [CrossRef]
39. Li, X.; Jiang, J.; Chen, Z.; Jackson, A. Transcriptomic, proteomic and metabolomic analysis of flavonoid biosynthesis during fruit maturation in *Rubus chingii Hu*. *Front. Plant Sci.* **2021**, *12*, 706667. [CrossRef]
40. Siebeneichler, T.J.; Crizel, R.L.; Reisser, P.L.; Perin, E.C.; Da Silva Messias, R.; Rombaldi, C.V.; Galli, V. Changes in the abscisic acid, phenylpropanoids and ascorbic acid metabolism during strawberry fruit growth and ripening. *J. Food Compos. Anal.* **2022**, *108*, 104398. [CrossRef]
41. Haure, M.; Nguyen, T.K.C.; Cendrès, A.; Perino, S.; Licandro, H.; Waché, Y. Glycosidically bound volatile profiles of green and roasted coffee beans and aromatic potential of the spent coffee ground. *Eur. Food Res. Technol.* **2022**, *248*, 2125–2134. [CrossRef]
42. Cheng, B.; Furtado, A.; Smyth, H.E.; Henry, R.J. Influence of genotype and environment on coffee quality. *Trends Food Sci. Technol.* **2016**, *57*, 20–30. [CrossRef]
43. Santos, M.O.; de Oliveira Silveira, H.R.; de Souza, K.R.D.; Lima, A.A.; Boas, L.V.V.; Barbosa, B.C.F.; Barreto, H.G.; Alves, J.D.; Chalfun-Junior, A. Antioxidant system differential regulation is involved in coffee ripening time at different altitudes. *Trop. Plant Biol.* **2018**, *11*, 131–140. [CrossRef]

MDPI AG
Grosspeteranlage 5
4052 Basel
Switzerland
Tel.: +41 61 683 77 34

Foods Editorial Office
E-mail: foods@mdpi.com
www.mdpi.com/journal/foods

www.ingramcontent.com/pod-product-compliance
Lightning Source LLC
Chambersburg PA
CBHW040753250926
51850CB00007BB/1483
* 9 7 8 3 7 2 5 8 7 4 4 8 4 *